Anti-disturbance Intrusion- and Fault-Tolerant Control of Multi-agent Systems

Xiang-Gui Guo · Pei-Ming Liu · Dong-Yu Zhang ·
Jian-Liang Wang · Lei Guo

Anti-disturbance Intrusion- and Fault-Tolerant Control of Multi-agent Systems

Springer

Xiang-Gui Guo
Beijing Engineering Research Center
of Industrial Spectrum Imaging
University of Science and Technology
Beijing
Beijing, China

Hangzhou Innovation Institute
Beihang University
Hangzhou, China

Dong-Yu Zhang
Northeastern University
Shenyang, China

Lei Guo
Beihang University
Beijing, China

Pei-Ming Liu
University of Science and Technology
Beijing
Beijing, China

Jian-Liang Wang
Hangzhou Innovation Institute
Beihang University
Hangzhou, China

ISBN 978-981-96-9613-0 ISBN 978-981-96-9614-7 (eBook)
https://doi.org/10.1007/978-981-96-9614-7

This Springer imprint is published by the registered company Springer Nature Singapore Pte Ltd.
The registered company address is: 152 Beach Road, #21-01/04 Gateway East, Singapore 189721, Singapore

Preface

Multi-agent systems (MASs) have gained significant attention and recognition in various fields due to their ability to model and control complex systems with distributed and interacting components. The application of MAS covers almost all engineering fields, including transportation, intelligent agriculture, disaster relief, and delivery services, and has gradually become a valuable and widely used tool. As an important branch of the multi-agent field, collaborative control has been deeply studied by researchers, and different control schemes are proposed from different dimensions to achieve the desired complex multi-agent behavior, such as formation of vehicle platoons and consensus of robots. In the collaborative control of MASs, consensus control forces the states of the agents in the system to converge to desired trajectories. Consensus control problem encompasses *complete consensus* control and *cluster consensus* control. In *complete consensus* problems, all agents in the system share the same control objective, whereas in *cluster consensus* problems, the agents are divided into different clusters, each with distinct control objectives, and interactions occur among different clusters.

On the other hand, as the representative of the intelligent systems, MASs are designed to incorporate certain biological functions during its development, granting it adaptive capabilities and resilience similar to those of living organisms when operating in complex environments. Inspired by biology, these unmanned systems convert fundamental biological principles into engineering guidelines, endowing them with immune intelligence for effective anti-disturbance, fault tolerance, and intrusion tolerance. H_∞ methods, sliding mode control techniques, and adaptive compensation methods are known for their robustness against disturbances. Since multi-source, heterogeneous disturbances are inevitable in the operation of MASs, this book is interested in investigating for the refined anti-disturbance consensus control problem by combining adaptive feedback control method with H_∞ methods, sliding mode control techniques, disturbance observer-based methods or adaptive compensation methods.

MASs operate in open network environments. While the introduction of network communication enhances system performance, it also brings about information security concerns, which may pose significant risks to the secure and stable operation of the system. Similar to the viral invasions faced by biological organisms, MASs are susceptible to various types of cyber attacks, such as denial-of-service (DoS) attacks, false data injection attacks (FDIAs), and so on. For different types of attacks, this book proposes different *intrusion-tolerant* control strategies to attenuate the negative effect of malicious attacks on the system. In addition, a hierarchical control approach is employed to avoid the dimension explosion problem that may occur when using linear matrix inequality (LMI) techniques to address multiple cyber attack issues. Besides, actuator faults may occur due to human factors or mechanical wear, aging, etc., in the operation of the system. These faults are similar to physical injuries in biological organisms, as they not only degrade system performance but may also lead to serious accidents. Different from the actuator fault in a single system, one agent's fault in MAS may degrade the performance of the whole system due to the cascade and coupling effect. *Fault-tolerant* control plays a crucial role in the stable operation of MAS. Furthermore, in some occasions, such as the chemical processes and industrial processes, circuits, and sensors, excessive faults and reverse faults may occur. Therefore, this book also investigates the *fault-tolerant* consensus control problem of MASs with time-varying unknown nonidentical direction (TVUND) faults.

In MASs, effective communication and interaction is the linchpin of collaborative control among the members of system. However, with the increase of the number of agents and the complexity of the system, the amount of data that needs to be transmitted among agents has grown significantly, which will pose a considerable challenge to the communication capabilities of the system. Event-triggered mechanisms, by setting trigger thresholds, can significantly reduce the transmission of redundant data. Therefore, event-triggered collaborative control has garnered widespread attention from researchers and has led to many meaningful achievements. Unfortunately, the reduction of transmission data will inevitably lead to the decline of system performance. Consequently, how to set the trigger threshold is of great theoretical and practical significance for the tradeoff between the amount of data transmission and system performance. For the case that there are multiple leaders and clusters in the system, due to the complexity of the communication topology and the interaction among different clusters, the restricted communication topology such as in-degree balance condition increases the conservativeness of the *cluster consensus* control strategy. This book also proposes some pinning control strategies to eliminate restrictions on communication topology of MAS. In addition, the pinning control technology achieves a good performance while saving system common resources.

In this monograph, the aim is to present our recent research results in investigating anti-disturbance *intrusion-* and *fault-tolerant* consensus control for MASs. The main feature of this book is the successful introduction of event-triggered distributed adaptive feedback control methodology, observer technology and sliding mode technology, giving MAS immune intelligence in anti-disturbance, *intrusion-tolerant*, and *fault-tolerant* capabilities. Meanwhile, pinning control strategies are proposed to realize the *cluster consensus* control of MAS. Moreover, based on Nussbaum function, the *fault-tolerant* control problems of different types of MASs subjected to time-varying unknown nonidentical direction (TVUND) faults are investigated. This monograph provides a coherent approach and contains valuable reference materials for researchers wishing to explore the area of MAS. Researchers, graduate students, and engineers in the fields of electrical engineering, control, applied mathematics, computer science and others will benefit from this book. Its contents are also suitable for a one-semester graduate course.

We are grateful to University of Science and Technology Beijing and Hangzhou Innovation Institute of Beihang University for providing resources for our research work. We also gratefully acknowledge the support of our research the National Natural Science Foundation of China under Grant 62573043, Grant 62173028, Grant 62233015, and Grant 62173024; in part by the Beijing Natural Science Foundation under Grant 4232060 and Grant IS23065; in part by the Science, Technology and Innovation Project of Xiongan New Area under Grant 2023XAGG0062; and in part by the High-End Foreign Experts Recruitment Plan of the Ministry of Human Resources and Social Security of China under Grant S20240349.

Finally, we would like to thank the entire team of Springer Group for their cooperation in bringing out the work in the form of a monograph.

Beijing, China/Hangzhou, China — Xiang-Gui Guo
Beijing, China — Pei-Ming Liu
Shenyang, China — Dong-Yu Zhang
Hangzhou, China — Jian-Liang Wang
Beijing, China — Lei Guo

Competing Interests The authors have no competing interests to declare that are relevant to the content of this manuscript.

Contents

About the Authors

Xiang-Gui Guo received the B.S. degree in Electrical Engineering and Automation from the College of Electrical Engineering, Northwest University for Nationalities, Lanzhou, China, in 2005, the M.S. degree in Control Theory and Engineering from the College of Electrical Engineering and Automation, Fuzhou University, Fuzhou, China, in 2008, and the Ph.D. degree in Control Science and Engineering from Northeastern University, Shenyang, China, in 2012. From 2014 to 2016, he was a Postdoctoral Fellow with the School of Electrical and Electronic Engineering, Nanyang Technological University, Singapore. From July to August 2018, he was a Visiting Scholar with the Department of Mechanical Engineering, University of Victoria, Victoria, BC, Canada. Since 2019, he has been a Professor with the School of Automation and Electrical Engineering, University of Science and Technology Beijing, Beijing, China. His research interest includes multi-agent systems, fuzzy systems, vehicular platoon control, and fault-tolerant Control.

Pei-Ming Liu received the B.S. degree in Electrical Engineering and Automation from the College of Engineering, Shanxi Agricultural University, Taigu, China, in 2018. He is currently working toward the Ph.D. degree in Instruments Science and Technology with the School of Automation and Electrical Engineering, University of Science and Technology Beijing, Beijing, China. His research interests include intrusion-tolerant and fault-tolerant control for multi-agent systems.

Dong-Yu Zhang received the B.S. degree in Measurement and Control Technology and Instrument from North University of China, Taiyuan, China, in 2019, and the M.S. degree in Control Science and Engineering from University of Science and Technology Beijing, Beijing, China, in 2022. He is currently pursuing the Ph.D. degree in Control Science and Engineering with Northeastern University, Shenyang, China. His research interests include multi-agent systems, anti-disturbance control, and security of cyber-physical systems.

Jian-Liang Wang received the B.E. degree in Electrical Engineering from the Beijing Institute of Technology, Beijing, China, in 1982, and the M.S.E. and Ph.D. degrees in Electrical Engineering from the Johns Hopkins University, Baltimore, MD, USA, in 1985 and 1988, respectively. From 1988 to 1990, he was a Lecturer with Beihang University, Beijing, China. From 1990 to 2019, he was a tenured Associate Professor with the Nanyang Technological University, Singapore. Since 2020, he has been a Senior Researcher/Professor with the Hangzhou Innovation Institute of Beihang University, China. His current research interests include multi-agent systems, UAVs and UAMs, vehicular platoon systems, and fault-tolerant control.

Lei Guo received the Ph.D. degree in Control Engineering from Southeast University, Nanjing, China, in 1997. He is currently a Professor of the School of Automation Science and Electronic Engineering, Beihang University. He is an Academician of the Chinese Academy of Sciences (CAS), a Fellow of IEEE and IET, and a Member of the Chinese Association of Automation and the China Association of Inventions. He servers as director of the Navigation, Guidance, and Control Professional Committee of the Chinese Association of Automation. He is actively involved in the research of anti-disturbance control theory and its applications. He has developed advanced theoretical approaches, such as refined anti-disturbance control methods for complex systems with multi-source heterogeneous disturbances. He has published more than 480 papers and seven monographs and served as an editor for several international journals. He has received a number of awards, including the second prize of National Natural Science Award, the second prize of National Technical Inventions Award, the first prize of the Provincial and Ministerial-Level Technology Invention Award, the first prize of the Natural Science Award from the Ministry of Education, the first prize of the Technical Inventions Award from the Ministry of Education, and the National Pioneer Innovation Award of China.

Symbols Description

$\mathbb{R}$	Real number
$\mathbb{Z}$	Integer number
$\mathbb{N}$	Natural number
$\mathbb{R}^n$	The n-dimensional real vector
$\mathbb{R}^{n\times m}$	The $n \times m$ real matrices
$\|x\|$	The Euclidean norm of a vector x
$\lvert \cdot \rvert$	The absolute value of real numbers
$D^+(f)$	The right-hand derivative of a function f
$\cup$	The union of sets
$\cap$	The intersection of sets
$\otimes$	The Kronecker product
I_N	The identity matrix with N dimensions
$\mathbf{1}_p$	$p \times 1$ column vector of all ones
$\mathbf{0}_p$	$p \times 1$ column vector of all zeros
$\mathbf{0}_{N\times N}$	$N \times N$ matrix of all zeros
max	Maximum
min	Minimum
sup	Supremum, the least upper bound
inf	Infimum, the greatest lower bound
sin	sin function
cos	cos function
sinh	The hyperbolic sine function
cosh	The hyperbolic cosine function
tanh	The hyperbolic tangent function
$P = P^T > 0$	Symmetric positive-definite matrix
$P = P^T \geq 0$	Symmetric nonnegative matrix or positive semi-definite
$\lambda_i(A)$	The ith eigenvalue set of A
$\lambda_{\min}(A)$	The minimum eigenvalue of a real symmetric matrix A
$\lambda_{\max}(A)$	The maximum eigenvalue of a real symmetric matrix A
A^{-1}	The inverse matrix of matrix A
A^T	The transpose matrix of matrix A

$\|A\|$	The induced 2-norm of a real matrix A
$*$	The symmetric entries of a matrix
$\mathrm{He}\{M\}$	$\mathrm{He}\{M\} \triangleq M + M^T$
$\mathrm{col}\{x_1, \ldots, x_N\}$	$\mathrm{col}\{x_1, \ldots, x_N\} = [x_1^T, \ldots, x_N^T]^T$
$\mathbb{E}\{\cdot\}$	The mathematical expectation of a stochastic variable
$\mathrm{prob}\{E\}$	The probability of event E happening
$\mathrm{diag}\{A_1, \cdots, A_n\}$	A block-diagonal matrix with matrices A_i, $i = 1, \cdots, n$
$\mathrm{sgn}(\cdot)$	The signum function
f^z	The zth-order time derivative of $f(t)$
$z!$	The factorial of z
$\mathscr{V}_N = \{1, 2, \cdots, N\}$	The integer set from 1 to N containing 1 and N
$\mathscr{V}_N \backslash \{N\}$	The relative complement of $\{N\}$ in $\mathscr{V}_N$
$\mathscr{V}_N \cup \{0\}$	The union of $\mathscr{V}_N$ and $\{0\}$
$\ln(x)$	The natural logarithm of x

Acronyms

AMETM	Adaptive Memory Event-Triggered Mechanism
ANAS	Asymmetric Nonlinear Actuator Saturation
DO	Disturbance Observer
DoS	Denial-of-Service
ETM	Event-Triggered Mechanism
FDIA	False Data Injection Attack
FLS	Fuzzy-Logic System
IQC	Integral Quadratic Constraint
ISS	Input-to-State Stability
LMI	Linear Matrix Inequality
MAS	Multi-Agent System
PID	Proportional-Integral-Derivative
SMC	Sliding-Mode Control
TVUND	Time-Varying Unknown Nonidentical Direction
UUB	Uniformly Ultimately Bounded
ZOH	Zero-Order Hold

Chapter 1
Introduction

Abstract With the advancement of network technology and intelligent sensing, the collaborative control of *multi-agent systems* (MASs) has been increasingly prevalent, spanning from smart agriculture to disaster relief and rescue operations. Consequently, the issues of consensus collaboration and self-tolerance of MASs have become prominent research topics in the academic community in recent years. This chapter provides an overview of the latest research findings in the realm of consensus control and control strategies for MASs.

1.1 Multi-agent Systems

1.1.1 Consensus of MASs

Multi-agent systems (MASs) refer to a field of study on the interactions, cooperation, and problem-solving capabilities of multiple agents, which can be robots, unmanned aerial vehicles (UAVs), autonomous vehicles, virtual entities, and more [1–4]. Inspired by biological organisms, an MAS consists of multiple autonomous agents, each of which has the capabilities of perception, decision-making, and action [5, 6]. These agents communicate and collaborate with each other to achieve individual and collective goals [7]. Research in MASs aims to understand and design these systems to operate efficiently in complex and dynamic environments [8]. The research on collaborative control of MAS focused by researchers encompasses several categories such as consensus problems [3, 9–16], aggregation behavior [17, 18], and formation control [19, 20]. Notably, consensus problems represent a fundamental concern in the study of MASs. As a result, the design of consensus strategies serves as a conceptual framework that can offer insights and methodologies applicable to the resolution of various other coordination control problems [18]. Based on different network topologies and application scenarios, consensus can be classified into two categories: *complete consensus* and *cluster consensus*, where *complete consensus* can be regarded as a special case of *cluster consensus* [1, 21].

X.-G. Guo et al., *Anti-disturbance Intrusion- and Fault-Tolerant Control of Multi-agent Systems*, https://doi.org/10.1007/978-981-96-9614-7_1

The *complete consensus* of MAS means that all agents in MAS tend to consensus or converge to a common goal or state by communicating and cooperating with each other [18, 22–27]. This *complete consensus* can cover different levels, including position consensus [12, 28], speed consensus [29], voltage consensus [8], and so on. In recent years, many valuable researches have been developed from different perspectives to ensure the complete consensus stability and performance of MASs, such as adaptive *complete consensus* [9, 12, 30, 31], leader-following *complete consensus* [10, 32–37], leaderless *complete consensus* [3, 38], quantized *complete consensus* [27, 39–41], *complete consensus* with time delays [42, 43], *complete consensus* over switching communication topologies [13, 32, 44, 45], and *complete consensus* with multiple leaders [46, 47]. However, these *complete consensus* control strategies are no longer applicable when there are multiple clusters in the MASs. Therefore, researchers put forward the concept of *cluster consensus*.

Cluster consensus of MAS means that the agents in the MAS are divided into different clusters, and the agents in the same cluster have the same consensus goal, while the agents in different clusters have different consensus goals [1, 8, 19, 48, 49]. Compared with *complete consensus*, *cluster consensus* is more challenging due to the existence of inter-cluster interaction [50]. As a special form of *cluster consensus*, bipartite consensus divides agents into two clusters, and this division is determined according to the cooperation and competition among agents [51]. More generally, in the case of multiple leaders in the system, pinning control technology is an effective way to achieve *cluster consensus*, which has been widely studied in recent years [1, 49]. However, the above research results have strict requirements on the communication topology of the system. Bipartite consensus requires the communication topology of the system to be structurally balanced [51], and for multi-leader *cluster consensus*, the communication topology of the system is constrained by the in-degree balance condition [19, 52, 53]. Meanwhile, in [54], the topological graph of the MAS requires each agent in a cluster with respect to other clusters having the same in-degree. These constraints on the communication topology weaken the scope of application of the control strategy and increase conservatism.

1.1.2 Communication Constraint

Compared to a single agent, MAS involves a larger number of agents (represented by nodes) and high information transfer volume [43, 55–58]. Therefore, reducing communication and energy consumption among agents is a crucial issue [1]. Unlike the traditional time-triggered mechanism, the event-triggered mechanism (ETM) operates based on predefined conditions to decide whether transmission and control tasks should be executed, thereby reducing the collection and transmission of redundant data and conserving network resources [59]. As a result, the ETM-based consensus problem for MASs has garnered significant attention [10, 16, 35, 55, 60–63]. Typically, [10] introduces an event-based consensus control scheme for MASs under *cyber attacks* by employing fixed threshold parameters. While effective, this approach still requires the use of a large amount of communication resources. A centralized ETM is proposed in [35], where all agents are triggered simultaneously. Although this simplifies design, it depends on global information, resulting in unnecessary communication and computation overhead. Alternatively, distributed adaptive ETMs, as explored in [16, 60], adjust thresholds dynamically based on the system state, requiring only local and neighbor information. This approach strikes a balance between performance and resource conservation, making it an effective way to improve event-triggered strategies [64]. For MASs with multiple clusters, the design of ETMs is more complex than for *complete consensus* due to the need for simultaneous intra-cluster and inter-cluster event triggering [1]. However, centralized ETMs in [65] and asynchronous ETMs in [66], where intra-cluster triggers are synchronous but inter-cluster ones are not, limit their applicability. Additionally, [67] implements an ETM requiring periodic sampling of neighbors' information to evaluate triggering conditions. It is worth noting that reducing communication burdens often degrades system performance [10, 15, 68]. Thus, a critical challenge lies in dynamically adjusting the number of trigger points to enhance system performance while reducing communication loads.

1.2 Self-tolerance of MASs

In practical engineering applications, the operating environment of MAS is complex and changing, and the main threats to MAS include *cyber attacks* [35, 68–70], actuator faults [71–74], disturbances [75–78] and so on. These adverse threats may lead to performance degradation and even instability of MASs. Therefore, it is necessary to empower MASs with "immune intelligence" for effective anti-disturbance, fault tolerance, and intrusion tolerance [79]. Anti-disturbance control, *intrusion-tolerant* control and fault-tolerant control, as the effective ways to deal with the above threats, have been widely studied in recent years.

1.2.1 Intrusion-Tolerant Control

With the expansion of information flow between agents and the challenges posed by open network environments, MASs are under growing threat from *cyber attacks* that can erode the integrity and veracity of data [3, 80]. These threats can significantly impair the operational effectiveness of MASs, and in the most severe instances, lead to system destabilization [81, 82]. Consequently, it is essential to confront the cyber security challenges inherent in MASs. The main type of *cyber attacks* that MASs must be defended against include false data injection attacks (FDIAs), denial-of-service (DoS) attacks, and so on [68, 83–86]. Over the past few years, the issue of MAS facing DoS attacks has garnered significant interest and seen substantial progress. This is largely because DoS attacks are not only relatively simple to execute but also have the potential to cause severe damage [1, 6, 20, 29, 87–89]. For DoS attacks, the switching input method is usually used to realize *intrusion-tolerant* control, which ensures that the MAS can achieve the control objective when the attack satisfies the limits about frequency and duration [1]. Meanwhile, FDIAs compel the controller to receive incorrect sensor signals or force actuators to receive incorrect control signals [90–93]. The attack data includes both continuous signals and discrete signals following a Bernoulli distribution. *Intrusion-tolerant* control methods for these attacks include adaptive compensation methods [90] and pulse control methods [92]. The FDIA model that obeys the Bernoulli distribution is more common in discrete systems [94]. However, continuous systems are also affected by discrete FDIAs. Furthermore, in response to multiple *cyber attacks*, the works proposed in [10, 95] have employed linear matrix inequality (LMI) methods to explore *intrusion-tolerant* schemes in MASs and networked control systems, respectively. Nevertheless, the scalability of these LMI-based control solutions is hindered when applied to extensive MASs due to the problem of exponential growth in dimensions. As a result, the development of *intrusion-tolerant* approaches for MASs under multiple of *cyber attacks* necessitates additional investigation.

1.2.2 Fault-Tolerant Control

In practical application, the actuator fault of the agent is inevitable, which may lead to the decline of system performance and even serious accidents [12, 16, 31, 73, 96–100]. *Fault-tolerant* control plays an important role in ensuring the reliable operation of the system with actuator faults. In response to multiplicative and additive faults, researchers have proposed numerous *fault-tolerant* methods, including *sliding mode control* method [101], adaptive control method [102–107], fault estimation method [108], and fuzzy control method [109]. However, the requirement for upper and lower bounds on fault factors is a major drawback. Furthermore, in some scenarios, such as chemical processes and industrial processes, excessive faults and reverse faults are possible, and fault factors are no longer limited to (0, 1], which makes all the above *fault-tolerant* methods inapplicable [2, 8, 90]. Therefore, more efficient methods should be designed to solve the problem of time-varying unknown direction of actuator efficiency factors.

1.2.3 Anti-disturbance Control

Besides *cyber attacks* and actuator faults, *disturbance* is a main factor that must be considered in practical systems [91, 110–116]. In practice, different types of *disturbances* in MASs are physically multi-source and mathematically heterogenous [117]. In addition, compared with the *disturbances* of a single agent, the interactions among *disturbances* originating from distinct agents in MASs cannot be ignored [91]. How to characterize the properties and rejection/attenuate the negative effects of *disturbances* is a critical problem [113, 118–120]. Roughly speaking, the practical *disturbances* can be classified into three types: the unknown dynamic signal [1, 21, 91, 121], the norm-bounded variable [9, 90, 122], and the stochastic variables [117, 123, 124]. Accordingly, different disturbance types are addressed using various anti-disturbance methods, including anti-disturbance observer-based methods [125, 126], adaptive compensation methods [21, 114], H_∞ attenuation methods [91, 127], and Kalman-type attenuation methods [117]. However, agents may encounter multiple disturbances simultaneously (multi-source heterogeneous *disturbances*), which will lead to the above anti-disturbance methods being ineffective in handling the disturbances according to the characteristics of the disturbances, and may even result in the decline of system control accuracy or even instability [128]. For the anti-disturbance problem under multi-source heterogeneous *disturbances*, composite hierarchical anti-disturbance control (CHADC) with *disturbance* observer plus controller structure is an effective refined control methodology [111, 128–131]. Furthermore, the coupling of actuator fault and disturbance aggravates the difficulty of controller design [99, 108, 132, 133], and disturbance separation is more challenging in this case. However, there are few refined anti-disturbance control methods for MASs with multi-source heterogeneous *disturbances* and actuator faults. Therefore, it is of interest and theoretical significance to construct effective refined anti-disturbance control schemes for MASs with actuator faults under *cyber attacks*.

1.3 Outline of the Book

This book is divided into 13 chapters. This section outlines a map for the book, guiding readers through the various chapters where the claims and concepts presented earlier are validated. Each part of the book addresses specific aspects, providing evidence, examples, or further elaboration to support the theoretical framework made in earlier sections.

This chapter presents the background of anti-disturbance *intrusion-* and *fault-tolerant* consensus control for MASs. It contains an overview of the essential concepts and comprehensive literature surveys within the closely allied research areas.

Chapter 2 develops the mathematical model of MASs that will be used throughout the book. Several preliminaries and basic lemmas associated with the MASs are also reviewed in this chapter.

Chapter 3 presents the combined anti-disturbance and *complete consensus* control of an MAS. Two distinct *disturbance* observation-based control methods is proposed with an memory ETM.

Chapter 4 extends the anti-disturbance *complete consensus* results to situations where the MAS is attacked. The state observer-based anti-disturbance strategy is applied to MAS under DoS attacks.

Chapter 5 considers the H_∞ *complete consensus* control problems of an MAS with norm-bounded disturbances under FDIAs. A new fault detection scheme is also designed to identify the actuator faults.

Chapter 6 develops a CHADC scheme for a class of MASs subjected to stochastic FDIAs and multiple disturbances. A novel adaptive disturbance compensation technique is detailed with estimation of the upper bounds of the unknown but bounded disturbance.

Chapter 7 aims to solve the *fault-tolerant* control problem of MAS with asymmetric nonlinear actuator saturation (ANASs) and IQCs. A pseudo-PID sliding-mode control (SMC) scheme is introduced to improve the *complete consensus* performance.

Chapter 8 investigates the *cluster consensus* control for heterogeneous MAS with actuator faults and IQCs. Two pinning control-based *fault-tolerant* controller with fixed and adaptive pinning gains respectively are proposed.

Chapter 9 studies the *cluster consensus* problem for nonlinear MASs with directed communication graph under periodic DoS attacks. A state tracking error systems id developed with an input switching method and the event-triggered controller design.

Chapter 10 extends the *cluster consensus* results to cases where MASs suffer from actuator faults and aperiodic DoS attacks. A event-triggered input switching scheme is proposed to handle the *intrusion-* and *fault-tolerant* control.

Chapter 11 concerns the *intrusion-* and *fault-tolerant cluster consensus* problems of MASs time-varying unknown nonidentical direction (TVUND) faults and stochastic FDIAs. The new *Nussbaum functions* and *Nussbaum functions*-based controller designs are detailed.

Chapter 12 investigates *cluster consensus* tracking problem for a class of high-order MASs with TVUND faults under multiple *cyber attacks*. A hierarchical framework is developed with virtual cyber layer and physical layer.

Chapter 13 summarizes the key findings and contributions of this book, discusses the challenges of *intrusion-tolerant, fault-tolerant*, and anti-disturbance control, and outlines potential future research directions.

References

1. X.G. Guo, P.M. Liu, J.L. Wang, C.K. Ahn, Event-triggered adaptive fault-tolerant pinning control for cluster consensus of heterogeneous nonlinear multi-agent systems under aperiodic DoS attacks. IEEE Trans. Netw. Sci. Eng. **8**(2), 1941–1956 (2021)
2. X.G. Guo, W.D. Xu, J.L. Wang, J.H. Park, Distributed neuroadaptive fault-tolerant sliding-mode control for 2-D plane vehicular platoon systems with spacing constraints and unknown direction faults. Automatica **129**, Art. No. 109675 (2021)
3. X.G. Guo, D.Y. Zhang, J.L. Wang, C.K. Ahn, Adaptive memory event-triggered observer-based control for nonlinear multi-agent systems under DoS attacks. IEEE/CAA J. Autom. Sinica **8**(10), 1644–1656 (2021)
4. H.Y. Yang, L. Guo, B.H. Xu, J.Z. Gu, Collaboration control of fractional-order multiagent systems with sampling delay. Math. Probl. Eng. Art. ID. 854960 (2013)
5. Y. Cui, Y. Jia, $L_2 - L_\infty$ consensus control for high-order multi-agent systems with switching topologies and time-varying delays. IET Control Theory Appl. **6**(12), 1933–1940 (2012)

6. C. Deng, C. Wen, MAS-based distributed resilient control for a class of cyber-physical systems with communication delays under DoS attacks. IEEE Trans. Cybern. **51**(5), 2347–2358 (2021)
7. H.Y. Yang, L. Guo, H.L. Zou, Robust consensus of multi-agent systems with time-delays and exogenous disturbances. Int. J. Control Autom. Syst. **10**(4), 797–805 (2012)
8. P.M. Liu, X.G. Guo, J.L. Wang, X.P. Xie, F.W. Yang, Fully distributed hierarchical ET intrusion- and fault-tolerant group control for MASs with application to robotic manipulators. IEEE Trans. Autom. Sci. Eng. **21**(3), 2868–2881 (2024)
9. G. Li, Y. Wu, X. Liu, Adaptive fixed-time consensus tracking control method for second-order multi-agent systems with disturbances. J. Franklin Inst. **357**(3), 1516–1531 (2020)
10. J. Liu, T. Yin, D. Yue, H.R. Karimi, J. Cao, Event-based secure leader-following consensus control for multiagent systems with multiple cyber attacks. IEEE Trans. Cybern. **51**(1), 162–173 (2021)
11. M. Meng, G. Xiao, B. Li, Adaptive consensus for heterogeneous multi-agent systems under sensor and actuator attacks. Automatica **122**, Art. No. 109242 (2020)
12. C.L. Wang, C.Y. Wen, L. Guo, Adaptive consensus control for nonlinear multi-agent systems with unknown control directions and time-varying actuator faults. IEEE Trans. Autom. Control **66**(9), 4222–4229 (2021)
13. P.J. Wang, G.H. Wen, X.H. Yu, W.W. Yu, Y.Z. Lv, Consensus disturbance rejection for linear multiagent systems with directed switching communication topologies. IEEE Trans. Control Netw. Syst. **7**(1), 254–265 (2020)
14. H.Y. Yang, L. Guo, X.L. Zhu, K.C. Cao, H.L. Zou, Consensus of compound-order multi-agent systems with communication delays. Cent. Eur. J. Phys. **11**(6), 806–812 (2013)
15. Y. Yang, Y. Li, D. Yue, Y.C. Tian, X. Ding, Distributed secure consensus control with event-triggering for multiagent systems under DoS attacks. IEEE Trans. Cybern. **51**(6), 2916–2928 (2021)
16. D. Ye, M.M. Chen, H.J. Yang, Distributed adaptive event-triggered fault-tolerant consensus of multiagent systems with general linear dynamics. IEEE Trans. Cybern. **49**(3), 757–767 (2019)
17. Z. Chen, T. Chu, Aggregation and pattern formation of multi-agent systems, in *2007 Chinese Control Conference*, pp. 606–610 (2007)
18. J. Qin, Q. Ma, Y. Shi, L. Wang, Recent advances in consensus of multi-agent systems: a brief survey. IEEE Trans. Ind. Electron. **64**(6), 4972–4983 (2017)
19. L. Ma, Y.L. Wang, Q.L. Han, H_∞ cluster formation control of networked multiagent systems with stochastic sampling. IEEE Trans. Cybern. **51**(12), 5761–5772 (2021)
20. Y. Tang, D. Zhang, P. Shi, W. Zhang, F. Qian, Event-based formation control for nonlinear multiagent systems under DoS attacks. IEEE Trans. Autom. Control **66**(1), 452–459 (2021)
21. X.G. Guo, P.M. Liu, H.J. Li, J.L. Wang, C.K. Ahn, Cluster synchronization of heterogeneous nonlinear multi-agent systems with actuator faults and IQCs through adaptive fault-tolerant pinning control. Inf. Sci. **575**, 289–305 (2021)
22. X.G. Guo, B.Q. Wang, J.L. Wang, C.K. Ahn, Z.G. Wu, Edge-event-triggered encryption-decryption observer-based control of multiagent systems for privacy protection under multiple cyber attacks. Inf. Sci. **642**, Art. No.119128 (2023)
23. Y. Liu, X. Hou, Event-triggered consensus control of disturbed multi-agent systems using output feedback. ISA Trans. **91**, 166–173 (2019)
24. C. Ma, H. Qiao, Distributed asynchronous event-triggered consensus of nonlinear multi-agent systems with disturbances: An extended dissipative approach. Neurocomputing **243**, 103–114 (2017)
25. Y.Y. Qian, L. Liu, G. Feng, Distributed event-triggered adaptive control for consensus of linear multi-agent systems with external disturbances. IEEE Trans. Cybern. **50**(5), 2197–2208 (2020)
26. J. Sun, J. Yang, S. Li, X. Wang, G. Li, Event-triggered output consensus disturbance rejection algorithm for multi-agent systems with time-varying disturbances. J. Franklin Inst. **357**(17), 12870–12885 (2020)

27. Z.G. Wu, Y. Xu, Y.J. Pan, H. Su, Y. Tang, Event-triggered control for consensus problem in multi-agent systems with quantized relative state measurements and external disturbance. IEEE Trans. Circuits Syst. I, Reg. Papers **65**(7), 2232–2242 (2018)
28. M. Zhao, C. Peng, W.L. He, Y. Song, Event-triggered communication for leader-following consensus of second-order multiagent systems. IEEE Trans. Cybern. **48**(6), 1888–1897 (2018)
29. T. Dong, Y.L. Gong, Leader-following secure consensus for second-order multi-agent systems with nonlinear dynamics and event-triggered control strategy under DoS attack. Neurocomputing **416**, 95–102 (2020)
30. C.L.P. Chen, G.X. Wen, Y.J. Liu, F.Y. Wang, Adaptive consensus control for a class of nonlinear multiagent time-delay systems using neural networks. IEEE Trans. Neural Netw. Learn. Syst. **25**(6), 1217–1226 (2014)
31. L. Zhao, G.H. Yang, Adaptive fault-tolerant control for nonlinear multi-agent systems with DoS attacks. Inf. Sci. **526**, 39–53 (2020)
32. S. Chen, H. Jiang, Z. Yu, Observer-based event-triggered consensus of leader-following linear multi-agent systems with input saturation and switching topologies. Neurocomputing **364**, 138–151 (2019)
33. S.L. Du, T. Liu, D.W.C. Ho, Dynamic event-triggered control for leader-following consensus of multiagent systems. IEEE Trans. Syst. Man Cybern. **50**(9), 3243–3251 (2020)
34. X. Ruan, J. Feng, C. Xu, J. Wang, Observer-based dynamic event-triggered strategies for leader-following consensus of multi-agent systems with disturbances. IEEE Trans. Netw. Sci. Eng. **7**(4), 3148–3158 (2020)
35. W. Xu, D.W.C. Ho, J. Zhong, B. Chen, Event/self-triggered control for leader-following consensus over unreliable network with dos attacks. IEEE Trans. Neural Netw. Learn. Syst. **30**(10), 3137–3149 (2019)
36. Y. Yang, D. Yue, C. Dou, Output-based event-triggered schemes on leader-following consensus of a class of multi-agent systems with lipschitz-type dynamics. Inf. Sci. **459**, 327–340 (2018)
37. S. Yazdani, H. Su (2018) Adaptive leader-follower flocking for uncertain Lagrange systems with input saturation and external disturbances, in *2018 37th Chinese Control Conference (CCC)*, pp. 7210–7214 (2018)
38. X.G. Guo, B.Q. Wang, J.L. Wang, Z.G. Wu, L. Guo, Adaptive event-triggered PIO-based anti-disturbance fault-tolerant control for MASs with process and sensor faults. IEEE Trans. Netw. Sci. Eng. **11**(1), 77–88 (2024)
39. D.V. Dimarogonas, K.H. Johansson, Stability analysis for multi-agent systems using the incidence matrix: quantized communication and formation control. Automatica **46**(4), 695–700 (2010)
40. J.J. Fu, J.Z. Wang, Adaptive coordinated tracking of multi-agent systems with quantized information. Syst. Control Lett. **74**, 115–125 (2014)
41. L.Y. Hao, Y. Yu, T.S. Li, H. Li, Quantized output-feedback control for unmanned marine vehicles with thruster faults via sliding-mode technique. IEEE Trans. Cybern. **52**(9), 9363–9376 (2022)
42. X. Dong, J. Xi, G. Lu, Y. Zhong, Formation control for high-order linear time-invariant multi-agent systems with time delays. IEEE Trans. Control Netw. Syst. **1**(3), 232–240 (2014)
43. Y. Luo, X. Xiao, J. Cao, A. Li, Event-triggered guaranteed cost consensus for uncertain nonlinear multi-agent systems with time delay. Neurocomputing **394**, 13–26 (2020)
44. Y.H. Cheng, L. Shi, J.L. Shao, W.X. Zheng, Sampled-data scaled group consensus for second-order multi-agent systems with switching topologies and random link failures. J. Frankl. Inst. **357**(5), 2868–2881 (2020)
45. D.M. Xie, L. Shi, F.C. Jiang, Group tracking control of second-order multi-agent systems with fixed and Markovian switching topologies. Neurocomputing **281**, 37–46 (2018)
46. Z.K. Li, Z.S. Duan, W. Ren, G. Feng, Containment control of linear multi-agent systems with multiple leaders of bounded inputs using distributed continuous controllers. Int. J. Robust Nonlinear Control **25**(13), 2101–2121 (2015)

47. W. Xiao, H. Ma, L. Zhou, H. Li, Adaptive fuzzy fixed-time formation-containment control for Euler-Lagrange systems. IEEE Trans. Fuzzy Syst. **31**(10), 3700–3709 (2023)
48. P.M. Liu, Z.Z Huang, X.G. Guo, Event-triggered secure group consensus of second-order multi-agent systems under periodic DoS attacks, in *2021 IEEE 10th Data Driven Control and Learning Systems Conference*, pp. 1007–1012 (2021)
49. Q. Ma, J.W. Lu, Cluster synchronization for directed complex dynamical networks via pinning control. Neurocomputing **101**, 354–360 (2013)
50. Z. Yaghoubi, Robust cluster consensus of general fractional-order nonlinear multi agent systems via adaptive sliding mode controller. Math. Comput. Simul. **172**, 15–32 (2020)
51. K. Li, C.C. Hua, X. You, C.K. Ahn, Output feedback predefined-time bipartite consensus control for high-order nonlinear multiagent systems. IEEE Trans. Circuits Syst. Regul. Pap. **68**(7), 3069–3078 (2021)
52. F. Li, Z. Ma, Q. Duan, Clustering component synchronization in a class of unconnected networks via pinning control. Phys. A **525**, 394–401 (2019)
53. L. Liu, X. Ding, W. Zhou, Prescribed-time cluster synchronization of uncertain complex dynamical networks with switching via pinning control. Neurocomputing **419**, 136–147 (2021)
54. Z. Liu, W.S. Wong, Group consensus of linear multiagent systems under nonnegative directed graphs. IEEE Trans. Autom. Control **67**(11), 6098–6105 (2022)
55. C. Deng, M.J. Er, G.H. Yang, N. Wang, Event-triggered consensus of linear multiagent systems with time-varying communication delays. IEEE Trans. Cybern. **50**(7), 2916–2925 (2020)
56. L. Ding, Q.L. Han, X.H. Ge, X.M. Zhang, An overview of recent advances in event-triggered consensus of multiagent systems. IEEE Trans. Cybern. **48**(4), 1110–1123 (2018)
57. W. Hu, L. Liu, G. Feng, Consensus of linear multi-agent systems by distributed event-triggered strategy. IEEE Trans. Cybern. **46**(1), 148–157 (2016)
58. W.D. Xu, X.G. Guo, J.L. Wang, H.C. Yan, Z.G. Wu, ETM-based fault-tolerant and intrusion-tolerant control for 2-D planar vehicular platoon with actual traffic scenarios. IEEE Trans. Intell. Transp. Syst. **25**(11), 18884–18898 (2024)
59. X.G. Guo, W.D. Xu, J.L. Wang, J.H. Park, H. Yan, BLF-based neuroadaptive fault-tolerant control for nonlinear vehicular platoon with time-varying fault directions and distance restrictions. IEEE Trans. Intell. Transp. Syst. **23**(8), 12388–12398 (2022)
60. B. Cheng, Z. Li, Fully distributed event-triggered protocols for linear multiagent networks. IEEE Trans. Autom. Control **64**(4), 1655–1662 (2019)
61. C. Du, X. Liu, W. Ren, P. Lu, H. Liu, Finite-time consensus for linear multiagent systems via event-triggered strategy without continuous communication. IEEE Trans. Control Netw. Syst. **7**(1), 19–29 (2020)
62. Z. Feng, G. Hu, Secure cooperative event-triggered control of linear multiagent systems under DoS attacks. IEEE Trans. Control Syst. Technol. **28**(3), 741–752 (2020)
63. T.Y. Zhang, D. Ye, Distributed event-triggered control for multi-agent systems under intermittently random denial-of-service attacks. Inf. Sci. **542**, 380–390 (2021)
64. Y. Xu, M. Fang, Z.G. Wu, Y.J. Pan, M. Chadli, T.W. Huang, Input-based event-triggering consensus of multiagent systems under denial-of-service attacks. IEEE Trans. Syst. Man Cybern. Syst. **50**(4), 1455–1464 (2020)
65. L. Liu, W. Zhou, X. Li, Y. Sun, Dynamic event-triggered approach for cluster synchronization of complex dynamical networks with switching via pinning control. Neurocomputing **340**, 32–41 (2019)
66. W.Y. Xu, D.W.C. Ho, Clustered event-triggered consensus analysis: An impulsive framework. IEEE Trans. Ind. Electron. **63**(11), 7133–7143 (2016)
67. A. Hu, J. Cao, M. Hu, L. Guo, Event-triggered group consensus for multi-agent systems subject to input saturation. J. Franklin Inst. **355**(15), 7384–7400 (2018)
68. S. Hu, D. Yue, X. Xie, X. Chen, X. Yin, Resilient event-triggered controller synthesis of networked control systems under periodic DoS jamming attacks. IEEE Trans. Cybern. **49**(12), 4271–4281 (2019)

69. X. Huang, J.X. Dong, Reliable leader-to-follower formation control of multiagent systems under communication quantization and attacks. IEEE Trans. Syst. Man Cybern. Syst. **50**(1), 89–99 (2020)
70. Y. Xu, M. Fang, P. Shi, Z.G. Wu, Event-based secure consensus of mutiagent systems against DoS attacks. IEEE Trans. Cybern. **50**(8), 3468–3476 (2020)
71. M. Davoodi, N. Meskin, K. Khorasani, Simultaneous fault detection and consensus control design for a network of multi-agent systems. Automatica **66**, 185–194 (2016)
72. Y. Wu, Z. Wang, Z. Huang, Distributed fault detection for nonlinear multi-agent systems under fixed-time observer. J. Franklin Inst. **356**(13), 7515–7532 (2019)
73. Y. Xu, Z.G. Wu, Distributed adaptive event-triggered fault-tolerant synchronization for multiagent systems. IEEE Trans. Ind. Electron. **68**(2), 1537–1547 (2021)
74. L. Zhao, G.H. Yang, Cooperative adaptive fault-tolerant control for multi-agent systems with deception attacks. J. Franklin Inst. **357**(6), 3419–3433 (2020)
75. B. Cheng, Z. Li, Consensus disturbance rejection with event-triggered communications. J. Franklin Inst. **356**(2), 956–974 (2019)
76. M. Ran, Q. Wang, C. Dong, L.H. Xie, Active disturbance rejection control for uncertain time-delay nonlinear systems. Automatica **112**, Art. No. 108692 (2020)
77. L. Rong, X. Liu, G.P. Jiang, S. Xu, Event-driven multiagent consensus disturbance rejection with input uncertainties via adaptive protocols. IEEE Trans. Syst. Man Cybern. Syst. **52**(5), 2911–2919 (2022)
78. Y. Zhu, J. Qiao, L. Guo, Adaptive sliding mode disturbance observer-based composite control with prescribed performance of space manipulators for target capturing. IEEE Trans. Ind. Electron. **66**(3), 1973–1983 (2019)
79. L. Guo, Y. Yuan, J. Qiao, X. Yu, Immune intelligence of unmanned system. Acta Aeronautica et Astronautica Sinica **41**(11), Art. No. 024618 (2020). (in Chinese)
80. Y. Yuan, H. Yang, L. Guo, F. Sun, *Analysis and Design of Networked Control Systems Under Attacks* (CRC Press, 2018)
81. Y.P. Gu, K.X. Guo, L. Guo, J.Z. Qiao, J.D. Jia, X. Yu, L.H. Xie. An enhanced UAV safety control scheme against attacks on desired trajectory. Aerosp. Sci. Technol. **119**, Art. No. 107212 (2021)
82. A.Y. Lu, G.H. Yang, Stability analysis for cyber-physical systems under denial-of-service attacks. IEEE Trans. Cybern. **51**(11), 5304–5313 (2021)
83. G. Franze, F. Tedesco, D. Famularo, Resilience against replay attacks: a distributed model predictive control scheme for networked multi-agent systems. IEEE-CAA J. Automatica Sin. **8**(3), 628–640 (2021)
84. W. Qi, Y. Hou, G. Zong, C.K. Ahn, Finite-time event-triggered control for Semi-Markovian switching cyber-physical systems with FDI attacks and applications. IEEE Trans. Circuits Syst. I Reg. Papers **68**(6), 2665–2674 (2021)
85. Y. Yuan, H. Yuan, D.W. Ho, L. Guo, Resilient control of wireless networked control system under denial-of-service attacks: a cross-layer design approach. IEEE Trans. Cybern. **50**(1), 48–60 (2018)
86. D. Zhang, G. Feng, Y. Shi, D. Srinivasan, Physical safety and cyber security analysis of multi-agent systems: a survey of recent advances. IEEE-CAA J. Automatica Sin. **8**(2), 319–333 (2021)
87. Z. Gu, C.K. Ahn, D. Yue, X.P. Xie, Event-triggered H_∞ filtering for T-S fuzzy-model-based nonlinear networked systems with multisensors against DoS attacks. IEEE Trans. Cybern. **52**(6), 5311–5321 (2020)
88. Y. Gu, X. Yu, K. Guo, J. Qiao, L. Guo, Detection, estimation, and compensation of false data injection attack for UAVs. Inf. Sci. **546**, 723–741 (2021)
89. D. Zhang, G. Feng, A new switched system approach to leader-follower consensus of heterogeneous linear multiagent systems with DoS attack. IEEE Trans. Syst. Man Cybern. Syst. **51**(2), 1258–1266 (2019)
90. X.G. Guo, P.M. Liu, Z.G. Wu, D. Zhang, C.K. Ahn, Hybrid event-triggered group consensus control for heterogeneous multi-agent systems with TVNUD faults and stochastic FDI attacks. IEEE Trans. Autom. Control **68**(12), 8013–8020 (2023)

91. X.G. Guo, D.Y. Zhang, J.L. Wang, J.H. Park, L. Guo, Observer-based event-triggered composite anti-disturbance control for multi-agent systems under multiple disturbances and stochastic FDIAs. IEEE Trans. Autom. Sci. Eng. **20**(1), 528–540 (2023)
92. W.L. He, X.Y. Gao, W.M. Zhong, F. Qian, Secure impulsive synchronization control of multi-agent systems under deception attacks. Inf. Sci. **459**, 354–368 (2018)
93. P.M. Liu, X.G. Guo, J.L. Wang, D. Coutinho, Z.G. Wu, Preset-time and preset-accuracy human-in-the-loop cluster consensus control for MASs under stochastic actuation attacks. IEEE Trans. Autom. Control **69**(3), 1675–1688 (2024)
94. L. Hu, Z. Wang, Q.L. Han, X. Liu, State estimation under false data injection attacks: Security analysis and system protection. Automatica **87**, 176–183 (2018)
95. L. Zha, R. Liao, J. Liu, X. Xie, E. Tian, J. Cao, Dynamic event-triggered output feedback control for networked systems subject to multiple cyber attacks. IEEE Trans. Cybern. **52**(12), 13800–13808 (2022)
96. L. Guo, H. Wang, Fault detection and diagnosis for general stochastic systems using B-spline expansions and nonlinear filters. IEEE Trans. Circuits Syst. Regul. Pap. **52**(8), 1644–1652 (2005)
97. L. Guo, L.P. Yin, H. Wang, T.Y. Chai, Entropy optimization filtering for fault isolation of nonlinear non-gaussian stochastic systems. IEEE Trans. Autom. Control **54**(4), 804–810 (2009)
98. Q.L. Hu, X.D. Shao, L. Guo, Adaptive fault-tolerant attitude tracking control of spacecraft with prescribed performance. IEEE/ASME Trans. Mechatron. **23**(1), 331–341 (2017)
99. X.M. Yao, L.G. Wu, L. Guo, Disturbance-observer-based fault tolerant control of high-speed trains: a Markovian jump system model approach. IEEE Trans. Syst. Man Cybern. Syst. **50**(4), 1476–1485 (2018)
100. W. Zou, C.K. Ahn, Z. Xiang, Fuzzy-approximation-based distributed fault-tolerant consensus for heterogeneous switched nonlinear multiagent systems. IEEE Trans. Fuzzy Syst. **29**(10), 2916–2925 (2021)
101. J. Qin, G. Zhang, W.X. Zheng, Y. Kang, Adaptive sliding mode consensus tracking for second-order nonlinear multiagent systems with actuator faults. IEEE Trans. Cybern. **49**(5), 1605–1615 (2019)
102. S. Chen, D.W.C. Ho, L. Li, M. Liu, Fault-tolerant consensus of multi-agent system with distributed adaptive protocol. IEEE Trans. Cybern. **45**(10), 2142–2155 (2015)
103. C. Deng, G.H. Yang, Distributed adaptive fault-tolerant control approach to cooperative output regulation for linear multi-agent systems. Automatica **103**, 62–68 (2019)
104. X. Jin, S. Wang, J. Qin, W.X. Zheng, Y. Kang, Adaptive fault-tolerant consensus for a class of uncertain nonlinear second-order multi-agent systems with circuit implementation. IEEE Trans. Circuits Syst. Regul. Pap. **65**(7), 2243–2255 (2018)
105. H.J. Ma, G.H. Yang, Adaptive fault tolerant control of cooperative heterogeneous systems with actuator faults and unreliable interconnections. IEEE Trans. Autom. Control **61**(11), 3240–3255 (2016)
106. X. Wang, G.H. Yang, Adaptive reliable coordination control for linear agent networks with intermittent communication constraints. IEEE Trans. Control Netw. Syst. **5**(3), 1120–1131 (2018)
107. Z. Zuo, J. Zhang, Y. Wang, Adaptive fault-tolerant tracking control for linear and Lipschitz nonlinear multi-agent systems. IEEE Trans. Ind. Electron. **62**(6), 3923–3931 (2015)
108. S.Y. Cao, L. Guo, X.Y. Wen, Fault tolerant control with disturbance rejection and attenuation performance for systems with multiple disturbances. Asian J. Control **13**(6), 1056–1064 (2011)
109. C. Deng, G.H. Yang, Distributed adaptive fuzzy control for nonlinear multiagent systems under directed graphs. IEEE Trans. Fuzzy Syst. **26**(3), 1356–1366 (2018)
110. L. Guo, S. Cao, *Anti-Disturbance Control for Systems With Multiple Disturbances* (CRC Press, 2013)
111. L. Guo, S. Cao, Anti-disturbance control theory for systems with multiple disturbances: a survey. ISA Trans. **53**(4), 846–849 (2014)

112. J.T. Jia, K.X. Guo, W.S. Li, X. Yu, L. Guo, Composite filtering for UWB-based localization of quadrotor UAV with skewed measurements and uncertain dynamics. IEEE Trans. Instrum. Meas. **71**, Art. No. 1002313 (2022)
113. M. Mohammadi, M.M. Arefi, P. Setoodeh, O. Kaynak, Optimal tracking control based on reinforcement learning value iteration algorithm for time-delayed nonlinear systems with external disturbances and input constraints. Inf. Sci. **554**, 84–98 (2021)
114. C.H. Xie, G.H. Yang, Decentralized adaptive fault-tolerant control for large-scale systems with external disturbances and actuator faults. Automatica **85**, 83–90 (2017)
115. X. Yu, Y.K. Zhu, J.Z. Qiao, L. Guo, Antidisturbance controllability analysis and enhanced antidisturbance controller design with application to flexible spacecraft. IEEE Trans. Aerosp. Electron. Syst. **57**(5), 3393–3404 (2021)
116. D.Y. Zhang, D.W. Ding, Z.M. Zhao, X.G. Guo, Adaptive memory event-triggered control for multi-agent systems under external disturbance, in *2021 11th International Conference on Information Science and Technology*, pp. 195–200 (2021)
117. L. Guo, W.S. Li, Y.K. Zhu, X. Yu, Z.D. Wang, Composite disturbance filtering: a novel state estimation scheme for systems with multisource, heterogeneous, and isomeric disturbances. IEEE Open J. Ind. Electron. Soc. **4**, 387–400 (2023)
118. X.G. Guo, C.K. Ahn, Adaptive fault-tolerant pseudo-PID sliding-mode control for high-speed train with integral quadratic constraints and actuator saturation. IEEE Trans. Intell. Transp. Syst. **22**(12), 7421–7431 (2021)
119. X.G. Guo, J.L. Wang, F. Liao, R.S.H. Teo, Distributed adaptive sliding mode control strategy for vehicle-following systems with nonlinear acceleration uncertainties. IEEE Trans. Veh. Technol. **66**(2), 981–991 (2017)
120. W.C. Zou, Z.R. Xiang, C.K. Ahn, Mean square leader–following consensus of second-order nonlinear multiagent systems with noises and unmodeled dynamics. IEEE Trans. Syst. Man Cybern. Syst. **49**(12), 2478–2486 (2019)
121. H.Y. Yang, L. Guo, C. Han, Robust consensus of multi-agent systems with uncertain exogenous disturbances. Commun. Theor. Phys. **56**(6), 1161–1166 (2011)
122. Z.T. Ding, Consensus disturbance rejection with disturbance observers. IEEE Trans. Ind. Electron. **62**(9), 5829–5837 (2015)
123. L. Guo, H. Wang, T.Y. Chai, Fault detection for non-linear non-Gaussian stochastic systems using entropy optimization principle. Trans. Inst. Meas. Control. **28**(2), 145–161 (2006)
124. L. Guo, H. Wang, Minimum entropy filtering for multivariate stochastic systems with non-Gaussian noises. IEEE Trans. Autom. Control **51**(4), 695–700 (2006)
125. I. Furtat, E. Fridman, A. Fradkov, Disturbance compensation with finite spectrum assignment for plants with input delay. IEEE Trans. Autom. Control **63**(1), 298–305 (2018)
126. Y. Yuan, Z. Wang, L. Guo, Event-triggered strategy design for discrete-time nonlinear quadratic games with disturbance compensations: the noncooperative case. IEEE Trans. Syst. Man Cybern. Syst. **48**(11), 1885–1896 (2018)
127. X.G. Guo, D.Y. Zhang, J.L. Wang, J.H. Park, L. Guo, Event-triggered observer-based $\mathscr{H}_\infty$ consensus control and fault detection of multiagent systems under stochastic false data injection attacks. IEEE Trans. Netw. Sci. Eng. **9**(2), 481–494 (2022)
128. Y.Y. Cui, J.Z. Qiao, Y.K. Zhu, X. Yu, L. Guo, Velocity-tracking control based on refined disturbance observer for gimbal servo system with multiple disturbances. IEEE Trans. Ind. Electron. **69**(10), 10311–10321 (2022)
129. L. Guo, X.Y. Wen, Hierarchical anti-disturbance adaptive control for non-linear systems with composite disturbances and applications to missile systems. Trans. Inst. Meas. Control. **33**(8), 942–956 (2011)
130. X. Yao, J.H. Park, L. Wu, L. Guo, Disturbance-observer-based composite hierarchical antidisturbance control for singular Markovian jump systems. IEEE Trans. Autom. Control **64**(7), 2875–2882 (2019)
131. Y. Zhu, L. Guo, J. Qiao, W. Li, An enhanced anti-disturbance attitude control law for flexible spacecrafts subject to multiple disturbances. Control. Eng. Pract. **84**, 274–283 (2019)

132. S.Y. Cao, L. Guo, W.H. Chen, Anti-disturbance fault tolerant initial alignment for inertial navigation system subjected to multiple disturbances. Aerosp. Sci. Technol. **72**, 95–103 (2018)
133. L. Guo, Y. Zhang, Generalized robust H_∞ fault diagnosis filtering based on conditional stochastic distributions of system outputs. Proc. Inst. Mech. Eng. Part I J. Syst. Control Eng. **221**(6), 857–864 (2007)

Chapter 2
Preliminaries

Abstract The terminology of topological graph theory is helpful to discuss any branch of MASs. This chapter introduces the basic topological graph theory, matrix theory, and fuzzy-logic system theory. In addition, some definitions and lemmas involved in the book are also introduced in this chapter.

2.1 Basic Algebraic Graph Theory

Assuming a consortium of agents engages in mutual interactions through communication or sensing networks, or a blend of both, in order to attain collective goals, it is customary to represent the interplay between these agents through the utilization of directed or undirected graphs. A *directed graph* $\mathscr{G}$ is a pair $(\mathscr{V}, \mathscr{E})$, where $\mathscr{V} = \{1, 2, \cdots, N\}$ is a nonempty finite node set and $\mathscr{E} \subset \mathscr{V} \times \mathscr{V} = \{(i, j)|i, j \in \mathscr{V}\}$ is an edge set. The edge (i, j) in the edge set $\mathscr{E}$ denotes that agent j can obtain information from agent i, but not necessarily vice versa. For an edge (i, j), node i is called the parent node, j is the child node, and i is a neighbor of j. The set of neighbors of i is denoted by $\mathscr{N}_i$, whose cardinality is called the in-degree of node i.

A graph with the property that $(i, j) \in \mathscr{E}$ implies $(j, i) \in \mathscr{E}$ for any $i, j \in \mathscr{V}$ is said to be undirected.

A weighted graph imparts weights on each of its edges, and the weights is usually represented by an adjacency matrix. The adjacency matrix, denoted as $\mathscr{A}$ and with entries $[a_{ij}] \in \mathbb{R}^{N \times N}$, is established for a *directed graph* $\mathscr{G}$ in such a manner that the weight $a_{ij} > 0$ if the edge $(j, i) \in \mathscr{E}$, and $a_{ij} = 0$ if $(j, i) \notin \mathscr{E}$. Obviously $a_{ij} \neq a_{ji}$ for a *directed graph*. Additionally, our stipulation includes $a_{ii} = 0$ for all $i \in \mathscr{V}$, thereby excluding self-loops (i, i) from consideration within this book. Correspondingly, the adjacency matrix for an *undirected graph* is defined in a similar manner, except that $a_{ij} = a_{ji}$ for all $i \neq j$ because $(i, j) \notin \mathscr{E}$ implies $(j, i) \notin \mathscr{E}$. It is important to emphasize that the selection of a_{ij} is undertaken to encapsulate the potency of interactions amid nodes. Should the absolute weight magnitude be of minimal concern, a_{ij} is set to 1 when $(j, i) \in \mathscr{E}$.

X.-G. Guo et al., *Anti-disturbance Intrusion- and Fault-Tolerant Control of Multi-agent Systems*, https://doi.org/10.1007/978-981-96-9614-7_2

The *Laplacian matrix* $\mathscr{L} = [l_{ij}]_{N\times N}$ is defined as $l_{ij} = -a_{ij}, i \neq j$, and $l_{ii} = \sum_{j=1}^{N} a_{ij}$. For an *undirected graph*, the *Laplacian matrix* $\mathscr{L}$ is symmetric, while for a *directed graph*, $\mathscr{L}$ is not necessarily symmetric and is sometimes called the nonsymmetric *Laplacian matrix* [1] or directed *Laplacian matrix* [2]. The node i that obtains information directly from the leader is called the pinning node, and the weight of the edge between the leader and the pinning node is called pinning gain. For pinning node, the pinning gain $b_i > 0$. Denote the pinning gain matrix $\mathscr{B} = \text{diag}\{b_1, \ldots, b_N\}$. In addition, a directed spanning tree refers to a graph with at least one directed path from a vertex to all other vertexes.

If the leader node labeled 0 has no neighbors, the *Laplacian matrix* of the weighted graph $\bar{\mathscr{G}}$ can be expressed as

$$\bar{\mathscr{L}} = \begin{bmatrix} 0 & 0_{1\times N} \\ \mathscr{L}_2 & \mathscr{L}_1 \end{bmatrix}_{(N+1)\times(N+1)}$$

where $\mathscr{L}_2 \in \mathbb{R}^N$ and $\mathscr{L}_1 = \mathscr{L} + \mathscr{B} \in \mathbb{R}^{N\times N}$. Furthermore, the matrix $\mathscr{L}_1$ is nonsingular if the weighted graph $\bar{\mathscr{G}}$ contains a directed spanning tree.

For cluster consensus, the nodes in $\mathscr{V}$ are divided into s $(2 \leq s < N)$ clusters $\mathscr{V}_1, \mathscr{V}_2, \cdots, \mathscr{V}_s$ with $\mathscr{V}_p \cap \mathscr{V}_q = \varnothing$ $(p \neq q)$ and $\mathscr{V}_1 \cup \mathscr{V}_2 \cup \cdots \cup \mathscr{V}_s = \mathscr{V}$. Let $\sigma_i = p$, if the ith node belongs to the pth cluster. Define $\hat{\mathscr{V}}_z \subseteq \mathscr{V}_z$ $(\tilde{\mathscr{V}}_z \subseteq \mathscr{V}_z)$ as a vertex set that sends (receives) information to (from) nodes in other clusters.

To facilitate the control law design and consensus analysis, several properties of the *Laplacian matrix* $\mathscr{L}$ are recalled in the following lemmas.

Lemma 2.1 ([3, 4]) *Zero is an eigenvalue of* $\mathscr{L}$ *with* $\mathbf{1}$ *a right eigenvector, and all nonzero eigenvalues are real and positive. Furthermore, zero is a simple eigenvalue of* $\mathscr{L}$ *if and only if* $\mathscr{G}$ *has a directed spanning tree. Then, the non-zero eigenvalues of* $\mathscr{L}$ *can be given as* $0 < \lambda_2 < \lambda_3 < \cdots < \lambda_N$.

Lemma 2.2 ([5, 6]) *Consider a directed graph* $\mathscr{G}$ *that has a directed spanning tree, and the root node is able to receive information from one of the virtual leaders. There exists a positive definite matrix* $\Theta = \text{diag}\{\theta_1, \cdots, \theta_N\}$ *such that the nonsingular* M*-matrix* $\mathscr{L}_1$ *satisfies*

$$\tilde{\mathscr{L}} = \Theta\mathscr{L}_1 + \mathscr{L}_1^T\Theta \geq \lambda_0 I_N > 0$$

where $\mathscr{L}_1 = \mathscr{L} + \mathscr{B}$, λ_0 *is the minimum eigenvalue of* $\tilde{\mathscr{L}}$, *and* $[\theta_1, \cdots, \theta_N]^T = (\mathscr{L}_1^T)^{-1}1_N$. *Furthermore, as* Θ *is a diagonal matrix, its maximum and minimum eigenvalues can be represented by* $\theta_{\max} = \max\{\theta_i\}$ *and* $\theta_{\min} = \min\{\theta_i\}$.

2.2 Consensus Theory

Consensus, as an important topic for MASs, implies that all subsystems (agents) of MAS can achieve dynamical objective through the designed control law. In a broad sense, consensus can be classified into two types, i.e., *complete consensus* and *cluster consensus*. Consider a second-order MAS consists of $N \geq 2$ follower agents. Let $x_i(t)$ and $v_i(t)$, $i = 1, \cdots, N$ denote the states of the ith follower agent of the MAS.

Definition 2.1 ([7] *Complete Consensus*) For any initial states $x_i(0)$ and $v_i(0)$, the leaderless MASs reach complete consensus if

$$\begin{aligned} &\lim_{t\to\infty} \|x_i(t) - x_j(t)\| = 0 \\ &\lim_{t\to\infty} \|v_i(t) - v_j(t)\| = 0, \quad \forall i, j = 1, \ldots, N. \end{aligned}$$

Further, the leader-follower MASs reach complete consensus if

$$\begin{aligned} &\lim_{t\to\infty} \|x_i(t) - x_0(t)\| = 0 \\ &\lim_{t\to\infty} \|v_i(t) - v_0(t)\| = 0, \quad \forall i = 1, \ldots, N \end{aligned}$$

where $x_0(t)$ and $v_0(t)$ are the states of the leader agent.

Definition 2.2 ([7, 8], *Cluster Consensus*) For any initial states $x_i(0)$ and $v_i(0)$), the second-order MASs reach cluster consensus if

$$\begin{aligned} &\lim_{t\to\infty} \|x_i(t) - x^h_{\sigma_i}(t)\| = 0, \quad \forall i = 1, \ldots, N \\ &\lim_{t\to\infty} \|v_i(t) - v^h_{\sigma_i}(t)\| = 0, \quad \sigma_i = 1, \ldots, s \end{aligned}$$

where $s \in \mathbb{N}_+$ represents the number of the clusters and leaders; $x^h_{\sigma_i}(t)$ and $v^h_{\sigma_i}(t)$ are the states of the leader agents.

2.3 Fuzzy Logic Systems

Fuzzy logic systems (FLSs) are rule-based systems. The components within the fuzzy rule base encompass offline membership functions, approximate inference algorithme, and defuzzification algorithms. The approximate inference algorithm is articulated through a collection of IF-THEN statements known as fuzzy rules, which can be structured as follows:

$$R^l: \text{ If } z_1 \text{ is } F_1^l \text{ and } z_2 \text{ is } F_2^l \text{ and } \cdots, \text{ and } z_n \text{ is } F_n^l$$
$$\text{Then } y \text{ is } G^l, l = 1, 2, \cdots, M$$

where F_j^l and G^l are fuzzy sets and $j = 1, 2, \cdots, n$. M denotes the number of IF-THEN rules. $Z = [z_1, \ldots, z_n]^T$ means the input of FLS, while y is the output.

The role of the fuzzification component is to translate a precise variable Z into a set of fuzzy values, which indicates the variable's affiliation with different fuzzy categories characterized by their respective membership functions. Subsequently, the fuzzy reasoning mechanism translates these fuzzy conditional statements into a mapping of sets in accordance with the principles of fuzzy logic. Following the application of the centroid-based defuzzification method, the output from the FLS can be articulated as

$$y(Z) = \frac{\sum_{l=1}^{M} \bar{y}_l \prod_{k=1}^{n} \mu_{F_k^l}(z_j)}{\sum_{l=1}^{M} [\prod_{k=1}^{n} \mu_{F_k^l}(z_j)]}$$

where $\bar{y}_l = \max_{y \in \mathscr{R}} \mu_{G^l}(y)$ is the center of the fuzzy set G_l. $\mu_{F_k^l}(z_j)$ is the membership function of z_j. By introducing vector $\xi(Z)$, $y(Z)$ is rewritten as

$$y(Z) = W^T \xi(Z)$$

where $W^T = [\bar{y}_1, \bar{y}_2, \cdots, \bar{y}_M]$ and $\xi(Z) = [\xi_1(Z), \xi_2(Z), \cdots, \xi_M(Z)]^T$. Then, the fuzzy basis function is defined as

$$\xi_l = \frac{\prod_{k=1}^{n} \mu_{F_k^l}(z_j)}{\sum_{l=1}^{M} [\prod_{k=1}^{n} \mu_{F_k^l}(z_j)]}.$$

Lemma 2.3 ([9]) *For any real function $f(Z)$ defined on a compact $\Omega \in \mathbb{R}^n$, when the number M of rules is large enough, the function $f(Z)$ can be expressed as follows:*

$$f(Z) = W^{*T} \xi(Z) + \varepsilon$$

where $W^* \in \mathbb{R}^M$ denotes the ideal approximation parameter vector and ε is the error of estimation bounded by constant ε^*, namely, $|\varepsilon| \leq \varepsilon^*$.

2.4 Some Other Definitions and Lemmas

Below are several supplementary definitions and lemmas that will be employed within this monograph.

Definition 2.3 ([10]) Let $\mathscr{I}$ be the infinitesimal operator, i.e., for a function $V(t)$,

$$\mathscr{I}V(t) = \lim_{\Delta \to 0} \frac{\mathbb{E}\{V(t+\Delta)|t\} - V(t)}{\Delta}.$$

Definition 2.4 The *Kronecker product* of matrices $A \in \mathbb{R}^{m \times n}$ and $B \in \mathbb{R}^{p \times q}$ is defined as

$$A \otimes B = \begin{bmatrix} a_{11}B & \cdots & a_{1n}B \\ \vdots & \ddots & \vdots \\ a_{m1}B & \cdots & a_{mn}B \end{bmatrix}$$

which has the following properties:

$$\begin{aligned} k(A \otimes B) &= (kA) \otimes B = A \otimes (kB) \\ A \otimes (B + C) &= A \otimes B + A \otimes C \\ (A + B) \otimes C &= A \otimes C + B \otimes C \\ (A \otimes B)(C \otimes D) &= (AC \otimes BD) \\ (A \otimes B)^T &= A^T \otimes B^T \\ (A \otimes B)^{-1} &= A^{-1} \otimes B^{-1}, \quad \text{if } A, B \text{ are invertible} \\ (A \otimes B) \otimes C &= A \otimes (B \otimes C) \end{aligned}$$

where k is a constant scalar, and A, B, C and D are matrices with appropriate dimensions.

Lemma 2.4 ([11]) *For a matrix equality constraint, i.e., $PB = BN$, it can be transformed into a standard linear matrix inequality and be rewritten approximately as*

$$\begin{bmatrix} -\nu I & PB - BN \\ * & -\nu I \end{bmatrix} < 0$$

where ν is an arbitrarily small positive constant.

Lemma 2.5 ([12]) *For any constant matrix $M_1 \in \mathbb{R}^{q \times l}$, symmetric positive definite matrix $M_2 \in \mathbb{R}^{l \times l}$ and scalar $\beta > 0$, the following inequality holds:*

$$2x^T M_1 y \le \beta x^T M_1 M_2 M_1^T x + \beta^{-1} y^T M_2^{-1} y$$

where $x \in \mathbb{R}^q$ and $y \in \mathbb{R}^l$.

Lemma 2.6 ([13]) *The following inequality is true for any scalars* $\ell > 0$ *and* $\varepsilon \in \mathbb{R}$*:*

$$0 < |\varepsilon| - \frac{\varepsilon^2}{\ell^2 + \varepsilon^2} < \ell.$$

Lemma 2.7 (Young's inequality [14, 15]) *For* $\forall(x, y) \in \mathbb{R}^2$*, the following inequality holds*

$$xy \leq \frac{\varepsilon^p}{p}|x|^p + \frac{1}{q\varepsilon^q}|y|^q$$

where $\varepsilon > 0$, $p > 1$, $q > 1$, *and* $(p-1)(q-1) = 1$.

Lemma 2.8 (Schur complement lemma, [16]) *For any symmetric matrix*

$$S = \begin{bmatrix} S_{11} & S_{12} \\ S_{21} & S_{22} \end{bmatrix}$$

with $S_{11} \in \mathbb{R}^{r \times r}$*, the following three conditions are equivalent.*

(1) $S < 0$
(2) $S_{11} < 0,\ S_{22} - S_{12}^T S_{11}^{-1} S_{12} < 0$
(3) $S_{22} < 0,\ S_{11} - S_{21}^T S_{22}^{-1} S_{21} < 0.$

Lemma 2.9 (Lyapunov stability theory, [17]) *Let function* $V(t) \geq 0$ *be a continuous function for* $\forall t \geq 0$ *and satisfying* $\dot{V}(t) \leq -\gamma V(t) + \varepsilon$*, where* $\gamma > 0$ *and* ε *are constants, then,*

$$V(t) \leq \left(V(0) - \frac{\varepsilon}{\gamma}\right) e^{-\gamma t} + \frac{\varepsilon}{\gamma}.$$

Lemma 2.10 (Differential mean value theorem, [18]) *In a region* $S \in \mathbb{R}^2$*,* $f(t, x(t), v(t))$ *is a continuously nonlinear differentiable function with* $(x(t_0), v(t_0)) \in S$*. For any* $(x(t), v(t)) \in S$*, the following equation holds:*

$$f(t, x(t), v(t)) - f(t, x(t_0), v(t_0)) = \frac{\partial F(t, x, v)}{\partial x}\Delta x + \frac{\partial F(t, x, v)}{\partial v}\Delta v$$

in which $F(t, x, v) = f(x(t_0) + \theta\Delta x, v(t_0) + \theta\Delta v)$, $\Delta x = (x(t) - x(t_0))$, $\Delta v = (v(t) - v(t_0))$*, and* $0 < \theta < 1$.

Lemma 2.11 ([19]) *For a symmetric matrix* $M \in \mathbb{R}^{N \times N}$ *and diagonal matrix* $D = \text{diag}\{d_1, d_2, \cdots, d_l, 0, \cdots, 0\}_{N \times N}$*, let*

$$M - D = \begin{bmatrix} B - \hat{D} & C \\ C^T & M_l \end{bmatrix}$$

where $d_l > 0$ $(l \leq N)$*;* M_l *is a nonsingular submatrix of* M*; namely,* M_l *is a matrix that removes the first* l *column-rows of* M*;* $\hat{D} = \text{diag}\{d_1, d_2, \cdots, d_l\}$*, and* B *and*

C are matrices with the corresponding dimensions. Thus, $M_l < 0$ *is equivalent to* $M - D < 0$ *if*

$$d_l > \lambda_{\max}(B - CM_l^{-1}C^T)$$

holds for all d_l $(l \le N)$.

Lemma 2.12 ([20]) *For a positive definite function* $V(t)$, *if it satisfies:*

a) $\dot{V}(t) \le -\alpha V(t)$, *for* $t_k \in \Psi(t)$
b) $\dot{V}(t) \le \rho \max\{V(t), V(t_{s+1})\}$, *for* $t_k \in \Phi(t)$

where α *and* ρ *are unknown positive constants,* $\Psi(t) \cap \Phi(t) = \varnothing$ *and* $\Psi(t) \cup \Phi(t) = [t_0, t)$, *and* t_{s+1} *is the next trigger instant with the latest successful trigger instant being* t_s *when* $t \in [t_k, t_{k+1})$ *and* $t_k \in \Phi(t)$. *Then, the following inequality holds:*

$$V(t) \le e^{[-\alpha(t-t_0-n(t_0,t)\sigma)+\rho n(t_0,t)\sigma]} V(t_0).$$

References

1. R. Agaev, P. Chebotarev, On the spectra of nonsymmetric Laplacian matrices. Linear Algebra Appl. **399**, 157–178 (2005)
2. G. Lafferriere, A. Williams, J. Caughman, J.J.P. Veerman, Decentralized control of vehicle formations. Syst. Control Lett. **54**(9), 899–910 (2005)
3. Z.K. Li, W. Ren, X.D. Liu, L.H. Xie, Distributed consensus of linear multi-agent systems with adaptive dynamic protocols. Automatica **49**(7), 1986–1995 (2013)
4. W. Ren, R. Beard, Consensus problems in networks of agents with switching toplogy and time-delays. IEEE Trans. Autom. Control **50**(5), 655–661 (2005)
5. J. Qin, H. Gao, W.X. Zheng, Exponential synchronization of complex networks of linear systems and nonlinear oscillators: a unified analysis. IEEE Trans. Neural Netw. Learn. **26**(3), 510–521 (2015)
6. W.W. Yu, G.R. Chen, M. Cao, J. Kurths, Second-order consensus for multiagent systems with directed topologies and nonlinear dynamics. IEEE Trans. Syst. Man Cybern. Part B Cybern. **40**(3), 881–891 (2010)
7. J.W. Kwon, D. Chwa, Adaptive bidirectional platoon control using a coupled sliding mode control method. IEEE Trans. Intell. Transp. Syst. **15**(5), 2040–2048 (2014)
8. G. Guo, W. Yue, Autonomous platoon control allowing range-limited sensors. IEEE Trans. Veh. Technol. **61**(7), 2901–2912 (2012)
9. W. Xiao, H. Ma, L. Zhou, H. Li, Adaptive fuzzy fixed-time formation-containment control for Euler-Lagrange systems. IEEE Trans. Fuzzy Syst. **31**(10), 3700–3709 (2023)
10. W.L. He, X.Y. Gao, W.M. Zhong, F. Qian, Secure impulsive synchronization control of multi-agent systems under deception attacks. Inf. Sci. **459**, 354–368 (2018)
11. X.G. Guo, J.L. Wang, F. Liao, Adaptive quantised H_∞ observer-based output feedback control for non-linear systems with input and output quantisation. IET Control Theory Appl. **11**(2), 263–272 (2017)
12. L.R. Huang, X.R. Mao, Delay-dependent exponential stability of neutral stochastic delay systems. IEEE Trans. Autom. Control **54**(1), 147–152 (2009)
13. C.L. Wang, C.Y. Wen, L. Guo, Adaptive consensus control for nonlinear multi-agent systems with unknown control directions and time-varying actuator faults. IEEE Trans. Autom. Control **66**(9), 4222–4229 (2021)

14. H. Deng, M. Krstić, Stochastic nonlinear stabilization, part I: A backstepping design. Syst. Control Lett. **32**(3), 143–150 (1997)
15. R. Horn, C. Johnson, *Matrix Analysis* (Cambridge University Press, New York, 1990)
16. S. Boyd, L. El Ghaoui, E. Feron, V. Balakrishnan, *Linear matrix inequalities in system and control theory*. SIAM Stud Appl Math (1994)
17. F.W. Lewis, S. Jagannathan, A. Yesildirak, *Neural Network Control of Robot Manipulators and Non-Linear Systems*. Taylor & Francis, New York (1998)
18. M.A. Shayman, *Geometry of the algebraic Riccati equation*, part I. SIAM J. Control. Optim. **21**(3), 375–394 (1983)
19. Q. Ma, J.W. Lu, Cluster synchronization for directed complex dynamical networks via pinning control. Neurocomputing **101**, 354–360 (2013)
20. W. Xu, D.W.C. Ho, J. Zhong, B. Chen, Event/self-triggered control for leader-following consensus over unreliable network with dos attacks. IEEE Trans. Neural Netw. Learn. Syst. **30**(10), 3137–3149 (2019)

Chapter 3
Observer-Based Adaptive Anti-disturbance Control of MASs by State Feedback

Abstract This chapter investigates the event-triggered problem of nonlinear *multi-agent systems* (MASs) satisfying Lipschitz condition under external *disturbance* in an undirected communication topology. To reduce the transmission of redundant data, an adaptive memory event-triggered mechanism (AMETM) is proposed, whose threshold can change dynamically with the state of the system and exclude *Zeno behavior*. Drawing upon this mechanism, two distinct anti-disturbance control strategies rooted in the mechanisms of *disturbance* observation and *disturbance* compensation mechanisms are proposed, respectively. In addition, the existence of a buffer can maintain a certain performance when the control channel is blocked.

3.1 Introduction

In recent years, multi-agent systems (MASs) have gained significant attention in the control field due to their extensive applications in unmanned aerial vehicles [1], intelligent cities [2, 3], power systems [4], etc. However, in practical MASs, due to the existence of *disturbance*, they may not be able to achieve the desired task or even become unstable [5]. Therefore, the anti-disturbance control problem for MASs is of great importance. To solve the *complete consensus* problem for MASs with external *disturbance*, studies have been carried out in [5–10]. It is discernible that countering the impact of disturbances contains two methodologies, prominently involving the construction of *disturbance* observers and the application of feedback and feedforward control mechanisms. In [5, 6, 9], the *disturbance* observers are designed using the system's state or output to estimate disturbances and mitigate their adverse effects. The feedforward compensation approach based on *disturbance* observers has gained significant attention due to its precision and fast response. However, the *disturbance* considered by these results has a specific model, simplifying the design of *disturbance* observers. A method of using the upper bound of *disturbance* for anti-disturbance control is proposed in [7, 10]. This method can easily make the derivative of the Lyapunov function negative, so that the MASs can reach *complete consensus*. Nevertheless, it often requires that the bound of the *disturbance* is known,

X.-G. Guo et al., *Anti-disturbance Intrusion- and Fault-Tolerant Control of Multi-agent Systems*, https://doi.org/10.1007/978-981-96-9614-7_3

so it is relatively conservative. Obviously, a more desirable alternative is to develop an anti-disturbance method that neither depends on a specific *disturbance* model nor requires the upper bound of the *disturbance*.

In addition, as the number of agent nodes and the amount of information transmission increase, energy consumption and network occupancy rise substantially. Consequently, optimizing network resource utilization has become a pressing issue. To reduce the need for continuous communication, an event-triggered mechanism (ETM) is proposed to replace the traditional time-triggered mechanism [11], and the *complete consensus* of MASs based on event-triggered control has been widely studied in recent years [12–16]. However, the fixed threshold ETM adopted in [12, 15, 16] will waste communication resources. To address this issue, [14] introduced an adaptive ETM with a threshold that varies according to the system state, but it is limited to linear systems and overlooks the existence of *disturbance*. Since nonlinear systems are more common in practical application and *disturbance* can not be neglected, it is of practical significance to study the anti-disturbance control for nonlinear MASs based on ETM.

Motivated by the above discussion, the *complete consensus* problem of nonlinear MASs satisfying Lipschitz condition based on event-triggered control under *disturbance* is studied in this chapter. Compared with the existing works, the main contributions are summarized as follows:

(1) Two anti-disturbance strategies based on ETM are proposed, which neither require a specific *disturbance* model nor a known upper bound. Moreover, the existence of a buffer can reduce the impact when the control channel is blocked.
(2) An adaptive memory event-triggered mechanism (AMETM) is introduced to save communication resources with the property that the threshold of the ETM can be changed dynamically and the *Zeno behavior* is excluded.

3.2 Problem Formulation

Considering a MAS consists of N agents, the dynamic of the ith agent is expressed as

$$\dot{x}_i(t) = Ax_i(t) + Bu_i(t) + Dw_i(t) + Ff(x_i(t), t) \tag{3.1}$$

where $x_i(t) \in \mathbb{R}^n$ and $u_i(t) \in \mathbb{R}^b$ are the state and input of the ith agent, respectively; $w_i(t) \in \mathbb{R}^d$ is the external *disturbance* which is bounded, i.e., $\|w_i(t)\| \leq \bar{w}_i$, where $\bar{w}_i$ is unknown; $f(x_i(t), t) = \mathrm{col}\{f_1(x_i(t), t), \cdots, f_n(x_i(t), t)\} \in \mathbb{R}^n$ is a nonlinear function vector. Moreover, A, B, D, and F are constant matrices with appropriate dimensions, and are assumed that $\mathrm{rank}(B) = \mathrm{rank}(B, D) = b$, which is equivalent to the *disturbance* matching condition.

Assumption 3.1 (A, B) is stabilizable.

Assumption 3.2 The derivative $\dot{w}_i(t)$ of external *disturbance* $w_i(t)$ is bounded, i.e., $\|\dot{w}_i(t)\| \leq \bar{w}_1$, where $\bar{w}_1$ is unknown.

Assumption 3.3 The nonlinear function $f(x_i(t), t)$ satisfies the Lipschitz condition and its upper bound is Δ, i.e., there exists a matrix Λ such that

$$\|f(x_2(t), t) - f(x_1(t), t)\| \leq \|\Lambda(x_2(t) - x_1(t))\| \tag{3.2}$$

where $x_1(t), x_2(t) \in \mathbb{R}^n$.

3.3 Adaptive Memory Event-Triggered Anti-disturbance Controller

3.3.1 Disturbance Observer Design

To reduce the influence of external *disturbance*, the *disturbance* observer is constructed as below

$$\begin{cases} \dot{z}_i(t) = -\Gamma D z_i(t) - \Gamma D \Gamma x_i(t) - \Gamma[A x_i(t) + B u_i(t) + F f(x_i(t), t)] \\ \hat{w}_i(t) = z_i(t) + \Gamma x_i(t) \end{cases} \tag{3.3}$$

where $\Gamma \in \mathbb{R}^{d \times n}$ is a matrix need to be designed, $\hat{w}_i(t)$ is the estimated value of *disturbance*, and $z_i(t) \in \mathbb{R}^d$ is the internal state of the *disturbance* observer.

Theorem 3.1 *For given positive scalars β and ρ, if there exists a symmetric positive definite matrix Q and a matrix Q_1 satisfy the following condition*

$$\begin{bmatrix} -D^T Q_1^T - Q_1 D + \rho Q & Q \\ * & -\beta \end{bmatrix} < 0 \tag{3.4}$$

then the disturbance observer error $e_{w_i}(t) = w_i(t) - \hat{w}_i(t)$ is uniformly ultimately bounded with Γ being designed as $\Gamma = Q^{-1} Q_1$.

Proof Based on the definition of $e_{w_i}(t)$, we can obtain

$$\begin{aligned} \dot{e}_{w_i}(t) &= \dot{w}_i(t) - \dot{\hat{w}}_i(t) = \dot{w}_i(t) - \dot{z}_i(t) - \Gamma \dot{x}_i(t) \\ &= \dot{w}_i(t) + \Gamma D z_i(t) + \Gamma D \Gamma x_i(t) - \Gamma D w_i(t) \\ &= -\Gamma D e_{w_i}(t) + \dot{w}_i(t). \end{aligned} \tag{3.5}$$

Consider the following Lyapunov function

$$W_i(t) = e_{w_i}^T(t)Qe_{w_i}(t). \tag{3.6}$$

Based on (3.5), the differential of $W_i(t)$ to time is given by

$$\begin{aligned}\dot{W}_i(t) =& - e_{w_i}^T(t)(D^T\Gamma^T Q + Q\Gamma D)e_{w_i}(t) + 2e_{w_i}^T(t)Q\dot{w}_i(t)\\ \leq& e_{w_i}^T(t)(\frac{1}{\beta}QQ^T - D^T\Gamma^T Q - Q\Gamma D)e_{w_i}(t) + \beta\dot{w}_i^T(t)\dot{w}_i(t).\end{aligned} \tag{3.7}$$

From (3.4), it is clear that

$$\begin{aligned}\dot{W}_i(t) \leq& e_{w_i}^T(t)(-\rho Q)e_{w_i}(t) + \beta\dot{w}_i^T(t)\dot{w}_i(t)\\ \leq& -\rho W_i(t) + \beta\bar{w}_1^2.\end{aligned} \tag{3.8}$$

Then, according to Assumption 3.2 and Lemma 2.9, it is easy to obtain that $e_{w_i}(t)$ is uniformly ultimately bounded. This completes the proof. □

Remark 3.1 It can be inferred from (3.8) that the boundary of $e_{w_i}(t)$ satisfies

$$\left\|e_{w_i}\right\|^2 \leq \frac{1}{\lambda_{\min}(Q)}[W_i(0)e^{-\rho t} + \frac{\beta\bar{w}_1^2}{\rho}(1 - e^{-\rho t})]$$

where $W_i(0)$ is the initial value of Lyapunov function $W_i(t)$. Furthermore, we can see that $\left\|e_{w_i}\right\|^2$ finally converges to the bound of $\frac{\beta\bar{w}_1^2}{\rho\lambda_{\min}(Q)}$ as time goes to $+\infty$. It is obvious that a small β and a large ρ can make the bound smaller. Therefore, we should try to reduce the value of β and increase the value of ρ on the basis of satisfying linear matrix inequality (3.4).

3.3.2 ETM-Based Controller Design

On the other hand, the limited communication bandwidth is also an important factor affecting multi-agent network communication. In order to save resources, based on *disturbance* observer (3.3), an adaptive memory event-triggered anti-disturbance controller is developed as

$$u_i(t) = -K\sum_{p=1}^{m}\varepsilon_p\xi_i(t_{k-p+1}^i) - B^\dagger D\hat{w}_i(t), t \in [t_k^i, t_{k+1}^i) \tag{3.9}$$

where $\xi_i(t) = \sum_{j\in\mathscr{N}_i}(x_i(t) - x_j(t))$, K is the gain matrix to be designed, and $B^\dagger$ is the solution of $(I_N - BB^\dagger)D = 0$. The index $p = \{1, 2, \cdots, m\}$ represents the number of data stored in buffer, and ε_p satisfies $0 \leq \varepsilon_p \leq 1$ and $\sum_{p=1}^{m} \varepsilon_p = 1$. In addition, the event-triggering time sequence set $\{t_k^i\}$ for the ith agent is given by

$$t_{k+1}^i = \inf\{t > t_k^i | g_i(t) > 0\} \tag{3.10}$$

where

$$g_i(t) = \left(\sum_{p=1}^{m} \varepsilon_p e_{i,t_{k-p+1}}(t)\right)^T \Theta \left(\sum_{p=1}^{m} \varepsilon_p e_{i,t_{k-p+1}}(t)\right) - c_i(t)\xi_i^T(t)\Theta\xi_i(t)$$

with

$$e_{i,t_{k-p+1}}(t) = \xi_i(t_{k-p+1}^i) - \xi_i(t)$$

$$\Theta = P^{-1}BB^TP^{-1}$$

and P is the positive definite matrix to be designed later. The triggering parameter $c_i(t) \in (0, \bar{c}] \subseteq (0, 1)$ is determined by the following adaptive law

$$\dot{c}_i(t) = \begin{cases} 0, & c_i(t) \geq \bar{c} \\ \chi_i \xi_i^T(t)PBB^TP\xi_i(t), & c_i(t) < \bar{c} \end{cases} \tag{3.11}$$

where $\chi_i > 0$, and $\bar{c}$ is the upper bound of $c_i(t)$.

Remark 3.2 In the controller (3.9), an AMETM is employed to optimize communication resource usage. The presence of the buffer achieves better performance, particularly in scenarios where the transmission channel is disrupted, such as during DoS attacks, and minimizes the occurrence of erroneous triggers caused by sudden changes in state. Moreover, unlike the fixed-threshold event-triggered mechanisms designed in [12, 15, 16], this method allows for dynamic adjustment of the threshold parameter, as $c_i(t)$ depends on the system state in (3.13), which can save resources more effectively in turn.

3.3.3 Complete Consensus Analysis

Recalling the following equation:

$$e_{i,t_{k-p+1}}(t) = \xi_i(t_{k-p+1}^i) - \xi_i(t). \tag{3.12}$$

Then, substituting (3.9) into (3.1), one has

$$\begin{aligned}\dot{x}_i(t) =&Ax_i(t) - BK\xi_i(t) - BK\sum_{p=1}^{m}\varepsilon_p e_{i,t_{k-p+1}}(t) + Dw_i(t)\\&- BB^{\dagger}D\hat{w}_i(t) + Ff(x_i(t), t).\end{aligned} \tag{3.13}$$

Let $x(t) = \text{col}\{x_1(t), \cdots, x_N(t)\}$, the compact form of (3.13) can be written as

$$\begin{aligned}\dot{x}(t) =&(I_N \otimes A)x(t) - (I_N \otimes BK)\xi(t) + (I_N \otimes D)w(t) - (I_N \otimes BB^{\dagger}D)\hat{w}(t)\\&- (I_N \otimes BK)\sum_{p=1}^{m}\varepsilon_p e_{t_{k-p+1}}(t) + (I_N \otimes F)f(x(t), t)\end{aligned} \tag{3.14}$$

where

$$\begin{aligned}&\xi(t) = \text{col}\{\xi_1(t), \cdots, \xi_N(t)\}\\&\hat{w}(t) = \text{col}\{\hat{w}_1(t), \cdots, \hat{w}_N(t)\}\\&w(t) = \text{col}\{w_1(t), \cdots, w_N(t)\}\\&f(x(t), t) = \text{col}\{f(x_1(t), t), \cdots, f(x_N(t), t)\}\\&\sum_{p=1}^{m}\varepsilon_p e_{t_{k-p+1}}(t) = \text{col}\left\{\sum_{p=1}^{m}\varepsilon_p e_{1,t_{k-p+1}}(t), \cdots, \sum_{p=1}^{m}\varepsilon_p e_{N,t_{k-p+1}}(t)\right\}.\end{aligned}$$

Define

$$\bar{x}(t) = \frac{1}{N}\sum_{i=1}^{N}x_i(t) \tag{3.15}$$

$$\delta_i(t) = x_i(t) - \bar{x}(t) \tag{3.16}$$

it is clear that

$$\delta(t) = (M \otimes I_n)x(t) \tag{3.17}$$

where

$$\delta(t) = \text{col}\{\delta_1(t), \cdots, \delta_N(t)\} \tag{3.18}$$

$$M = I_N - \frac{1}{N}1_N 1_N^T. \tag{3.19}$$

Besides, it is easy to obtain that

$$\mathscr{L}M = M\mathscr{L} = \mathscr{L} \tag{3.20}$$

$$\xi(t) = (\mathscr{L} \otimes I_n)\delta(t) \tag{3.21}$$

with $\xi(t) = \mathrm{col}\{\xi_1(t), \cdots, \xi_N(t)\}$. Then, it follows from the above definitions and (3.14) that

$$\begin{aligned}\dot{\delta}(t) =&(I_N \otimes A - \mathscr{L} \otimes BK)\delta(t) - (M \otimes BB^{\dagger}D)\hat{w}(t) + (M \otimes D)w(t)\\&- (M \otimes BK)\sum_{p=1}^{m} \varepsilon_p e_{t_{k-p+1}}(t) + (I_N \otimes F)\tilde{f}(x(t), t)\end{aligned} \tag{3.22}$$

where

$$\tilde{f}(x(t), t) = \mathrm{col}\left\{(f(x_1(t), t) - \frac{1}{N}\sum_{i=1}^{N} f(x_i(t), t)), \cdots, f(x_N(t), t) - \frac{1}{N}\sum_{i=1}^{N} f(x_i(t), t))\right\}.$$

Next, the *complete consensus* result is presented in the following theorem.

Theorem 3.2 *Assume that Assumptions 3.1–3.3 hold. Consider the MASs (3.1) with event-triggered anti-disturbance controller (3.9), the complete consensus error vector* δ *is uniformly bounded if for given positive scalars* β_1, β_2, *and* α, *there exists a symmetric positive definite matrix* P *such that*

$$\begin{bmatrix} \Pi_1 & P\Lambda^T \\ * & -\beta_1 \end{bmatrix} < 0 \tag{3.23}$$

where

$$\begin{aligned}\Pi_1 =&PA^T + AP + \beta_1 FF^T - \alpha_1 BB^T + \frac{1}{\beta_2}\lambda_N DD^T + \alpha P\\ \alpha_1 =&(1 - \bar{c})\lambda_2 > 0\end{aligned}$$

and the controller gain K *in (3.9) is designed as* $K = B^T P^{-1}$.

Proof Construct the Lyapunov function as

$$V(t) = \delta^T(t)(\mathscr{L} \otimes P^{-1})\delta(t). \tag{3.24}$$

From (3.22), the derivative of $V(t)$ is

$$\begin{aligned}\dot{V}(t) = &\delta^T(t)[\mathscr{L} \otimes (A^T P^{-1} + P^{-1}A) - \mathscr{L}^2 \otimes (K^T B^T P^{-1}\\&+ P^{-1}BK)]\delta(t) - 2\delta^T(t)(\mathscr{L} \otimes P^{-1}BB^{\dagger}D)\hat{w}(t)\\&+ 2\delta^T(t)(\mathscr{L} \otimes P^{-1}D)w(t) + 2\delta^T(t)(\mathscr{L} \otimes P^{-1}F)\tilde{f}(x(t), t)\\&- 2\delta^T(t)(\mathscr{L} \otimes P^{-1}BK)\sum_{p=1}^{m} \varepsilon_p e_{t_{k-p+1}}(t).\end{aligned} \tag{3.25}$$

Note that $D = BB^{\dagger}D$, according to Lemma 2.7 one can deduce that

$$\begin{aligned}&2\delta^T(t)(\mathscr{L} \otimes P^{-1}D)w(t) - 2\delta^T(t)(\mathscr{L} \otimes P^{-1}BB^{\dagger}D)\hat{w}(t)\\ \le&\frac{1}{\beta_2}\delta^T(t)(\mathscr{L}^2 \otimes P^{-1}DD^TP^{-1})\delta(t) + \beta_2 e_w^T(t)e_w(t).\end{aligned} \tag{3.26}$$

Furthermore, since

$$2\delta^T(t)(\mathscr{L} \otimes P^{-1}F)[1_N \otimes (f(\bar{x}(t), t) - \frac{1}{N}\sum_{i=1}^{N} f(x_i(t), t))] = 0 \tag{3.27}$$

holds, it is not difficult to obtain that

$$\begin{aligned}&2\delta^T(t)(\mathscr{L} \otimes P^{-1}F)\tilde{f}(x(t), t)\\ \le&\beta_1\delta^T(t)(\mathscr{L} \otimes P^{-1}FF^TP^{-1})\delta(t) + \frac{1}{\beta_1}\delta^T(t)(\mathscr{L} \otimes \Lambda^T\Lambda)\delta(t).\end{aligned} \tag{3.28}$$

Based on the fact that $K = B^TP^{-1}$ and $\xi(t) = (\mathscr{L} \otimes I_n)\delta(t)$, one has

$$\begin{aligned}&-2\delta^T(t)(\mathscr{L} \otimes P^{-1}BK)\sum_{p=1}^{m}\varepsilon_p e_{t_{k-p+1}}(t)\\ =&-2\sum_{i=1}^{N}\xi_i^T(t)P^{-1}BB^TP^{-1}\sum_{p=1}^{m}\varepsilon_p e_{i,t_{k-p+1}}(t)\\ \le&\left(\sum_{p=1}^{m}\varepsilon_p e_{t_{k-p+1}}(t)\right)^T(I_N \otimes P^{-1}BB^TP^{-1})\left(\sum_{p=1}^{m}\varepsilon_p e_{t_{k-p+1}}(t)\right)\\ &+\xi^T(t)(I_N \otimes P^{-1}BB^TP^{-1})\xi(t)\end{aligned} \tag{3.29}$$

and it follows from the event-triggered condition (3.10) that

$$\begin{aligned}&\left(\sum_{p=1}^{m}\varepsilon_p e_{t_{k-p+1}}(t)\right)^T(I_N \otimes \Theta)\left(\sum_{p=1}^{m}\varepsilon_p e_{t_{k-p+1}}(t)\right)\\ \le&c(t)\xi^T(t)(I_N \otimes \Theta)\xi(t) = c(t)\delta^T(t)(\mathscr{L}^2 \otimes \Theta)\delta(t)\end{aligned} \tag{3.30}$$

where $c(t) = \text{diag}\{c_1(t), \cdots, c_N(t)\}$.

Next, substituting (3.26)–(3.29) into (3.25) yields

$$
\begin{aligned}
\dot{V}(t) \leq & \delta^T(t)[\mathscr{L} \otimes (A^T P^{-1} + P^{-1}A + \beta_1 P^{-1} F F^T P^{-1} + \frac{1}{\beta_1}\Lambda^T \Lambda)]\delta(t) \\
& - (1-\bar{c})\delta^T(t)(\mathscr{L}^2 \otimes P^{-1} B B^T P^{-1})\delta(t) \\
& + \frac{1}{\beta_2}\delta^T(t)(\mathscr{L}^2 \otimes P^{-1} D D^T P^{-1})\delta(t) + \beta_2 e_w^T(t) e_w(t) \\
\leq & \delta^T(t)[\mathscr{L} \otimes (A^T P^{-1} + P^{-1}A + \beta_1 P^{-1} F F^T P^{-1} + \beta_2 e_w^T(t) e_w(t) \\
& + \frac{1}{\beta_1}\Lambda^T \Lambda - (1-\bar{c})\lambda_2 P^{-1} B B^T P^{-1} + \frac{1}{\beta_2}\lambda_N P^{-1} D D^T P^{-1})]. \quad (3.31)
\end{aligned}
$$

Using Lemma 2.8 for (3.23), then pre- and post-multiplying with P^{-1} and substituting it into (3.31), one gets

$$
\begin{aligned}
\dot{V}(t) & \leq \delta^T(t)[\mathscr{L} \otimes (-\alpha P^{-1})]\delta(t) + \beta_2 e_w^T(t) e_w(t) \\
& \leq -\alpha V(t) + \beta_2 e_w^T(t) e_w(t). \quad (3.32)
\end{aligned}
$$

Then, using Lemma 2.9 and Theorem 3.1, one can conclude that *complete consensus* error vector $\delta(t)$ is uniformly ultimately bounded. This completes the proof. □

From Theorems 3.1 and 3.2, when the *disturbance* is a constant or changes slowly, good performance can be achieved by using the controller (3.9). However, when the *disturbance* changes rapidly, the decrease of the performance of the *disturbance* observer (3.3) makes the system performance worse. To solve this problem, we give another anti-disturbance method. In addition, in order to facilitate the discussion, we assume that the *disturbance* is one-dimensional ($d = 1$).

Similar to (3.9), the controller is designed as

$$
u_i(t) = -K_1 \sum_{p=1}^{m} \varepsilon_p \xi_i(t_{k-p+1}^i) - B^\dagger D z_i(t), t \in [t_k^i, t_{k+1}^i) \quad (3.33)
$$

where K_1 is the controller gain, $z_i(t)$ is the *disturbance* compensation term and is given by

$$
z_i(t) = \frac{D^T P_1^{-1}(x_i - x_j)\tilde{w}_i^2(t)}{\|D^T P_1^{-1}(x_i - x_j)\| \, \|\tilde{w}_i(t)\| + \vartheta_i} \quad (3.34)
$$

where P_1 is a symmetric positive definite matrix to be designed, x_j is the state of a neighbor of the ith agent, ϑ_i is a positive number, and $\tilde{w}_i(t)$ satisfies

$$
\dot{\tilde{w}}_i(t) = -\eta_i \vartheta_i \tilde{w}_i(t) + \eta_i \, \|D^T P_1^{-1}(x_i - x_j)\| \quad (3.35)
$$

where η_i is a positive scalar. Besides, the new event-trigger mechanism is similar to (3.10)–(3.11), and it only needs to change P to P_1. Then, substituting (3.33) into (3.1) and according to the definition of $\delta_i(t)$ in (3.16), we have

$$\begin{aligned}\dot{\delta}(t) =&(I_N \otimes A - \mathscr{L} \otimes BK)\delta(t) - (M \otimes BB^{\dagger}D)z(t) + (M \otimes D)w(t) \\ &- (M \otimes BK)\sum_{p=1}^{m} \varepsilon_p e_{t_{k-p+1}}(t) + (I_N \otimes F)\tilde{f}(x(t), t) \end{aligned} \tag{3.36}$$

where $z(t) = \text{col}\{z_1(t), \cdots, z_N(t)\}$.

Next, based on the proof process of Theorem 3.2, we can easily deduce the following corollary.

Corollary 3.1 *Suppose that Assumptions 3.1–3.3 hold, under the action of the controller (3.33), if there exists a symmetric positive definite matrix P_1 and a scalar $\beta_3 > 0$ such that*

$$\begin{bmatrix} \Pi_2 & P_1\Lambda^T \\ * & -\beta_3 \end{bmatrix} < 0 \tag{3.37}$$

where

$$\Pi_2 = P_1A^T + AP_1 + \beta_3FF^T - \alpha_1BB^T$$

$$\alpha_1 = (1 - \bar{c})\lambda_2 > 0$$

and the controller gain K_1 in (3.33) is given by $K_1 = B^TP_1^{-1}$. Then, the complete consensus error vector δ is uniformly ultimately bounded.

Proof Define $e_{\tilde{w}_i}(t) = \bar{w}_i - \tilde{w}_i(t)$, and choose the Lyapunov function as

$$V_1(t) = \delta^T(t)(\mathscr{L} \otimes P_1^{-1})\delta(t) + \sum_{i=1}^{N}\sum_{j\in\mathscr{N}_i}\frac{1}{\eta_i}a_{ij}e^2_{\tilde{w}_i}(t). \tag{3.38}$$

The proof process is similar to that of Theorem 3.2 and hence omitted for brevity. □

3.3.4 Zeno Behavior Analysis

Theorem 3.3 *In consideration of the MASs (3.1) with AMETM (3.10), Zeno behavior does not occur due to the minimum time interval between any two event-triggered execution intervals is strictly positive, that is*

$$t^i_{k+1} - t^i_k \ge \frac{\Omega_i^2}{\|B^TP^{-1}\|[2\|A\|\Xi_i^2 + 2\Xi_i\Phi_i(t^i_k)]} > 0 \tag{3.39}$$

where

$$\Phi_i(t_k^i) = \|A\| \left\| \sum_{p=1}^{m} \varepsilon_p \xi_i(t_{k-p+1}^i) \right\| + \|B\| \left\| \sum_{j \in \mathscr{N}_i} (u_i(t) - u_j(t)) \right\|$$
$$+ 2\mathscr{N}_i^*(\|D\| \, \|\bar{w}\| + \|F\| \, \Delta) \tag{3.40}$$

$$\Omega_i^2 = \left\| B^T P^{-1} \sum_{p=1}^{m} \varepsilon_p e_{i,t_{k-p+1}}(t) \right\|^2, \quad \Xi_i^2 = \left\| \sum_{p=1}^{m} \varepsilon_p e_{i,t_{k-p+1}}(t) \right\|^2 \tag{3.41}$$

and $\mathscr{N}_i^*$ is the number of agents adjacent to the i-th agent.

Proof Differentiating Ξ_i^2 with respect to $t \in [t_k^i, t_{k+1}^i)$, one has

$$\begin{aligned}
\frac{d}{dt} \Xi_i^2 \leq & 2 \left\| \sum_{p=1}^{m} \varepsilon_p e_{i,t_{k-p+1}}(t) \right\| \left\| \sum_{j \in \mathscr{N}_i} (\dot{x}_i(t) - \dot{x}_j(t)) \right\| \\
\leq & 2 \left\| \sum_{p=1}^{m} \varepsilon_p e_{i,t_{k-p+1}}(t) \right\| \left(\|A\| \left\| \sum_{j \in \mathscr{N}_i} (x_i(t) - x_j(t)) \right\| + \left\| B \sum_{j \in \mathscr{N}_i} (u_i(t) - u_j(t)) \right\| \right. \\
& \left. + \left\| D \sum_{j \in \mathscr{N}_i} (w_i(t) - w_j(t)) \right\| + \left\| F \sum_{j \in \mathscr{N}_i} (f(x_i(t), t) - f(x_j(t), t)) \right\| \right) \\
\leq & 2\|A\| \left\| \sum_{p=1}^{m} \varepsilon_p e_{i,t_{k-p+1}}(t) \right\|^2 + 2 \left\| \sum_{p=1}^{m} \varepsilon_p e_{i,t_{k-p+1}}(t) \right\| \left(\|A\| \left\| \sum_{p=1}^{m} \varepsilon_p \xi_i(t_{k-p+1}^i) \right\| \right. \\
& \left. + \|B\| \left\| \sum_{j \in \mathscr{N}_i} (u_i(t) - u_j(t)) \right\| + 2\mathscr{N}_i^* (\|D\| \, \|\bar{w}\| + \|F\| \, \Delta) \right).
\end{aligned} \tag{3.42}$$

By using Ω^2 in (3.39) and based on (3.42), the following inequality holds:

$$\frac{d}{dt} \left(\Omega_i^2 \right) \leq \|B^T P^{-1}\|^2 [2\|A\| \, \Xi_i^2 + 2\Xi_i \Phi_i(t_k^i)]. \tag{3.43}$$

At the event-triggering instants $t_1^i, t_2^i, \cdots, t_k^i, \cdots$, by employing event-triggering condition (3.10), it holds that

$$\Omega_i^2 > c_i(t) \left\| B^T P^{-1} \xi_i(t_k^i) \right\|^2 \geq 0. \tag{3.44}$$

Together with (3.43) and (3.44), we can acquire

$$t_{k+1}^i - t_k^i \geq \epsilon_i \geq \frac{\Omega_i^2}{\|B^T P^{-1}\|^2[2\|A\|\Xi_i^2 + 2\Xi_i \Phi_i(t_k^i)]}. \tag{3.45}$$

Then, one can conclude that $\Xi_i > 0$ and $\Phi_i(t_k^i) > 0$, which imply that $\epsilon_i > 0$, thus ϵ_i is strictly positive. This completes the proof. □

3.4 Simulation Examples

In this section, we use the communication topology shown in Fig. 3.1 to verify the effectiveness of the proposed controller. The system matrices in (3.1) are selected as follows

$$A = \begin{bmatrix} -2.9 & 0.3 & 0.4 & 1.2 \\ -0.1 & -0.2 & 0.6 & 1.5 \\ 1.2 & 2.1 & -2.8 & 3.4 \\ 1 & -2 & -2.5 & -2.5 \end{bmatrix}, \quad B = \begin{bmatrix} 0 & 0 \\ 0 & 0 \\ -1 & 0.5 \\ -0.1 & 0.2 \end{bmatrix}$$
$$D = \begin{bmatrix} 0 & 0 & -1 & -0.1 \end{bmatrix}^T, \quad F = 0.01 I_4$$

and the nonlinear function $f(x_i(t), t)$ is selected as

$$f(x_i(t), t) = [0, 0, 0, 0.01\sin(x_{i1}(t))]^T.$$

Then, based on Assumption 3.3, one gets $\Lambda = \text{diag}\{0.01, \cdots, 0.01\}$. Moreover, the parameters for controller (3.9) and (3.33) are chosen as $m = 3$, $\varepsilon_1 = 0.7$, $\varepsilon_2 = 0.2$, $\varepsilon_3 = 0.1$. Triggering parameters $\bar{c} = 0.6$ and $c_i(0) = 0.1$.

Letting $\beta = 0.05$, $\rho = 0.1$, $\beta_1 = \beta_2 = 1$, $\alpha = 3.2$, and $\beta_3 = 0.01$, and solving LMIs (3.4), (3.23), and (3.37), one obtains

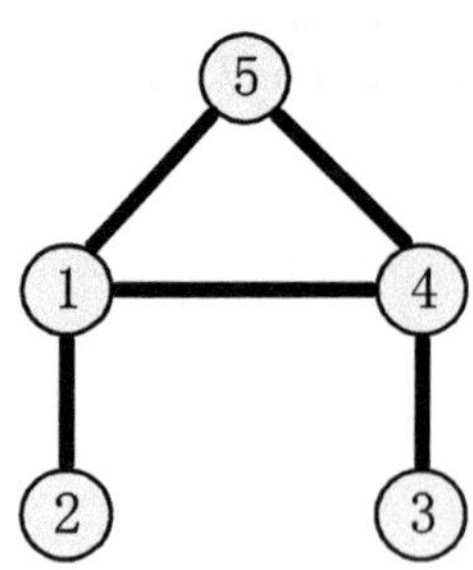

Fig. 3.1 Communication topology

$$\Gamma = \begin{bmatrix} 0 & 0 & -21.4632 & 181.4758 \end{bmatrix}$$

$$P = \begin{bmatrix} 16.1402 & -0.5310 & 19.8974 & 2.5482 \\ -0.5310 & 21.0103 & 4.3546 & -22.0028 \\ 19.8974 & 4.3546 & 42.6132 & 4.3881 \\ 2.5482 & -22.0028 & 4.3881 & 43.1258 \end{bmatrix}$$

$$K = \begin{bmatrix} 0.0792 & 0.0327 & -0.0655 & 0.0163 \\ -0.0366 & -0.0065 & 0.0294 & 0.0005 \end{bmatrix}$$

$$P_1 = \begin{bmatrix} 0.2564 & 0.0120 & 0.1428 & 0.0837 \\ 0.0120 & 0.6962 & 0.1662 & -0.3476 \\ 0.1428 & 0.1662 & 0.3437 & 0.0139 \\ 0.0837 & -0.3476 & 0.0139 & 0.5309 \end{bmatrix}$$

$$K_1 = \begin{bmatrix} 2.3174 & 1.2012 & -4.4671 & 0.3497 \\ -1.2704 & -0.3377 & 2.1336 & 0.3001 \end{bmatrix}.$$

Without loss of generality, consider that two kinds of *disturbances* act on agent 1, which are denoted by $w_1^1(t) = 6$, $t \geq 6s$ and $w_1^2(t) = \sin(2t)$, $t \geq 0$, respectively, and assume that there is no *disturbance* on other agent. Figure 3.2 shows the *disturbance* signals and their estimations, from which we can see that the designed *disturbance* observer (3.3) can accurately estimate the *disturbance* when $\dot{w}_i(t) = 0$, but if $\dot{w}_i(t) \neq 0$, the performance of the *disturbance* observer will degrade.

Define the complete consensus error of the whole system as follows:

$$J(t) = \frac{1}{N}\sqrt{\sum_{i=1}^{N} \|x_i(t) - \bar{x}(t)\|^2}.$$

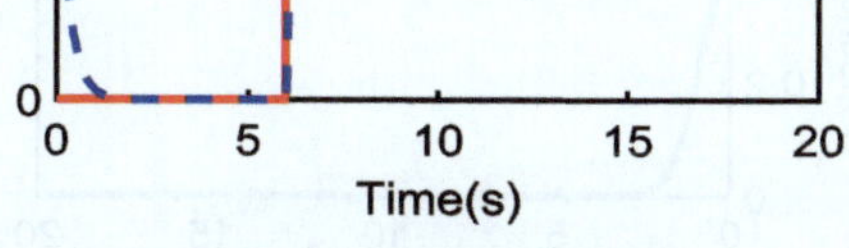

(a) $w_1^1(t)$ and $\hat{w}_1^1(t)$

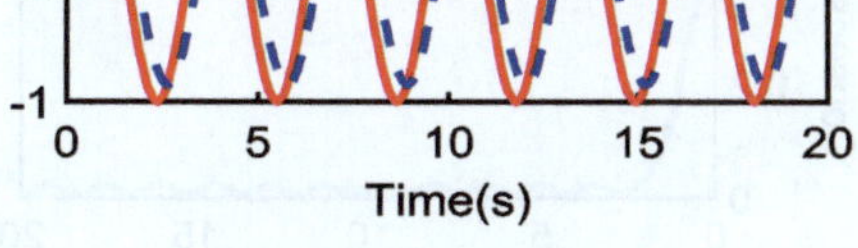

(b) $w_1^2(t)$ and $\hat{w}_1^2(t)$

Fig. 3.2 The disturbance signals and their estimations

Figure 3.3a exhibits the *complete consensus* error by using the controller (3.9) under $w_1^1(t)$, which shows that the system can achieve *complete consensus* even if the *disturbance* occurs. The adaptive parameters $c_i(t)$ are given in Fig. 3.3b, and Table 3.1 shows the number of triggers under two different controllers. From Table 3.1, one can see that the communication burden can be reduced. The comparison of *complete consensus* error $J(t)$ by using the controller (3.9) and the controller (3.33) under $w_1^2(t) = \sin 2t, t \geq 0$, is shown in Fig. 3.4. It can be seen that, due to the degradation of the performance of the *disturbance* observer (3.3) as shown in Fig. 3.2b, the *complete consensus* performance of the system becomes worse by

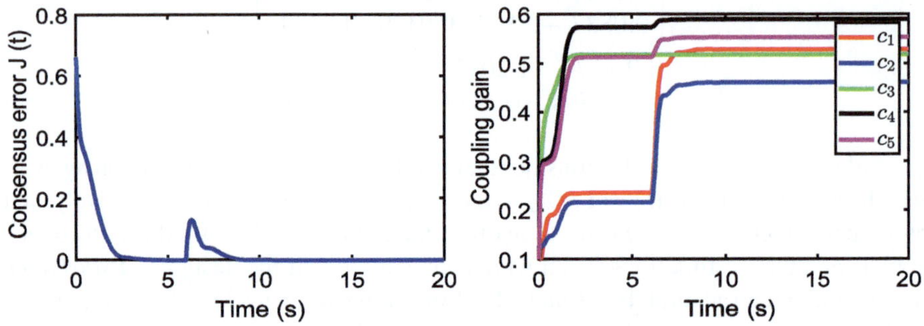

(a) Complete consensus error $J(t)$ (b) Adaptive parameters $c_i(t)$

Fig. 3.3 Simulation results by using the controller (3.9) in Theorem 2 under $w_1^1(t) = 6, t \geq 6s$

Table 3.1 Trigger numbers of each agent under different controllers

Controller	Agent 1	Agent 2	Agent 3	Agent 4	Agent 5
Controller (3.9)	91	96	66	76	74
Time-triggered Controller	1000	1000	1000	1000	1000

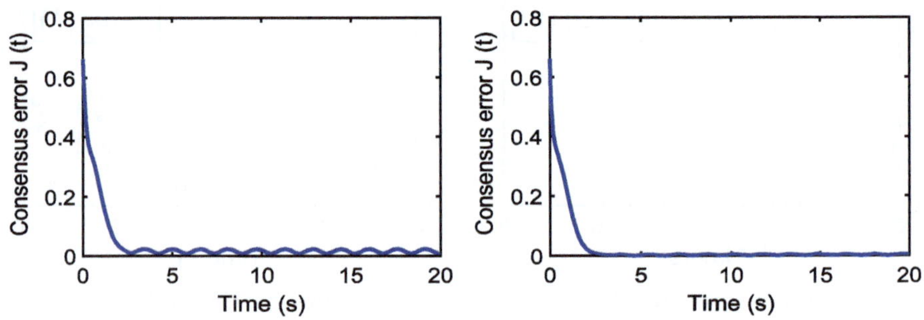

Fig. 3.4 Comparison of complete consensus error $J(t)$ by using the controller (3.9) in Theorem 2 (left) and using the controller (3.33) in Corollary 1 (right) under $w_1^2(t) = \sin 2t, t \geq 0s$

using the controller (3.9), while a smaller *complete consensus* error can be obtained by using the controller (3.33) based on *disturbance* compensation $z_i(t)$ in (3.34). Thus, the advantages of the proposed method are verified.

3.5 Conclusion

This chapter addresses the challenge of nonlinear MASs under external *disturbances* by introducing two types of anti-disturbance controllers. These controllers are designed based on *disturbance* observers and *disturbance* compensation methods. Additionally, an AMETM is proposed to alleviate communication load while ensuring the exclusion of *Zeno behavior*. Finally, simulation examples are provided to demonstrate the effectiveness of the proposed control strategy.

References

1. J.J. Cui, Y.W. Liu, A. Nallanathan, Multi-agent reinforcement learning-based resource allocation for UAV networks. IEEE Trans. Wirel. Commun. **19**(2), 729–743 (2020)
2. A. Colombo, G.R. de Campos, F.D. Rossa, Control of a city road network: Distributed exact verification of traffic safety. IEEE Trans. Autom. Control **62**(10), 4933–4948 (2017)
3. D.Y. Zhang, D.W. Ding, Z.M. Zhao, X.G. Guo. Adaptive memory event-triggered control for multi-agent systems under external disturbance, in *2021 11th International Conference on Information Science and Technology*, pp. 195–200 (2021)
4. S.R. Islam, D. Sutanto, K.M. Muttaqi, A distributed multi-agent based emergency control approach following catastrophic disturbances in interconnected power systems. IEEE Trans. Power Syst. **31**(4), 2764–2775 (2016)
5. Y. Yuan, Z. Wang, L. Guo, Event-triggered strategy design for discrete-time nonlinear quadratic games with disturbance compensations: the noncooperative case. IEEE Trans. Syst. Man Cybern. Syst. **48**(11), 1885–1896 (2018)
6. Z.T. Ding, Consensus disturbance rejection with disturbance observers. IEEE Trans. Ind. Electron. **62**(9), 5829–5837 (2015)
7. G. Li, Y. Wu, X. Liu, Adaptive fixed-time consensus tracking control method for second-order multi-agent systems with disturbances. J. Franklin Inst. **357**(3), 1516–1531 (2020)
8. Q. Liu, Z. Wang, X. He, D.H. Zhou, Event-based H_∞ consensus control of multi-agent systems with relative output feedback: the finite-horizon case. IEEE Trans. Autom. Control **60**(9), 2553–2558 (2015)
9. P.J. Wang, G.H. Wen, X.H. Yu, W.W. Yu, Y.Z. Lv, Consensus disturbance rejection for linear multiagent systems with directed switching communication topologies. IEEE Trans. Control Netw. Syst. **7**(1), 254–265 (2020)
10. S. Yazdani, H. Su, Adaptive leader-follower flocking for uncertain Lagrange systems with input saturation and external disturbances, in *2018 37th Chinese Control Conference (CCC)*, pp. 7210–7214 (2018)
11. X.G. Guo, X. Fan, C.K. Ahn, Adaptive event-triggered fault detection for interval type-2 T-S fuzzy systems with sensor saturation. IEEE Trans. Fuzzy Syst. **29**(8), 2310–2321 (2021)
12. C. Deng, M.J. Er, G.H. Yang, N. Wang, Event-triggered consensus of linear multiagent systems with time-varying communication delays. IEEE Trans. Cybern. **50**(7), 2916–2925 (2020)
13. L. Ding, Q.L. Han, X.H. Ge, X.M. Zhang, An overview of recent advances in event-triggered consensus of multiagent systems. IEEE Trans. Cybern. **48**(4), 1110–1123 (2018)

14. C. Du, X. Liu, W. Ren, P. Lu, H. Liu, Finite-time consensus for linear multiagent systems via event-triggered strategy without continuous communication. IEEE Trans. Control Netw. Syst. **7**(1), 19–29 (2020)
15. W. Hu, L. Liu, G. Feng, Consensus of linear multi-agent systems by distributed event-triggered strategy. IEEE Trans. Cybern. **46**(1), 148–157 (2016)
16. Y. Luo, X. Xiao, J. Cao, A. Li, Event-triggered guaranteed cost consensus for uncertain nonlinear multi-agent systems with time delay. Neurocomputing **394**, 13–26 (2020)

Chapter 4
Anti-disturbance Control of MASs Under DoS Attacks by Output Feedback

Abstract This chapter investigates the event-triggered anti-disturbance secure *complete consensus* problem for nonlinear multi-agent systems (MASs) subjected to denial-of-service (DoS) attacks over an *undirected graph.* A novel adaptive memory state observer-based anti-disturbance control scheme is presented to improve the observer accuracy by incorporating a buffer for the system output measurements. Meanwhile, this control strategy can offer more reasonable control signals in the event of DoS attacks. Compared with the disturbance observer-based controller in Chap. 3, this control scheme designs the state observer to remove the assumption that the state is measurable. It is worth mentioning that the updates of the adopted adaptive memory event-triggered mechanism (AMETM) do not require global information. Then, the observer and controller gains are obtained by using the LMI technique.

4.1 Introduction

In Chap. 3, an anti-disturbance strategy based on *disturbance* observer and *disturbance* compensation is proposed to eliminate the adverse effects of external disturbance. However, since multi-agent systems (MASs) operate in networked environments, they are vulnerable to *cyber attacks*, which may cause system performance degradation or even instability [1]. In multi-agent networks, agents may be subjected to denial-of-service (DoS) attacks, false data injection attacks (FDIAs), etc. [2, 3]. Among these attacks, DoS attacks can be categorized into two types: one targets the cyber-physical network (which is constituted of a physical layer with fixed physical links and a cyber layer with cyber control units) to disrupt timely updates of control inputs, while the other attacks the communication network within the cyber layer, affecting the communication topology and impeding information exchange among agents [4]. Thus, they are more destructive than other types of attacks [5]. To resist DoS attacks, there are existing results on solving security problems for MASs [6–12]. However, most of these works only focus on linear MASs, despite the fact that real-world systems are inherently nonlinear [13–15], making the study of nonlinear systems is more challenging.

X.-G. Guo et al., *Anti-disturbance Intrusion- and Fault-Tolerant Control of Multi-agent Systems*, https://doi.org/10.1007/978-981-96-9614-7_4

Moreover, although some results, such as those in [8, 12], consider the secure *complete consensus* of nonlinear MASs, they generally require that all system states are measurable as the case for Chap. 3. Nevertheless, in practical applications, obtaining full system states is often difficult [14, 16–18]. Therefore, the state observer-based feedback control is often used. Additionally, external disturbances are widely regarded as a primary cause of performance degradation and instability in systems [19]. To meet the increasing demand for high-accuracy performance under such disturbances, *disturbance* compensation control approaches have been widely used in the control community due to their strong disturbance rejection capabilities [20, 21]. The composition of the state observer with the *disturbance* compensation mechanism and the design of an observer-based anti-disturbance controller under DoS attacks make the secure *complete consensus* control problems of MASs highly complicated and challenging. Therefore, developing an observer-based anti-disturbance controller for nonlinear MASs under DoS attacks is the core focus of this article.

Considering the above discussion, how to design a more accurate state observer and a more effective controller under DoS attacks and *disturbance* to achieve *complete consensus* for nonlinear MASs is the problem that this chapter aims to solve. The main contributions are as follows:

(1) An anti-disturbance control scheme based on a novel state observer: Compared with the observer designed in [16, 18], the proposed method integrates a buffer into the state observer to store past system output measurements, enhancing observer accuracy during DoS attacks. Unlike [16, 22] where the control signal is set to zero when DoS attacks occur, the proposed observer-based controller can reduce the impact of packet loss caused by DoS attacks due to the existence of the buffer, which can provide more reliable control signals and improve the security and of the MAS.
(2) A novel adaptive memory event-triggered mechanism (AMETM): Similar to the event-triggered mechanism (ETM) in Chap. 3, the AMETM stores recent information to dynamically adjust trigger points, improving system performance and reducing erroneous triggers caused by sudden measurement changes. Meanwhile, the adjustable threshold adapts to state changes, saving network resources, with a simpler design than the memory ETM in [23, 24]. Moreover, a hybrid update mechanism is used to eliminate Zeno behaviors.
(3) The compensation mechanism for the observer and controller: In contrast to Chap. 3 in which a *disturbance* observer is used based on the state feedback control technique to resist disturbances, the proposed output feedback-based compensation mechanism reduces challenges associated with unmeasurable states and resists the influence of disturbances. In addition, the complete consensus performance of the system can be improved.

4.2 Problem Formulation and Preliminaries

Figure 4.1 depicts the structure of the ith agent under DoS attacks, which mainly consists of sensor, actuator, controller, buffers, zero-order hold (ZOH), AMETM, and observer. When the AMETM picks out state variables that need to be sampled, these data are transmitted over the network to the observers and the controllers. If DoS attacks occur, the data transmission will be interrupted, which will adversely affect the system stability and performance.

4.2.1 System Model

Consider a nonlinear MAS of N agents with the dynamics of the ith agent being given by

$$\begin{cases} \dot{x}_i(t) = (A + \Delta A)x_i(t) + Bu_i(t) + Dw_i(t) + Ff(x_i(t), t) \\ y_i(t) = Cx_i(t), \ i = 1, \cdots, N \end{cases} \tag{4.1}$$

where $x_i(t) \in \mathbb{R}^n$, $u_i(t) \in \mathbb{R}^b$, $y_i(t) \in \mathbb{R}^q$, and $w_i(t) \in \mathbb{R}^d$ are the state, input, output and external *disturbance* of the ith agent, respectively. A, B, C, D, and F are known matrices with appropriate dimensions, and ΔA is a time-varying uncertain matrix. Furthermore, the nonlinear dynamics of the ith agent is represented as

$$f(x_i(t), t) = \text{col}\{f_1(x_i(t), t), \cdots, f_n(x_i(t), t)\} \in \mathbb{R}^n. \tag{4.2}$$

Assumption 4.1 The pair (A, B) is stabilizable, the pair (A, C) is observable, matrix C is of full row rank, and $\text{rank}(B) = \text{rank}(B, D) = b$.

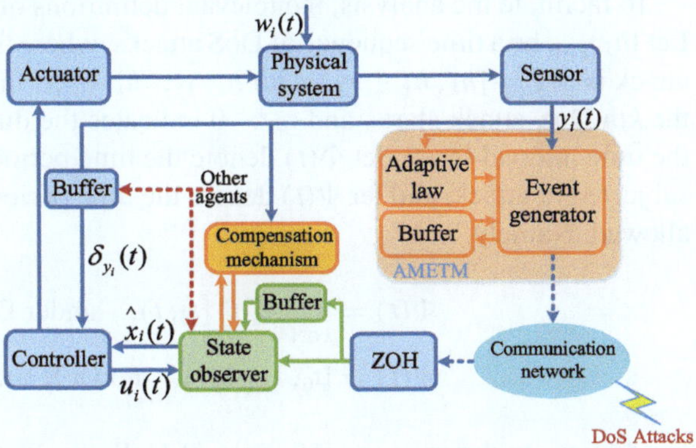

Fig. 4.1 Framework of the ith agent under DoS attacks

Assumption 4.2 There exists a positive constant ϖ and a matrix $E(t)$, such that

$$\Delta A(t) = \varpi A E(t) \tag{4.3}$$

where time-varying matrix $E(t)$ satisfies $E^T(t)E(t) \leq I_N$.

Assumption 4.3 There exists a matrix Λ such that $\forall x_1(t), x_2(t) \in \mathbb{R}^n$, the nonlinear function $f(x_i(t), t)$ satisfies

$$\|f(x_2(t), t) - f(x_1(t), t)\| \leq \|\Lambda(x_2(t) - x_1(t))\|. \tag{4.4}$$

Assumption 4.4 The *disturbance* signal $w_i(t)$ is bounded, i.e., $\|w_i(t)\| \leq \bar{w}$, where $\bar{w}$ is a known upper bound of the *disturbance* signal.

Remark 4.1 Assumption 4.1 is common in many existing results [16, 25, 26]. Assumption 4.2 serves as a constraint for uncertainties arising from mismatched parameters, model inaccuracies, and similar factors [27]. The mild Lipschitz condition in Assumption 4.3 applies to a range of nonlinear systems, including robotic systems. In fact, most linear and piecewise time-invariant continuous functions also meet this assumption [28, 29].

4.2.2 DoS Attack Model and AMETM

In this chapter, as illustrated in Fig. 4.1, we focus on DoS attacks that disrupt the controller's updating process, specifically the first type of DoS attack described in the introduction. These attacks block the transmission of sensor data by damaging the communication link between sensors and observers, which in turn impacts the observer and controller updates. Furthermore, under DoS attacks, each agent may be targeted at any time. However, in reality, it is often difficult for an attacker to maintain a continuous assault, as they must accumulate energy for subsequent attacks.

To facilitate the analysis, the relevant definitions of DoS attacks are given below. Let $\{h_k\}_{k\geq 0}$ be a time sequence of DoS attacks, where the time interval of the kth DoS attack is $\mathscr{H}_k = [h_k, h_k + \eta_k)$ with $h_{k+1} > h_k + \eta_k$, in which h_k is the instant when the kth DoS attack starts, and $\eta_k > 0$ indicates the duration of the kth attack. Over the time interval $[t_0, t)$, let $\Phi(t)$ denote the time period set where communication is subjected to attack, and let $\Psi(t)$ denote the time period set where communication is allowed. Namely,

$$\begin{aligned} \Phi(t) &= \bigcup_{k\in\mathbb{N}} \mathscr{H}_k \cap [t_0, t), \quad \text{under DoS attack} \\ \Psi(t) &= [t_0, t) \backslash \Phi(t), \quad \text{no DoS attack.} \end{aligned} \tag{4.5}$$

In addition, to describe the influence of DoS attacks on trigger points, we give the following definition.

Definition 4.1 (*Communication Failure Frequency* [30]) For $\forall t > t_0$, let $n(t_0, t)$ denote the total number of trigger points under attack during $[t_0, t)$ for each agent. Then, the frequency of attack in teams of trigger points $\ell(t_0, t)$ over $[t_0, t)$ is defined as

$$\ell(t_0, t) = \frac{n(t_0, t)}{t - t_0}. \tag{4.6}$$

Remark 4.2 To differentiate between DoS attacks and packet loss, the distinctions can be summarized as follows: (a). Packet loss typically involves a finite, small number of consecutive losses, whereas DoS attacks can persist for extended durations. For instance, if the communication failure frequency (under a DoS attack) remains below a specified threshold, scenarios where the network is blocked for a period can be considered DoS attacks. Such cases generally do not qualify as packet loss, as the consecutive losses are fewer and within a small, unknown range. (b). Based on (a), the proposed method remains effective when the communication failure frequency is below a specific threshold. Consequently, it can also address certain types of packet loss that satisfy this condition. Notably, the presented mathematical framework for handling packet loss is novel and has not been explored in existing studies, making it a fresh approach in addressing packet loss challenges.

Assume that $\{t_k^i\}$ is the event time set for the ith agent. For convenience of statement, define

$$\begin{aligned} \xi_i(t) &= \sum_{j\in\mathcal{N}_i} (x_i(t) - x_j(t)) \\ e_{i,t_{k-p+1}}(t) &= \xi_i(t^i_{k-p+1}) - \xi_i(t) \end{aligned} \tag{4.7}$$

where $1 \le p \le m$ denotes the number of data $e_{i,t_{k-p+1}}(t)$ stored in a buffer. Then, in contrast to the traditional event triggering scheme, a novel AMETM is proposed as follows:

$$g_i(t) = \left(\sum_{p=1}^{m} \varepsilon_p e_{i,t_{k-p+1}}(t)\right)^T \Theta \left(\sum_{p=1}^{m} \varepsilon_p e_{i,t_{k-p+1}}(t)\right) - c_i(t)\xi_i^T(t)\Theta\xi_i(t) \tag{4.8}$$

where $\Theta = C^T C$, ε_p's are given positive constants with $\sum_{p=1}^{m} \varepsilon_p = 1$, and $c_i(t) \in (0, \bar{c}] \subseteq (0, 1)$ is the adaptive trigger parameter and obeys the following update law:

$$\dot{c}_i(t) = \begin{cases} 0, & c_i(t) \ge \bar{c} \\ \beta_i \xi_i^T(t)\Theta\xi_i(t), & c_i(t) < \bar{c} \end{cases} \tag{4.9}$$

with β_i being a positive constant and $\bar{c}$ representing the upper bound of $c_i(t)$.

Remark 4.3 In [30], a centralized event trigger condition with a fixed threshold σ is designed, which can be rewritten as

$$\sum_{i=1}^{N}[\xi_i(t_k)-\xi_i(t)]^T[\zeta_i(t_k)-\xi_i(t)]\leq\sigma\left\{\frac{1}{2}\sum_{i=1}^{N}\sum_{j\in\mathcal{N}_i}[x_i(t)-x_j(t)]^T[x_i(t)-x_j(t)]\right\}$$

if the consideration of leader is removed. Compared to this condition, this chapter adopts the event triggered condition (4.8) that requires neighboring agents only. This can effectively reduce event triggers and hence unnecessary communication. Different from the traditional event triggered condition in [8], recently triggered packets are also used by adding a buffer of size m to (4.8). Note that the latest released packet is more important, so the choice of ε_1 should be larger than the other ε_p's. By adjusting the values of ε_p, we can change the number of trigger points. A smaller ε_1 and a larger m may trigger more points to ensure system performance. When $\varepsilon_1 = 1$ and $m = 1$, the event triggering scheme degenerates to a normal adaptive trigger mechanism, as in [31]. In addition, compared with the fixed threshold σ in [30], the trigger speed of this paper depends on $c_i(t)$, which can be adjusted dynamically. Specifically, the smaller $c_i(t)$ is, the more triggers there are but the better the system performance is. Alternately, the larger $c_i(t)$ is, the fewer triggers there are but the worse the system performance is. Obviously, the larger the number of triggers, the heavier the communication burden. Moreover, when $c_i(t) = 0$, the event triggering scheme becomes a time triggering mechanism. Therefore, the choice of $c_i(t)$ should be a trade-off between system performance and communication burden.

To avoid *Zeno behavior*, similarly to [7, 30], the next triggered instant is determined as follows:

$$t_{k+1}^i=\begin{cases}\inf\{t>t_k^i|g_i(t)>0\}, & t_k^i\in\Psi(t)\\ t_k^i+\sigma, & t_k^i\in\Phi(t)\end{cases}\tag{4.10}$$

where σ is a given positive constant.

In this chapter, undirected communication networks are considered, and the objective is to achieve the control objective that the *multi-agent systems* (4.1) can achieve *complete consensus* in the presence of DoS attacks and *disturbance* $w_i(t)$, i.e.,

$$\lim_{t\to\infty}\left\|x_i(t)-\frac{1}{N}\sum\nolimits_{i=1}^{N}x_i(t)\right\|=0,\ i=1,2,\cdots,N.\tag{4.11}$$

4.3 State Observer-Based Anti-disturbance Control Strategy

This section presents an anti-disturbance control strategy based on a state observer, which enhances observer accuracy by utilizing a buffer to store past data and compensating for the uncertainty and *disturbance*. The secure *complete consensus* of the MAS under external disturbances and DoS attacks is analyzed, with conditions for achieving *complete consensus* provided. Additionally, the coupling issue between observer and controller gains with unknown matrix variables is addressed using the LMI technique. Finally, Zeno behavior is eliminated for all agents.

4.3.1 Design of Observer and Controller

The state observer for the ith agent is constructed as below:

$$\begin{cases} \dot{\hat{x}}_i(t)=\begin{cases} A\hat{x}_i(t)+Bu_i(t)+Ff(\hat{x}_i(t),t)+H_i(t)+DG_i(t) \\ +LC(\sum\limits_{p=1}^{m}\varepsilon_p\xi_i(t^i_{k-p+1})-\hat{\xi}_i(t)), \quad t^i_k\in\Psi(t) \\ A\hat{x}_i(t)+Bu_i(t)+Ff(\hat{x}_i(t),t)+H_i(t)+DG_i(t) \\ +LC(\sum\limits_{p=1}^{m}\varepsilon_p\xi_i(t^i_{s-p+1})-\hat{\xi}_i(t)), \quad t^i_k\in\Phi(t) \end{cases} \\ \hat{y}_i(t)=C\hat{x}_i(t),\ i\in\mathscr{V} \end{cases} \tag{4.12}$$

where $\hat{x}_i(t)\in\mathbb{R}^n$ and $\hat{y}_i(t)\in\mathbb{R}^q$ denote the estimated state and output of the ith agent, respectively; L is the observer gain to be designed; t^i_s is the latest successful trigger instant while t^i_k is the instant of DoS attack; and $H_i(t)$ and $G_i(t)$ are the compensation for the uncertainty and *disturbance* of the ith agent, respectively, and can be designed as follows

$$H_i(t)=\begin{cases}\chi_2\varpi^2(1+\chi_3)\frac{\hat{x}_i^T(t)A^TA\hat{x}_i(t)}{2e^T_{y_i}e_{y_i}}P_1^{-1}C^Te_{y_i}, & e_{y_i}\neq 0\\ 0, & e_{y_i}=0\end{cases} \tag{4.13}$$

$$G_i(t)=\begin{cases}\bar{w}\frac{We_{y_i}(t)}{\|We_{y_i}(t)\|+\hbar_1}, & e_{y_i}\neq 0\\ 0, & e_{y_i}=0\end{cases} \tag{4.14}$$

where $\chi_2>0$, $\chi_3>0$, $P_1\in\mathbb{R}^{n\times n}$ is symmetric positive definite and $W\in\mathbb{R}^{d\times q}$ is a matrix variable to be designed, $\hbar_1$ is a small positive constant, and the output estimation error $e_{y_i}(t)$ is defined as

$$e_{y_i}(t)=y_i(t)-\hat{y}_i(t).$$

Next, based on state observer (4.12), for each agent i, the observer-based controller is designed as

$$u_i(t) = -K\hat{x}_i(t) - B^{\dagger} D z_i(t) \tag{4.15}$$

where matrix $B^{\dagger}$ satisfies $(I_N - BB^{\dagger})D = 0$ and can be calculated by Lemma 2.4, K is the controller gain to be designed, and $z_i(t)$ obeys

$$z_i(t) = \begin{cases} \bar{w}\frac{W_1\delta_{y_i}(t)}{\|W_1\delta_{y_i}(t)\|+\hbar_2}, & \delta_{y_i}(t) \neq 0 \\ 0, & \delta_{y_i}(t) = 0 \end{cases} \tag{4.16}$$

with $\hbar_2$ being a small positive constant, $\bar{w}$ denoting the upper bound of the *disturbance* signal as in Assumption 4.4, $W_1 \in \mathbb{R}^{d\times q}$ being the matrix parameter to be designed, and the output *complete consensus* error $\delta_{y_i}(t)$ being defined as

$$\delta_{y_i}(t) = y_i(t) - \frac{1}{N}\sum_{i=1}^{N} y_i(t).$$

Remark 4.4 It is worth mentioning that the control strategy (4.15)–(4.16) removes the assumption that the state is measurable as in Chap. 3 and [12] and only needs to use state estimation information $\hat{x}_i(t)$ and output complete consensus error information $\delta_{y_i}(t)$, which can be obtained in practical applications. Therefore, the control strategy is more practical. In addition, in contrast to the observer designed in [16], by introducing a buffer in observer (4.12), not only the current released packet $\xi_i(t_k^i)$ but also some recent packets, $\{\xi_i(t_{k-1}^i), \xi_i(t_{k-2}^i), \cdots \xi_i(t_{k-m}^i)\}$, are used. Even in the presence of DoS attacks, due to the existence of the buffer, the state can also be accurately estimated. Moreover, unlike Chap. 3 in which a *disturbance* observer was used based on the state-feedback control technique to reduce the influence of external disturbances, the accuracy of the state observer (4.12) is improved by introducing the uncertainty compensation term $H_i(t)$ in (4.13) and the external *disturbance* compensation term $G_i(t)$ in (4.14) simultaneously. Furthermore, the difficulty caused by unmeasurable states can be avoided.

4.3.2 Complete Consensus Analysis

To simplify the notation, denote the state estimation error $e_{x_i}(t) = x_i(t) - \hat{x}_i(t)$. Then, it follows from (4.1) and (4.15) that

$$\begin{aligned} \dot{x}(t) = {} & [I_N \otimes (A - BK) + \Delta\tilde{A}]x(t) + (I_N \otimes BK)e_x(t) \\ & + (I_N \otimes F)f(x(t), t) + (I_N \otimes D)w(t) - (I_N \otimes BB^{\dagger}D)z(t) \end{aligned} \tag{4.17}$$

where

$$\begin{aligned}
&x(t) = \text{col}\{x_1(t), \cdots, x_N(t)\}, \quad z(t) = \text{col}\{z_1(t), \cdots, z_N(t)\} \\
&\Delta\tilde{A} = \text{diag}\{\Delta A, \cdots, \Delta A\}, \quad w(t) = \text{col}\{w_1(t), \cdots, w_N(t)\} \\
&e_x(t) = \text{col}\{e_{x_1}(t), \cdots, e_{x_N}(t)\} \\
&f(x(t), t) = \text{col}\{f(x_1(t), t), \cdots, f(x_N(t), t)\}.
\end{aligned}$$

Define the average state $\bar{x}(t)$ and complete consensus error $\delta_i(t)$ as

$$\begin{aligned}
\bar{x}(t) &= \frac{1}{N}\sum_{i=1}^{N} x_i(t) \\
\delta_i(t) &= x_i(t) - \bar{x}(t).
\end{aligned}$$

Then, we have

$$\delta(t) = (M \otimes I_n)x(t)$$

where $\delta(t) = \text{col}\{\delta_1(t), \cdots, \delta_N(t)\}$ and $M = I_N - \frac{1}{N}1_N 1_N^T$. Moreover, it is not difficult to see that

$$\mathscr{L}M = M\mathscr{L} = \mathscr{L} \tag{4.18}$$

$$\xi(t) = (\mathscr{L} \otimes I_n)\delta(t) \tag{4.19}$$

with $\xi(t) = \text{col}\{\xi_1(t), \cdots, \xi_N(t)\}$.

Then, based on the above definitions and (4.17), the dynamic of the complete consensus error δ is given by

$$\begin{aligned}
\dot{\delta}(t) = {}& [I_N \otimes (A - BK) + \Delta\tilde{A}]\delta(t) + (M \otimes BK)e_x(t) \\
&+ (I_N \otimes F)\tilde{f}(x(t), t) + (M \otimes D)w(t) - (M \otimes BB^{\dagger}D)z(t)
\end{aligned} \tag{4.20}$$

where

$$\begin{aligned}
\tilde{f}(x(t), t) = \text{col}\bigg\{&(f(x_1(t), t) - \frac{1}{N}\sum\nolimits_{i=1}^{N} f(x_i(t), t)), \cdots, \\
&(f(x_N(t), t) - \frac{1}{N}\sum\nolimits_{i=1}^{N} f(x_i(t), t))\bigg\}.
\end{aligned}$$

Next, the main results of this chapter are presented in the following theorem.

Theorem 4.1 *Suppose that Assumptions 4.1–4.4 hold. Under the observer-based anti-disturbance controller (4.15) and the triggering function (4.8) with the triggering sequence (4.10), MAS (4.1) achieves complete consensus and the designed observer (4.12) gives accurate estimate of the state in the presence of DoS attacks* *and external disturbances, if there exist symmetric positive definite matrices* P_1, P_2 *and matrices* L, K, W, *and* W_1 *satisfying*

$$\begin{bmatrix} \Pi_{11} & * \\ \lambda_{\max}(M)P_2BK & \Pi_{22} \end{bmatrix} < 0 \tag{4.21}$$

$$P_1D = C^TW^T \tag{4.22}$$

$$P_2D = C^TW_1^T \tag{4.23}$$

$$\begin{bmatrix} \bar{\Pi}_{11} & * \\ \lambda_{\max}(M)P_2BK & \bar{\Pi}_{22} \end{bmatrix} \leq \frac{\rho}{2}\bar{P} \tag{4.24}$$

$$\begin{bmatrix} 0 & 0 \\ * & \tilde{\Pi}_{22} \end{bmatrix} \leq \frac{\rho}{2}\bar{P} \tag{4.25}$$

$$\ell(t_0, t) \leq \frac{\alpha - \varsigma}{(\alpha + \rho)\sigma} \tag{4.26}$$

where

$$\begin{aligned}
\Pi_{11} &= A^TP_1 + P_1A + \frac{1}{\gamma}P_1FF^TP_1 + \gamma\Lambda^T\Lambda + \frac{1}{\chi_1}P_1LL^TP_1 \\
&\quad + \frac{1}{\chi_2}P_1P_1 + \chi_2\varpi^2(1 + \frac{1}{\chi_3})A^TA - \lambda_2(P_1LC + C^TL^TP_1) + \alpha P_1 \\
\Pi_{22} &= A^TP_2 + P_2A - P_2BK - K^TB^TP_2 + \varpi P_2AA^TP_2 \\
&\quad + \varpi I + \frac{1}{\gamma}P_2FF^TP_2 + \gamma\Lambda^T\Lambda + \chi_1\bar{c}\lambda_N^2\Theta + \alpha P_2 \\
\bar{\Pi}_{11} &= A^TP_1 + P_1A + \frac{1}{\gamma}P_1FF^TP_1 + \gamma\Lambda^T\Lambda + \theta_1P_1LL^TP_1 \\
&\quad + \frac{1}{\chi_2}P_1P_1 + \chi_2\varpi^2(1 + \frac{1}{\chi_3})A^TA - \lambda_2(P_1LC + C^TL^TP_1) \\
\bar{\Pi}_{22} &= A^TP_2 + P_2A - P_2BK - K^TB^TP_2 + \varpi P_2AA^TP_2 \\
&\quad + \varpi I + \frac{1}{\gamma}P_2FF^TP_2 + \gamma\Lambda^T\Lambda + 2\bar{\theta}_2\lambda_N^2\Theta \\
\tilde{\Pi}_{22} &= (\bar{\theta}_1\bar{c} + 2\bar{\theta}_2)\lambda_N^2\Theta
\end{aligned}$$

with $\bar{\theta}_1 = (1 + 1/\theta_2)/\theta_1$, $\bar{\theta}_2 = (1 + \theta_2)/\theta_1$, *and* $\Theta = C^TC$. *Besides,* α, χ_1, χ_2, χ_3, θ_1, θ_2, γ, *and* ρ *are positive constants,* ς *satisfies* $0 < \varsigma < \alpha$, *and matrix* $\bar{P} = \text{diag}\{P_1, P_2\}$.

Proof Choose the Lyapunov function candidate as follows

$$V(t) = \sum_{i=1}^{N} V_i(t) = \sum_{i=1}^{N} \zeta_i^T(t)\bar{P}\zeta_i(t) \tag{4.27}$$

where $\zeta_i(t) = \text{col}\{e_{x_i}(t), \delta_i(t)\}$. Then, according to whether t_k^i is an under attack instant, we divide the proof process into two parts. First, consider $t \in [t_k^i, t_{k+1}^i)$.

(1) If $t_k^i \in \Psi(t)$ with $\Psi(t)$ as in (4.5), this means that the latest sampling information can be collected, transmitted, and properly received.

Recalling that $e_{x_i}(t) = x_i(t) - \hat{x}_i(t)$, one can deduce that

$$\begin{aligned}\dot{e}_x(t) &= (I_N \otimes A - \mathscr{L} \otimes LC)e_x(t) + (I_N \otimes \Delta A)x(t) + (I_N \otimes D)w(t) \\ &\quad + (I_N \otimes F)\hat{f}(x(t), t) - (I_N \otimes I)H(t) - (I_N \otimes D)G(t) \\ &\quad - (I_N \otimes LC)\sum_{p=1}^{m} \varepsilon_p e_{t_{k-p+1}}(t)\end{aligned} \tag{4.28}$$

where

$$\begin{aligned}\hat{f}(x(t), t) &= \text{col}\{f(x_1(t), t) - f(\hat{x}_1(t), t), \cdots, f(x_N(t), t) - f(\hat{x}_N(t), t)\} \\ \sum_{p=1}^{m} \varepsilon_p e_{t_{k-p+1}}(t) &= \text{col}\left\{\sum_{p=1}^{m} \varepsilon_p e_{1,t_{k-p+1}}(t), \cdots, \sum_{p=1}^{m} \varepsilon_p e_{N,t_{k-p+1}}(t)\right\}.\end{aligned}$$

Based on (4.20) and (4.28), the time derivative of $V(t)$ is

$$\begin{aligned}\dot{V}(t) &= e_x^T(t)[I_N \otimes (A^T P_1 + P_1 A) - \mathscr{L} \otimes (P_1 LC + C^T L^T P_1)]e_x(t) \\ &\quad + 2e_x^T(t)(I_N \otimes P_1 \Delta A)x(t) + 2e_x^T(t)(I_N \otimes P_1 D)w(t) \\ &\quad + 2e_x^T(t)(I_N \otimes P_1 F)\hat{f}(x(t), t) - 2e_x^T(t)(I_N \otimes P_1)H(t) \\ &\quad - 2e_x^T(t)(I_N \otimes P_1 D)G(t) - 2e_x^T(t)(I_N \otimes P_1 LC)\sum_{p=1}^{m} \varepsilon_p e_{t_{k-p+1}}(t) \\ &\quad + \delta^T(t)[I_N \otimes (A^T P_2 + P_2 A - P_2 BK - K^T B^T P_2) + \Delta\tilde{A}^T(I_N \otimes P_2) \\ &\quad + (I_N \otimes P_2)\Delta\tilde{A}]\delta(t) + 2\delta^T(t)(M \otimes P_2 BK)e_x(t) \\ &\quad + 2\delta^T(t)(I_N \otimes P_2 F)\tilde{f}(x(t), t) + 2\delta^T(t)(M \otimes P_2 D)w(t) \\ &\quad - 2\delta^T(t)(M \otimes P_2 BB^\dagger D)z(t).\end{aligned} \tag{4.29}$$

From Assumption 4.3 and by using *Young's inequality* (Lemma 2.7) it is not difficult to derive that

$$2e_x^T(t)(I_N \otimes P_1 F)\hat{f}(x(t), t) \leq e_x^T(t)[I_N \otimes (\frac{1}{\gamma} P_1 F F^T P_1 + \gamma \Lambda^T \Lambda)]e_x(t). \quad (4.30)$$

Note that $\sum_{i=1}^{N} \delta_i = 0$, so we have

$$\sum_{i=1}^{N} \delta_i{}^T P_2 F[f(\bar{x}(t), t) - \frac{1}{N}\sum_{i=1}^{N} f(x_i(t), t)] = 0.$$

Then, one can obtain

$$\begin{aligned}
&2\delta^T(t)(I_N \otimes P_2 F)\tilde{f}(x(t), t) = 2\sum_{i=1}^{N} \delta_i{}^T P_2 F \tilde{f}(x_i(t), t) \\
&= 2\sum_{i=1}^{N} \delta_i{}^T P_2 F[f(x_i(t), t) - f(\bar{x}(t), t) + f(\bar{x}(t), t) - \frac{1}{N}\sum_{i=1}^{N} f(x_i(t), t)] \\
&\leq \delta^T(t)[I_N \otimes (\frac{1}{\gamma} P_2 F F^T P_2 + \gamma \Lambda^T \Lambda)]\delta(t). \qquad (4.31)
\end{aligned}$$

By letting $\tilde{E}(t) = \text{diag}\{E(t), \cdots, E(t)\}$, we obtain

$$\Delta\tilde{A} = \varpi(I_N \otimes A)\tilde{E}(t), \quad \Delta\tilde{A}^T = \varpi\tilde{E}^T(I_N \otimes A^T).$$

Thus,

$$\delta^T(t)[\Delta\tilde{A}^T(I_N \otimes P_2) + (I_N \otimes P_2)\Delta\tilde{A}]\delta(t) \leq \delta^T(t)[\varpi I_N \otimes (P_2 A A^T P_2 + I)]\delta(t). \quad (4.32)$$

Furthermore, it follows from *Young's inequality* (Lemma 2.7) that

$$\begin{aligned}
&- 2e_x^T(t)(I_N \otimes P_1 L C)\sum_{p=1}^{m} \varepsilon_p e_{t_{k-p+1}}(t) \\
&\leq \frac{1}{\chi_1} e_x^T(t)(I_N \otimes P_1 L L^T P_1)e_x(t) \\
&+ \chi_1 \left(\sum_{p=1}^{m} \varepsilon_p e_{t_{k-p+1}}(t)\right)^T (I_N \otimes C^T C)\left(\sum_{p=1}^{m} \varepsilon_p e_{t_{k-p+1}}(t)\right) \qquad (4.33)
\end{aligned}$$

$$
\begin{aligned}
&2e_x^T(t)(I_N \otimes P_1 \Delta A)x(t) \\
\leq &\frac{1}{\chi_2} e_x^T(t)(I_N \otimes P_1 P_1)e_x(t) + \chi_2 \varpi^2 x^T(t)(I_N \otimes A^T A)x(t) \\
\leq &\frac{1}{\chi_2} e_x^T(t)(I_N \otimes P_1 P_1)e_x(t) + \chi_2 \varpi^2 (1 + \frac{1}{\chi_3}) e_x^T(t)(I_N \otimes A^T A)e_x(t) \\
&+ \chi_2 \varpi^2 (1 + \chi_3)\hat{x}^T(t)(I_N \otimes A^T A)\hat{x}(t)
\end{aligned} \tag{4.34}
$$

and according to (4.8), we have

$$
\begin{aligned}
&\left(\sum_{p=1}^{m} \varepsilon_p e_{t_{k-p+1}}(t)\right)^T (I_N \otimes C^T C) \left(\sum_{p=1}^{m} \varepsilon_p e_{t_{k-p+1}}(t)\right) \\
&\leq c(t)\xi^T(t)(I_N \otimes C^T C)\xi(t) = c(t)\delta^T(t)(\mathscr{L}^2 \otimes C^T C)\delta(t)
\end{aligned}
$$

where $c(t) = \operatorname{diag}\{c_1(t), \cdots, c_N(t)\}$.

Since $P_1 D = C^T W^T$ and $P_2 D = C^T W_1^T$ hold in (4.22) and (4.23), respectively, we can easily deduce that

$$
\begin{aligned}
2e_x^T(t)(I_N \otimes P_1 D)w(t) - 2e_x^T(t)(I_N \otimes P_1 D)G(t) \leq 0 \\
2\delta^T(t)(M \otimes P_2 D)w(t) - 2\delta^T(t)(M \otimes P_2 B B^{\dagger} D)z(t) \leq 0.
\end{aligned} \tag{4.35}
$$

Then, substituting (4.30)–(4.35) into (4.29) yields

$$
\begin{aligned}
\dot{V}(t) \leq &e_x^T(t)[I_N \otimes (A^T P_1 + P_1 A + \frac{1}{\gamma} P_1 F F^T P_1 + \gamma \Lambda^T \Lambda + \frac{1}{\chi_2} P_1 P_1) \\
&- \mathscr{L} \otimes (P_1 L C + C^T L^T P_1) + \chi_2 \varpi^2 (1 + \frac{1}{\chi_3})(I_N \otimes A^T A) \\
&+ \frac{1}{\chi_1} I_N \otimes P_1 L L^T P_1] e_x(t) + \delta^T(t)[I_N \otimes (A^T P_2 + P_2 A - P_2 B K \\
&- K^T B^T P_2 + \frac{1}{\gamma} P_2 F F^T P_2 + \gamma \Lambda^T \Lambda + \varpi P_2 A A^T P_2 + \varpi I) \\
&+ \chi_1 c(t)(\mathscr{L}^2 \otimes \Theta)]\delta(t) + 2\delta^T(t)(M \otimes P_2 B K)e_x(t).
\end{aligned} \tag{4.36}
$$

By using condition (4.21), the following inequality can be determined

$$
\dot{V}_i(t) \leq \zeta_i^T(t)(-\alpha \bar{P})\zeta_i(t) = -\alpha V_i(t) \tag{4.37}
$$

where α is a positive constant.

(2) If $t_k^i \in \Phi(t)$ with $\Phi(t)$ as in (4.5), this means that the latest sampling information cannot be obtained. Moreover, it is assumed that DoS attacks block the network channel from t_{s+1}^i, so one obtains

$$\begin{aligned}\dot{e}_x(t) &= (I_N \otimes A - \mathscr{L} \otimes LC)e_x(t) + (I_N \otimes \Delta A)x(t) + (I_N \otimes D)w(t)\\&\quad + (I_N \otimes F)\hat{f}(x(t), t) - (I_N \otimes I)H(t) - (I_N \otimes D)G(t)\\&\quad - (I_N \otimes LC)\Big(\sum_{p=1}^{m} \varepsilon_p \xi(t_{s-p+1}) - \xi(t_{s+1})\Big)\\&\quad - (I_N \otimes LC)(\xi(t_{s+1}) - \xi(t)) \end{aligned} \tag{4.38}$$

where $\xi(t_{s+1}) - \xi(t) = \mathrm{col}\{\xi_1(t_{s+1}^1) - \xi_1(t), \cdots, \xi_N(t_{s+1}^N) - \xi_N(t)\}$. Then, we have

$$\begin{aligned}\dot{V}(t) &= e_x^T(t)[I_N \otimes (A^T P_1 + P_1 A) - \mathscr{L} \otimes (P_1 LC + C^T L^T P_1)]e_x(t)\\&\quad + 2e_x^T(t)(I_N \otimes P_1 \Delta A)x(t) + 2e_x^T(t)(I_N \otimes P_1 D)w(t)\\&\quad + 2e_x^T(t)(I_N \otimes P_1 F)\hat{f}(x(t), t) - 2e_x^T(t)(I_N \otimes P_1)H(t)\\&\quad - 2e_x^T(t)(I_N \otimes P_1 D)G(t) - 2e_x^T(t)(I_N \otimes P_1 LC)(\xi(t_{s+1}) - \xi(t))\\&\quad - 2e_x^T(t)(I_N \otimes P_1 LC)\left(\sum_{p=1}^{m} \varepsilon_p \xi(t_{s-p+1}) - \xi(t_{s+1})\right)\\&\quad + \delta^T(t)[I_N \otimes (A^T P_2 + P_2 A - P_2 BK - K^T B^T P_2) + \Delta\tilde{A}^T(I_N \otimes P_2)\\&\quad + (I_N \otimes P_2)\Delta\tilde{A}]\delta(t) + 2\delta^T(t)(M \otimes P_2 BK)e_x(t)\\&\quad + 2\delta^T(t)(I_N \otimes P_2 F)\tilde{f}(x(t), t) + 2\delta^T(t)(M \otimes P_2 D)w(t)\\&\quad - 2\delta^T(t)(M \otimes P_2 BB^\dagger D)z(t). \end{aligned} \tag{4.39}$$

According to *Young's inequality* (Lemma 2.7), we can obtain

$$\begin{aligned}&- 2e_x^T(t)(I_N \otimes P_1 LC)\left(\sum_{p=1}^{m} \varepsilon_p \xi(t_{s-p+1}) - \xi(t_{s+1})\right)\\&\le \bar{\theta}_1\Big(\sum_{p=1}^{m} \varepsilon_p \xi(t_{s-p+1}) - \xi(t_{s+1})\Big)^T (I_N \otimes C^T C)\Big(\sum_{p=1}^{m} \varepsilon_p \xi(t_{s-p+1}) - \xi(t_{s+1})\Big)\\&\quad + \frac{1}{\bar{\theta}_1} e_x^T(t)(I_N \otimes P_1 LL^T P_1)e_x(t) \end{aligned} \tag{4.40}$$

$$\begin{aligned}&- 2e_x^T(t)(I_N \otimes P_1 LC)(\xi(t_{s+1}) - \xi(t))\\&\le \frac{1}{\bar{\theta}_2} e_x^T(t)(I_N \otimes P_1 LL^T P_1)e_x(t) + \bar{\theta}_2(\xi(t_{s+1}) - \xi(t))^T (I_N \otimes C^T C)(\xi(t_{s+1}) - \xi(t)). \end{aligned} \tag{4.41}$$

From the event-triggered condition (4.8), we have

$$\left(\sum_{p=1}^{m} \varepsilon_p \xi(t_{s-p+1}) - \xi(t_{s+1})\right)^T (I_N \otimes C^T C)\left(\sum_{p=1}^{m} \varepsilon_p \xi(t_{s-p+1}) - \xi(t_{s+1})\right)$$
$$\leq c(t_{s+1})\xi^T(t_{s+1})(I_N \otimes C^T C)\xi(t_{s+1}). \tag{4.42}$$

Additionally, it is easy to conclude that

$$\begin{aligned}&[(\xi(t_{s+1}) - \xi(t)]^T (I_N \otimes C^T C)[\xi(t_{s+1}) - \xi(t)]\\&\leq 2\xi^T(t_{s+1}) \times (I_N \otimes C^T C)\xi(t_{s+1}) + 2\xi^T(t)(I_N \otimes C^T C)\xi(t).\end{aligned} \tag{4.43}$$

Next, using the analysis methods in Part 1), combining (4.40) and (4.42) with (4.43), and according to LMIs (4.24) and (4.25), we can deduce from (4.39) that

$$\begin{aligned}\dot{V}_i(t) &\leq \zeta_i^T(t)\frac{\rho\bar{P}}{2}\zeta_i(t) + \zeta_i^T(t_{s+1}^i)\frac{\rho\bar{P}}{2}\zeta_i(t_{s+1}^i)\\&\leq \rho\max\{V_i(t), V_i(t_{s+1}^i)\}.\end{aligned} \tag{4.44}$$

Then, it follows from (4.37), (4.44) and Lemma 2.12 that

$$V_i(t) \leq e^{[-\alpha(t-t_0-n(t_0,t)\sigma)+\rho n(t_0,t)\sigma]}V_i(t_0). \tag{4.45}$$

According to (4.26), is equivalent to

$$V_i(t) \leq e^{-\varsigma(t-t_0)}V_i(t_0). \tag{4.46}$$

Thus,

$$V(t) \leq e^{-\varsigma(t-t_0)}V(t_0), \tag{4.47}$$

which means that $V(t)$ is bounded. Therefore, the secure average *complete consensus* of MASs (4.1) is guaranteed. □

Remark 4.5 To simplify solving LMIs (4.21)–(4.25), we can set parameters of the same type, such as χ_1, χ_2, and χ_3, to equal values and then adjust them using a trial-and-error approach. Moreover, for conditions (4.24) and (4.25) in Theorem 1, since (4.21)–(4.23) have already determined P_1, P_2, L, and K, we only need to find an appropriate ρ to satisfy these conditions, rather than solving for L and K again using (4.21)–(4.25). This can simplify the solution process.

Remark 4.6 In many existing results (such as [7, 10, 11]) on the secure *complete consensus* of MASs under DoS attacks, time is divided into two periods based on whether t is during the interval of DoS attacks, so the duration and frequency of DoS attacks are restricted [7]. In contrast, this chapter focuses on whether trigger times occur during DoS attacks, eliminating the need to separately constrain their duration and frequency. Instead, we only restrict the frequency of communication failures, which indirectly limits the occurrence of DoS attacks. Additionally, these constraints must be satisfied by all agents.

Remark 4.7 In previous works [7, 9–11, 30, 32], algebraic Riccati equation (ARE) and algebraic Riccati inequality (ARI) are used to solve for the control gain. However, these approaches are very conservative when the matrix dimension is large or there are many parameters. In this chapter, the observer and controller are designed simultaneously using the LMI technique, without requiring prior adjustment of parameters or symmetric positive definite matrices. Moreover, the derived conditions avoid the use of the *Kronecker product*, simplifying the solution process.

4.3.3 Joint Design of Observer and Controller

The conditions in Theorem 4.1 guarantee that the MAS achieves *complete consensus* and that observers can reconstruct the system state under external disturbances and DoS attacks. However, since L and K in (4.21) are coupled with unknown matrices, and (4.22), (4.23) are equality constraints, they cannot be solved directly using the LMI toolbox in MATLAB. To facilitate finding a solution, the following theorem introduces a design method for determining the observer and controller gains.

Theorem 4.2 *For given positive scalars α, γ, χ_1, χ_2, χ_3, ν and ϖ, if there exist symmetric positive definite matrices P_1 and P_2, and matrices Q_1, N, $\bar{K}$, W and W_1 satisfying the following conditions:*

$$\begin{bmatrix} \Pi'_{11} & Q_1 & P_1 & P_1F & \Pi'_{15} & 0 & 0 \\ * & -\chi_1 & 0 & 0 & 0 & 0 & 0 \\ * & * & -\chi_2 & 0 & 0 & 0 & 0 \\ * & * & * & -\gamma & 0 & 0 & 0 \\ * & * & * & * & \Pi'_{55} & \sqrt{\varpi}P_2A & P_2F \\ * & * & * & * & * & -I & 0 \\ * & * & * & * & * & * & -\gamma \end{bmatrix} < 0 \tag{4.48}$$

$$\begin{bmatrix} -\nu I & P_2B - BN \\ * & -\nu I \end{bmatrix} < 0 \tag{4.49}$$

$$\begin{bmatrix} -\nu I & P_1D - C^TW^T \\ * & -\nu I \end{bmatrix} < 0 \tag{4.50}$$

$$\begin{bmatrix} -\nu I & P_2D - C^TW_1^T \\ * & -\nu I \end{bmatrix} < 0 \tag{4.51}$$

where

$$\Pi_{11}^{'} = A^T P_1 + P_1 A + \gamma\Lambda^T\Lambda + \chi_2\varpi^2(1+\frac{1}{\chi_3})A^T A - \lambda_2(Q_1 C + C^T Q_1^T) + \alpha P_1$$
$$\Pi_{15}^{'} = \lambda_{\max}(M)\bar{K}^T B^T$$
$$\Pi_{55}^{'} = A^T P_2 + P_2 A - B\bar{K} - \bar{K}^T B^T + \varpi I + \gamma\Lambda^T\Lambda + \chi_1\bar{c}\lambda_N^2\Theta + \alpha P_2.$$

Then, the observer and the controller gains are given by $L = P_1^{-1}Q_1$ *and* $K = N^{-1}\bar{K}$, *respectively.*

Proof Let $Q_1 = P_1 L$, $P_2 B = BN$, and $NK = \bar{K}$. Then, (4.21) can be written as

$$\begin{bmatrix} \Pi_{11} & * \\ \lambda_{\max}(M)B\bar{K} & \Pi_{22} \end{bmatrix} < 0$$

where

$$\begin{aligned}\Pi_{11} &= A^T P_1 + P_1 A + \frac{1}{\gamma}P_1 F F^T P_1 + \gamma\Lambda^T\Lambda + \frac{1}{\chi_1}Q_1 Q_1^T + \frac{1}{\chi_2}P_1 P_1 \\ &\quad + \chi_2\varpi^2(1+\frac{1}{\chi_3})A^T A - \lambda_2(Q_1 C + C^T Q_1^T) + \alpha P_1 \\ \Pi_{22} &= A^T P_2 + P_2 A - B\bar{K} - \bar{K}^T B^T + \varpi P_2 A A^T P_2 + \varpi I + \frac{1}{\gamma}P_2 F F^T P_2 \\ &\quad + \gamma\Lambda^T\Lambda + \chi_1\bar{c}\lambda_N^2\Theta + \alpha P_2.\end{aligned}$$

Next, using Lemmas 2.8 and 2.4, we can obtain (4.48)–(4.51). This completes the proof. □

4.3.4 Zeno Behavior Analysis

Zeno behavior is a critical issue in event-triggering strategies, as it leads to infinite triggers in *finite time*, which is physically impossible and contradicts the goal of reducing trigger frequency. Therefore, it is important to prevent it.

In this chapter, a mixed event-triggering approach is used to determine the trigger times. When DoS attacks occur, a fixed time σ is used to determine the next triggered moment, ensuring that *Zeno behavior* is avoided during this period. If DoS attacks do not occur, the trigger function $g_i(t)$ in (4.8) decides the next trigger time. The following theorem then shows that there is a positive lower bound between any two consecutive event moments, proving that *Zeno behavior* is avoided.

Theorem 4.3 *Under the event triggering strategy (4.10), Zeno behavior will not be exhibited in MAS (4.1) and the minimum time interval between any inter-event instants is strictly positive when DoS attacks do not occur. That is,*

$$t_{k+1}^{i}-t_{k}^{i}\geq\frac{\Omega_{i}^{2}}{\|C\|[2\left\|\bar{A}\right\|\Xi_{i}^{2}+2\Xi_{i}\Gamma_{i}(t_{k}^{i})]}>0 \tag{4.52}$$

where

$$\Xi_{i}^{2}=\left\|\sum_{p=1}^{m}\varepsilon_{p}e_{i,t_{k-p+1}}(t)\right\|^{2} \tag{4.53}$$

$$\Omega_{i}^{2}=\left\|C\sum_{p=1}^{m}\varepsilon_{p}e_{i,t_{k-p+1}}(t)\right\|^{2} \tag{4.54}$$

$$\|\bar{A}\|=\|A\|+\|\varpi A\|+\|F\Lambda\| \tag{4.55}$$

$$\Gamma_{i}(t_{k}^{i})=\|\bar{A}\|\left\|\sum_{p=1}^{m}\varepsilon_{p}\xi_{i}(t_{k-p+1}^{i})\right\|+2|\mathscr{N}_{i}|\,\|D\|\,\|\bar{w}\|+\|B\|\left\|\sum_{j\in\mathscr{N}_{i}}(u_{i}(t)-u_{j}(t))\right\| \tag{4.56}$$

and $|\mathscr{N}_i|$ represents the number of neighbors of the agent i.

Proof For any $t\in[t_k^i,t_{k+1}^i)$, we have

$$\begin{aligned}
&\frac{d}{dt}\left\|\sum_{p=1}^{m}\varepsilon_{p}e_{i,t_{k-p+1}}(t)\right\|^{2}\\
&\leq 2\left\|\sum_{p=1}^{m}\varepsilon_{p}e_{i,t_{k-p+1}}(t)\right\|\left\|\sum_{j\in\mathscr{N}_{i}}(\dot{x}_{i}(t)-\dot{x}_{j}(t))\right\|\\
&\leq 2\left\|\sum_{p=1}^{m}\varepsilon_{p}e_{i,t_{k-p+1}}(t)\right\|\left(\|\bar{A}\|\left\|\sum_{p=1}^{m}\varepsilon_{p}e_{i,t_{k-p+1}}(t)\right\|+\|\bar{A}\|\left\|\sum_{p=1}^{m}\varepsilon_{p}\xi_{i}(t_{k-p+1}^{i})\right\|\right.\\
&\quad\left.+\left\|\sum_{j\in\mathscr{N}_{i}}(Bu_{i}(t)-Bu_{j}(t))+\sum_{j\in\mathscr{N}_{i}}(Dw_{i}(t)-Dw_{j}(t))\right\|\right)\\
&\leq 2\left\|\bar{A}\right\|\left\|\sum_{p=1}^{m}\varepsilon_{p}e_{i,t_{k-p+1}}(t)\right\|^{2}+2\left\|\sum_{p=1}^{m}\varepsilon_{p}e_{i,t_{k-p+1}}(t)\right\|\left(\|\bar{A}\|\left\|\sum_{p=1}^{m}\varepsilon_{p}\xi_{i}(t_{k-p+1}^{i})\right\|\right.\\
&\quad\left.+2|\mathscr{N}_{i}|\,\|D\|\,\|\bar{w}\|+\|B\|\left\|\sum_{j\in\mathscr{N}_{i}}(u_{i}(t)-u_{j}(t))\right\|\right).
\end{aligned} \tag{4.57}$$

By using the definition Ω_i^2 in (4.54), the inequality (4.57) can be rewritten as

$$\frac{d}{dt}\left(\Omega_i^2\right) \leq \|C\| \left[2 \left\|\bar{A}\right\| \Xi_i^2 + 2\Xi_i\Gamma_i(t_k^i)\right] \tag{4.58}$$

where $\|\bar{A}\|$ and $\Gamma_i(t_k^i)$ are defined in (4.52). At the event-triggering instants $t_1^i, t_2^i, \cdots, t_k^i, \cdots$, it can be seen from (4.8) that the following inequality holds:

$$\Omega_i^2 > c_i(t)\left\|C\xi_i(t_k^i)\right\|^2 \geq 0. \tag{4.59}$$

Then, by combining (4.58) with (4.59), we obtain

$$t_{k+1}^i - t_k^i \geq \epsilon_i > \frac{\Omega_i^2}{\|C\|[2\left\|\bar{A}\right\| \Xi_i^2 + 2\Xi_i\Gamma_i(t_k^i)]} \tag{4.60}$$

and can conclude that $\Xi_i > 0$ and $\Gamma_i(t_k^i) > 0$, which implies that $\epsilon_i > 0$. Hence, we have $t_{k+1}^i - t_k^i \geq \min\{\epsilon_i, \sigma\} > 0$, and the proof is completed. □

Remark 4.8 It should be pointed out that the proof of Theorem 4.3 is inspired by [32]. Note that the interval ϵ_i between two events is influenced by $c_i(t)$, so choosing an appropriate $c_i(t)$ is crucial for balancing performance and communication load. Furthermore, the presence of a hybrid update mechanism ensures that a lower bound always exists during DoS attacks, effectively preventing continuous triggering.

4.4 Simulation Examples

In this section, we demonstrate the validity of our method by using the communication topology shown in Fig. 3.1. Consider each agent with the dynamic as (4.1) and system matrices as follows:

$$A = \begin{bmatrix} -2.9 & 0.3 & 0.4 & 1.2 \\ -0.1 & -0.2 & 0.6 & 1.5 \\ 1.2 & 2.1 & -2.8 & 3.4 \\ 1 & -2 & -2.5 & -2.5 \end{bmatrix}, \quad B = \begin{bmatrix} 0 & 0 \\ 0 & 0 \\ -1 & 0.5 \\ -0.1 & 0.2 \end{bmatrix}$$

$$C = \begin{bmatrix} 1 & 0 & 0 & 0 \\ 0 & 1 & 0 & 0 \\ 0 & 0 & 1 & 1 \end{bmatrix}, \quad D = \begin{bmatrix} 0 \\ 0 \\ -0.1 \\ -0.1 \end{bmatrix}, \quad E(t) = \sin(t)I_4$$

$$F = 0.01I_4.$$

In addition, the function $f(x_i(t), t)$ is denoted as $f(x_i(t), t) = [0, 0, 0, 0.01\sin(x_{i1}(t))]^T$. Then, according to Assumption 4.3, one has $\Lambda =$

diag{0.01, $\cdots$, 0.01}. Moreover, from Fig. 3.1, it is easy to obtain the *Laplacian matrix*

$$\mathscr{L} = \begin{bmatrix} 3 & -1 & 0 & -1 & -1 \\ -1 & 1 & 0 & 0 & 0 \\ 0 & 0 & 1 & -1 & 0 \\ -1 & 0 & -1 & 3 & -1 \\ -1 & 0 & 0 & -1 & 2 \end{bmatrix}$$

whose minimum non-zero eigenvalue and maximum eigenvalue are 0.6972 and 4.3028, respectively.

In the following simulation, we consider two different types of disturbances affecting the first agent to demonstrate the effectiveness of the proposed method, i.e.,

$$w_1^1(t) = 0.7\cos(t), \quad t \geq 0s$$

$$w_1^2(t) = \begin{cases} 0.5\cos(2t+1), & 0s \leq t < 9s \\ 0.4, & t \geq 9s \end{cases}$$

and there are no disturbances to other agents.

Then, based on Theorems 4.1 and 4.2, we choose parameters $\varpi = 0.0001, \gamma = 1$, $\chi_1 = 0.01$, $\chi_2 = 4.5$, $\chi_3 = 0.01$, $\nu = 0.001$, $\alpha = 0.25$, $\rho = 4$, $\varsigma = 0.001$ and $\sigma = 0.04$. By solving LMIs (4.48)–(4.51), the observer gain L, the controller gain K and matrices P_1, W, and W_1 are obtained as follows:

$$L = \begin{bmatrix} 0.0006 & 0.0000 & 0.0007 \\ 0.0000 & 0.0023 & -0.0010 \\ 0.0005 & 0.0005 & 0.0014 \\ 0.0002 & -0.0015 & 0.0020 \end{bmatrix}$$

$$K = \begin{bmatrix} 0.1435 & -2.5926 & 0.2402 & -1.9829 \\ 0.5258 & -2.2773 & -0.1509 & -1.5060 \end{bmatrix}$$

$$P_1 = \begin{bmatrix} 15.4884 & -0.1750 & -5.4519 & -1.6695 \\ -0.1750 & 7.5176 & -3.1816 & 6.2076 \\ -5.4519 & -3.1816 & 8.9619 & -2.4892 \\ -1.6695 & 6.2076 & -2.4892 & 8.9626 \end{bmatrix}$$

$$W = \begin{bmatrix} 0.7121 & -0.3026 & -0.6473 \end{bmatrix}$$

$$W_1 = \begin{bmatrix} 0.0001 & 0.0004 & -5.3476 \end{bmatrix}.$$

To achieve a trade-off between the system performance and communication burden, the parameters and initial coupling gains in the trigger function (4.8) and the adaptive law (4.9) are selected as $m = 3$, $\varepsilon_1 = 0.8$, $\varepsilon_2 = 0.15$, $\varepsilon_3 = 0.05$, $\beta_i = 0.1$, $\bar{c} = 0.3$, $c_1(0) = 0.27$, $c_2(0) = 0.28$, $c_3(0) = 0.28$, $c_4(0) = 0.1$ and $c_5(0) = 0.2$. Furthermore, the initial states of each agent are given in Table 4.1.

Table 4.1 Initial states of each agent

Initial states	Agent 1	Agent 2	Agent 3	Agent 4	Agent 5
$x_{i1}/\hat{x}_{i1}$	0.05/-0.15	0.15/–0.05	0.21/0.11	–0.1/0.18	0.1/0.1
$x_{i2}/\hat{x}_{i2}$	–0.1/0.24	0.21/0.16	0.08/0.02	0.31/0.15	–0.25/0.1
$x_{i3}/\hat{x}_{i3}$	0.14/–0.2	0.07/0.21	–0.21/0.23	0.11/–0.24	0.13/0.14
$x_{i4}/\hat{x}_{i4}$	0.21/0.17	–0.12/0.1	0.19/-0.1	0.32/0.02	–0.07/–0.08

It is assumed that the *disturbance* signal $w_1^1(t)$ occurs for the first agent. Define the complete consensus error of the whole system as

$$J(t) = \frac{1}{N}\sqrt{\sum_{i=1}^{N} \|x_i(t) - \bar{x}(t)\|^2}.$$

The simulation results are shown in Figs. 4.2, 4.3, 4.4, 4.5, 4.6. Figure 4.2 displays the state trajectories of the five agents, with gray bars indicating DoS attacks that

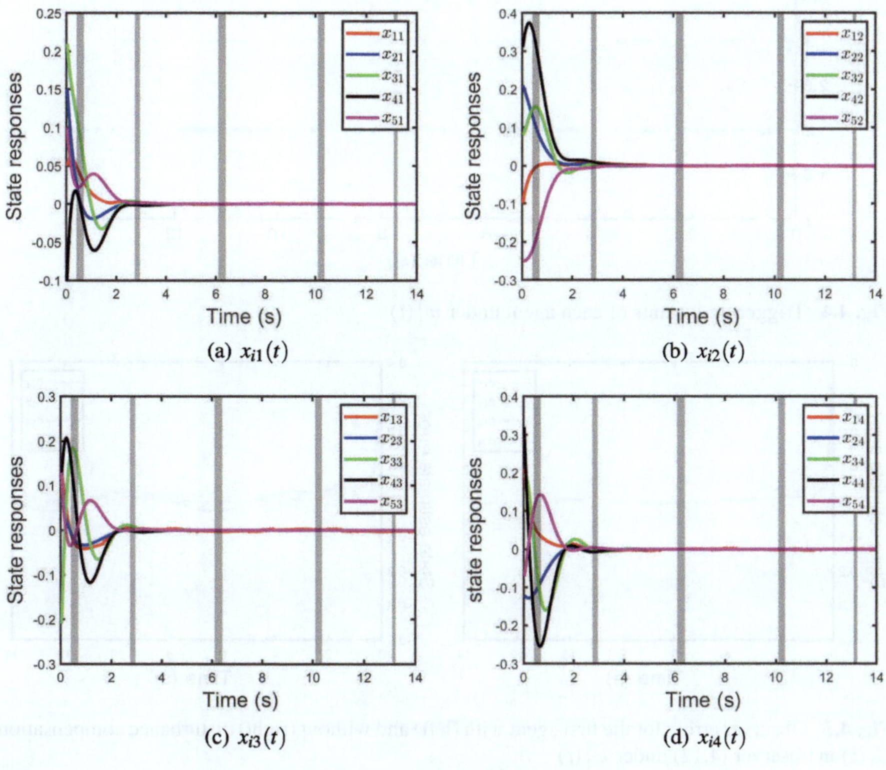

Fig. 4.2 State trajectories of five agents under DoS attacks and $w_1^1(t)$

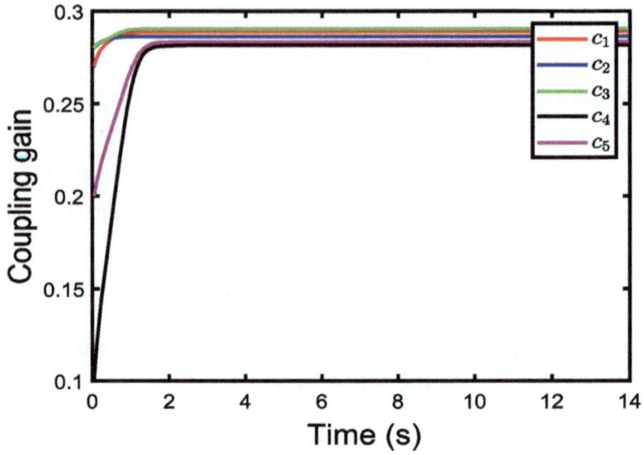

Fig. 4.3 Adaptive coupling gains $c_i(t)$ under $w_1^1(t)$

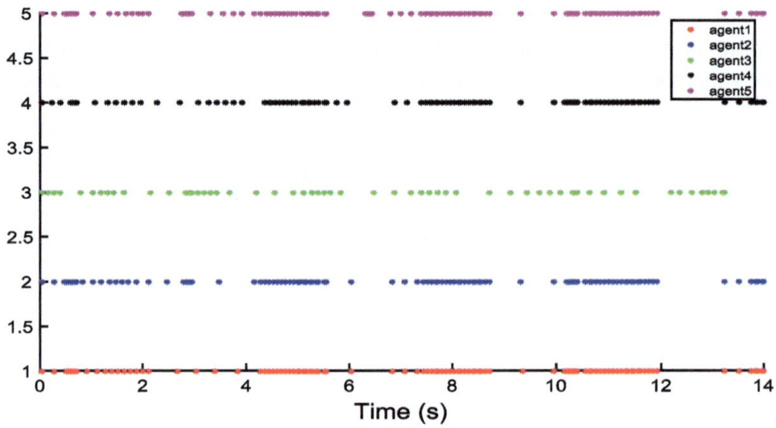

Fig. 4.4 Triggering instants of each agent under $w_1^1(t)$

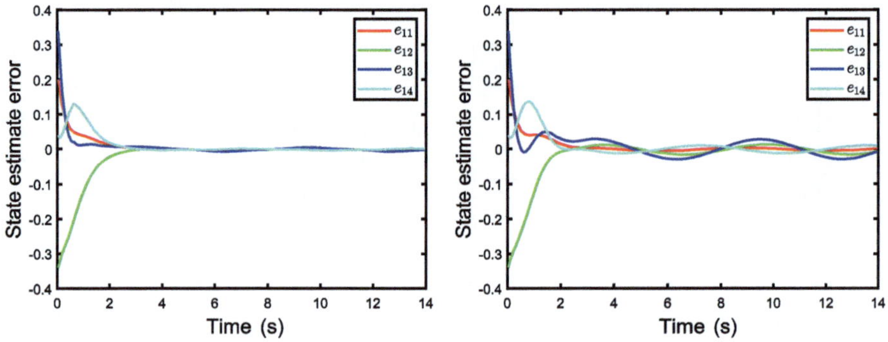

Fig. 4.5 Observer errors for the first agent with (left) and without (right) disturbance compensation $G_i(t)$ in observer (4.12) under $w_1^1(t)$

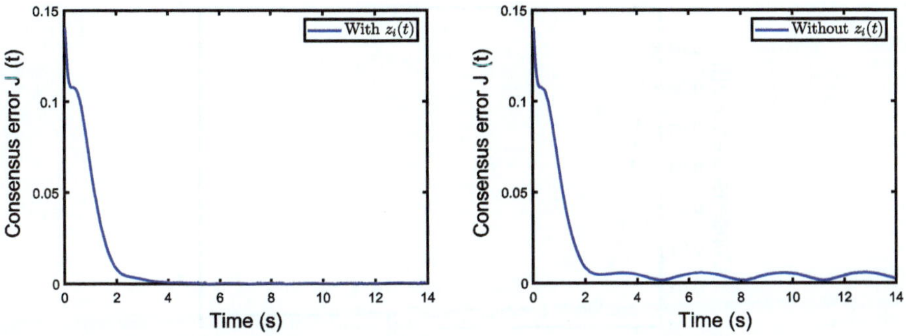

Fig. 4.6 The complete consensus error of five agents with (left) and without (right) disturbance compensation $z_i(t)$ in control scheme (4.15) under $w_1^1(t)$

Table 4.2 Trigger numbers of each agent under different control schemes

Control scheme	Agent 1	Agent 2	Agent 3	Agent 4	Agent 5
Control scheme (4.15)	69	74	68	71	75
Method in [33]	72	104	104	98	96

act on all agents when the attacks occur. Note that our proposed control scheme can achieve better *complete consensus* even in the presence of external disturbances and DoS attacks. The adaptive coupling gains $c_i(t)$ and triggering instants of each agent are shown in Figs. 4.3 and 4.4, respectively. From Fig. 4.4 and through calculation, we can find that the communication failure frequency of each agent satisfies (4.26) and prevents *Zeno behavior*. Additionally, we observe that the periods with dense trigger points correspond to times when the *disturbance* signal experiences the most rapid changes, i.e., when its first derivative is the largest. To achieve the ideal performance in the presence of disturbances, more packets are required to adapt to these disturbances. This requires the event-triggered mechanism to generate more trigger points during the *disturbance*, which aligns with the theoretical analysis. The observer errors for the first agent with or without *disturbance* compensation $G_i(t)$ in observer (4.12) under $w_1^1(t)$ are shown in Fig. 4.5. These results demonstrate that the observer errors gradually approach zero with the inclusion of $G_i(t)$. The complete consensus errors for the five agents with and without *disturbance* compensation $z_i(t)$ in control scheme (4.15) are displayed in Fig. 4.6. From Figs. 4.5 and 4.6, it is clear that introducing *disturbance* compensation $G_i(t)$ and $z_i(t)$ leads to better state estimation and complete consensus performance, validating the proposed method.

To further highlight the advantages of our design method, comparisons with existing methods are made in next. To make the comparisons more obvious, suppose that the *disturbance* signal $w_1^2(t)$ occurs for the first agent, and the initial states are reset as $x_1(0) = [-0.05\ -0.05\ -0.05\ -0.05]^T, x_2(0) = [0.05\,0.05\,0.05\,0.05]^T, x_3(0) = [0.1\ 0.1\ 0.1\ 0.1]^T$, $x_4(0) = [0.15\ 0.15\ 0.15\ 0.15]^T$, $x_5(0) = [0.2\ 0.2\ 0.2\ 0.2]^T$. Figure 4.7 displays the comparison between two different methods, one of which

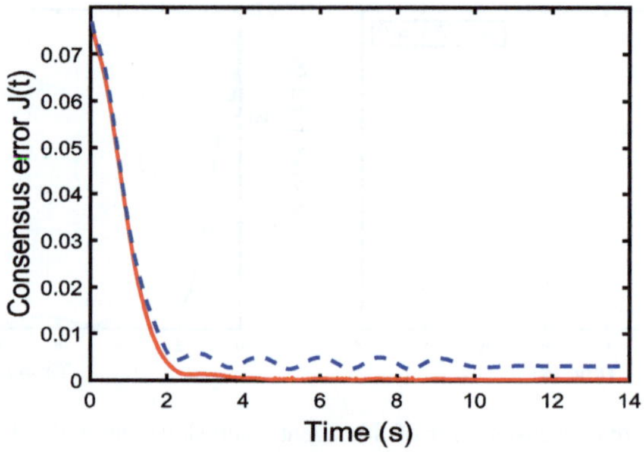

(a) Complete consensus error $J(t)$

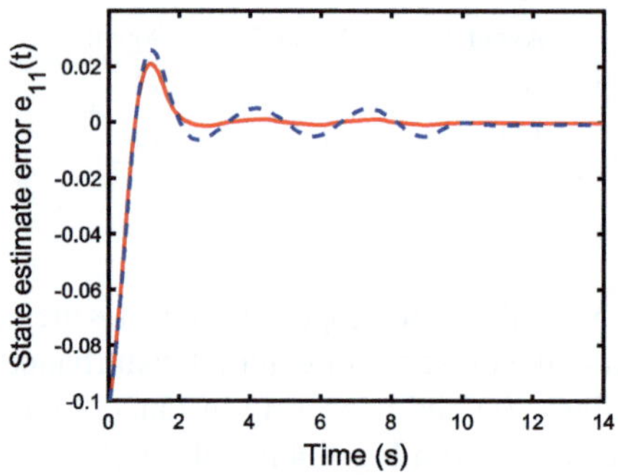

(b) State estimate error $e_{11}(t)$

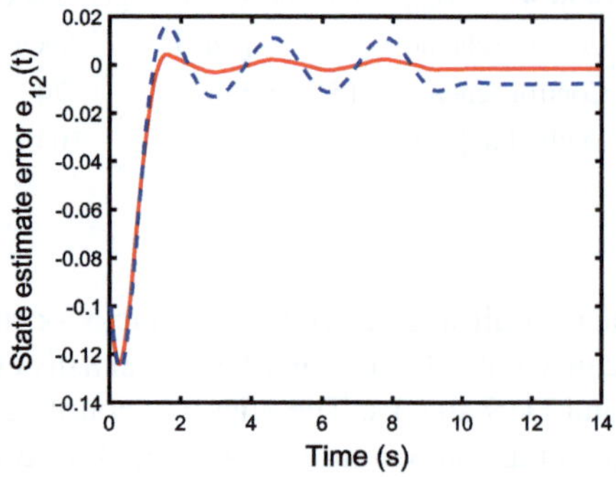

(c) State estimate error $e_{12}(t)$

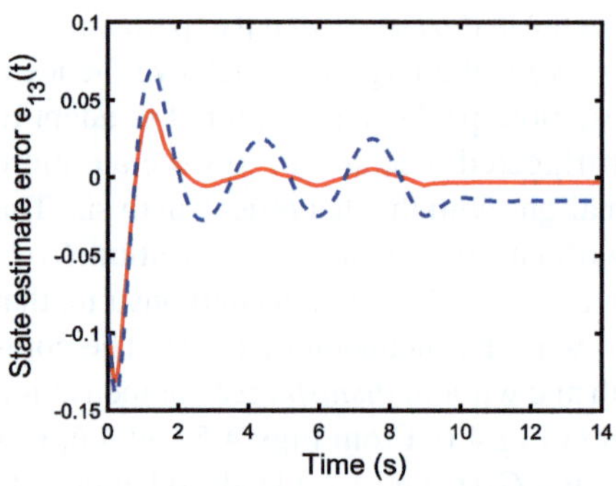

(d) State estimate error $e_{13}(t)$

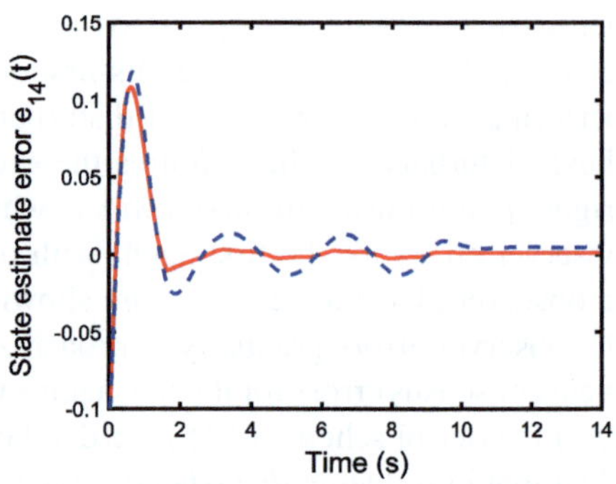

(e) State estimate error $e_{14}(t)$

Fig. 4.7 Comparison between proposed observer-based control scheme (4.15) (solid) in Theorem 1 and the method in [33] (dashed) under $w_1^2(t)$

uses the AMETM (4.8) and the observer-based controller (4.15) proposed in this chapter. The other uses the method of [33]. From Fig. 4.7, it is clear that the control strategy in [33] fails to achieve *complete consensus* due to external disturbances, and the observer cannot accurately track the states. In contrast, our method incorporates a compensation mechanism for both the observer and controller, resulting in better performance. Additionally, Table 4.2 presents the number of triggers for both control schemes. A comparison shows that the method proposed in this chapter provides superior complete consensus and state estimation performance, while also reducing the communication burden.

4.5 Conclusion

This chapter addresses the *complete consensus* problem for nonlinear MASs under DoS attacks, considering external disturbances and uncertainties. An observer-based controller design method is proposed to ensure the *complete consensus* of MAS under DoS attacks and *disturbance* as well as uncertainties signals. The controller gain and observer gain are obtained through a joint design method. Moreover, to reduce the communication burden, an AMETM for MAS is designed, which updates based on both current states and stored information, and *Zeno behavior* is avoided. Simulation results demonstrate the effectiveness of the proposed method. It is worth noting that this chapter focuses on undirected network communication in the cyber layer. Extending these findings to directed network communication remains an interesting future research direction, which will be explored in later chapters.

References

1. D. Zhang, G. Feng, Y. Shi, D. Srinivasan, Physical safety and cyber security analysis of multi-agent systems: a survey of recent advances. IEEE-CAA J. Automatica Sin. **8**(2), 319–333 (2021)
2. G. Franze, F. Tedesco, D. Famularo, Resilience against replay attacks: a distributed model predictive control scheme for networked multi-agent systems. IEEE-CAA J. Automatica Sin. **8**(3), 628–640 (2021)
3. C. Zhou, B. Hu, Y. Shi, Y.C. Tian, X. Li, Y. Zhao, A unified architectural approach for cyberattack-resilient industrial control systems. Proc. IEEE **109**(4), 517–541 (2021)
4. Y. Wan, G. Wen, X. Yu, T. Huang, Distributed consensus tracking of networked agent systems under denial-of-service attacks. IEEE Trans. Syst. Man Cybern. Syst. **51**(10), 6183–6196 (2021)
5. E. Mousavinejad, F.W. Yang, Q.L. Han, X.H. Ge, L. Vlacic, Distributed cyber attacks detection and recovery mechanism for vehicle platooning. IEEE Trans. Intell. Transp. Syst. **21**(9), 3821–3834 (2020)
6. C. De Persis, P. Tesi, Input-to-state stabilizing control under denial-of-service. IEEE Trans. Autom. Control **60**(11), 2930–2944 (2015)
7. Z. Feng, G. Hu, Secure cooperative event-triggered control of linear multiagent systems under DoS attacks. IEEE Trans. Control Syst. Technol. **28**(3), 741–752 (2020)

8. J. Liu, T. Yin, D. Yue, H.R. Karimi, J. Cao, Event-based secure leader-following consensus control for multiagent systems with multiple cyber attacks. IEEE Trans. Cybern. **51**(1), 162–173 (2021)
9. Y. Xu, M. Fang, Z.G. Wu, Y.J. Pan, M. Chadli, T.W. Huang, Input-based event-triggering consensus of multiagent systems under denial-of-service attacks. IEEE Trans. Syst. Man Cybern. Syst. **50**(4), 1455–1464 (2020)
10. Y. Yang, Y. Li, D. Yue, Y.C. Tian, X. Ding, Distributed secure consensus control with event-triggering for multiagent systems under DoS attacks. IEEE Trans. Cybern. **51**(6), 2916–2928 (2021)
11. T.Y. Zhang, D. Ye, Distributed event-triggered control for multi-agent systems under intermittently random denial-of-service attacks. Inf. Sci. **542**, 380–390 (2021)
12. L. Zhao, G.H. Yang, Adaptive fault-tolerant control for nonlinear multi-agent systems with DoS attacks. Inf. Sci. **526**, 39–53 (2020)
13. X.G. Guo, X. Fan, J.L. Wang, J.H. Park, Event-triggered switching-type fault detection and isolation for fuzzy control systems under DoS attacks. IEEE Trans. Fuzzy Syst. **29**(11), 3401–3414 (2021)
14. X.G. Guo, J.L. Wang, F. Liao, Adaptive quantised H_∞ observer-based output feedback control for non-linear systems with input and output quantisation. IET Control Theory Appl. **11**(2), 263–272 (2017)
15. W. Zou, C.K. Ahn, Z. Xiang, Fuzzy-approximation-based distributed fault-tolerant consensus for heterogeneous switched nonlinear multiagent systems. IEEE Trans. Fuzzy Syst. **29**(10), 2916–2925 (2021)
16. S. Hu, D. Yue, Q.L. Han, X. Xie, X. Chen, C. Dou, Observer-based event-triggered control for networked linear systems subject to denial-of-service attacks. IEEE Trans. Cybern. **50**(5), 1952–1964 (2020)
17. Y. Wu, H. Liang, Y. Zhang, C.K. Ahn, Cooperative adaptive dynamic surface control for a class of high-order stochastic nonlinear multiagent systems. IEEE Trans. Cybern. **51**(11), 5214–5224 (2021)
18. Z. Wu, H. Mo, J. Xiong, M. Xie, Adaptive event-triggered observer-based output feedback $\mathscr{L}_\infty$ load frequency control for networked power systems. IEEE Trans. Ind. Inf. **16**(6), 3952–3962 (2020)
19. X.G. Guo, J.L. Wang, F. Liao, R.S.H. Teo, Distributed adaptive sliding mode control strategy for vehicle-following systems with nonlinear acceleration uncertainties. IEEE Trans. Veh. Technol. **66**(2), 981–991 (2017)
20. I. Furtat, E. Fridman, A. Fradkov, Disturbance compensation with finite spectrum assignment for plants with input delay. IEEE Trans. Autom. Control **63**(1), 298–305 (2018)
21. Y. Yuan, Z. Wang, L. Guo, Event-triggered strategy design for discrete-time nonlinear quadratic games with disturbance compensations: the noncooperative case. IEEE Trans. Syst., Man Cybern. Syst. **48**(11), 1885–1896 (2018)
22. S. Hu, D. Yue, X. Xie, X. Chen, X. Yin, Resilient event-triggered controller synthesis of networked control systems under periodic DoS jamming attacks. IEEE Trans. Cybern. **49**(12), 4271–4281 (2019)
23. E. Tian, K. Wang, X. Zhao, S. Shen, J. Liu, An improved memory-event-triggered control for networked control systems. J. Franklin Inst. **356**(13), 7210–7223 (2019)
24. K. Wang, E. Tian, S. Shen, L. Wei, J. Zhang, Input-output finite-time stability for networked control systems with memory event-triggered scheme. J. Franklin Inst. **356**(15), 8507–8520 (2019)
25. X. Huang, J.X. Dong, Reliable leader-to-follower formation control of multiagent systems under communication quantization and attacks. IEEE Trans. Syst. Man Cybern. Syst. **50**(1), 89–99 (2020)
26. X. Wang, Z. Fei, T. Wang, L. Yang, Dynamic event-triggered actuator fault estimation and accommodation for dynamical systems. Inf. Sci. **525**, 119–133 (2020)
27. B. Wang, W. Chen, J. Wang, B. Zhang, Z. Zhang, X. Qiu, Accurate cooperative control for multiple leaders multiagent uncertain systems: a two-layer node-to-node communication framework. IEEE Trans. Ind. Inf. **14**(6), 2395–2405 (2018)

28. Y. Gao, J. Liu, G. Sun, M. Liu, L. Wu, Fault deviation estimation and integral sliding mode control design for Lipschitz nonlinear systems. Syst. Control Lett. **123**, 8–15 (2019)
29. A.U. Rehman, M. Rehan, N. Iqbal, C.K. Ahn, LPV scheme for robust adaptive output feedback consensus of Lipschitz multiagents using Lipschitz nonlinear protocol. IEEE Trans. Syst. Man Cybern. Syst. **51**(11), 7040–7050 (2021)
30. W. Xu, D.W.C. Ho, J. Zhong, B. Chen, Event/self-triggered control for leader-following consensus over unreliable network with dos attacks. IEEE Trans. Neural Netw. Learn. Syst. **30**(10), 3137–3149 (2019)
31. D. Ye, M.M. Chen, H.J. Yang, Distributed adaptive event-triggered fault-tolerant consensus of multiagent systems with general linear dynamics. IEEE Trans. Cybern. **49**(3), 757–767 (2019)
32. Y. Xu, M. Fang, P. Shi, Z.G. Wu, Event-based secure consensus of mutiagent systems against DoS attacks. IEEE Trans. Cybern. **50**(8), 3468–3476 (2020)
33. R. Yang, W.X. Zheng, Output-based event-triggered predictive control for networked control systems. IEEE Trans. Ind. Electron. **67**(12), 10631–10640 (2020)

Chapter 5
H_∞ Anti-disturbance Control and Fault Detection of MASs Under FDIAs

Abstract This chapter addresses the event-triggered observer-based security *complete consensus* and fault detection problem for nonlinear multi-agent systems (MASs) under external disturbances and stochastic false data injection attacks (FDIAs) in a directed communication network. The randomly occurring FDIAs are modeled by Bernoulli-distributed random variables. Then, an observer-based event-triggered control strategy is developed, relying on local measurements and information from neighboring agents, while ensuring that *Zeno behavior* is avoided. Interestingly, the observer errors are first regarded as *disturbance*, and then attenuated by H_∞ norm bounds together with the external disturbances. Additionally, the same measurement information used by the state observers is employed to construct residuals with adaptive thresholds to detect faults in any agent. The accuracy of the observer and the fault detection performance are enhanced by a *disturbance* compensation mechanism. Finally, simulation results demonstrate the effectiveness and advantages of the proposed strategy.

5.1 Introduction

In Chap. 4, the event-triggered secure *complete consensus* problem for nonlinear multi-agent systems (MASs) with uncertainty under denial-of-service (DoS) attacks over an *undirected graph* is studied. Compared with DoS attacks, false data injection attacks (FDIAs) destroy the integrity and accuracy of the data by tampering with the transmitted information, and have stronger concealment ability [1]. However, research on the security of MAS under FDIA is still in the initial stage [2–8]. In [5, 7], a control strategy with an adaptive attack compensator is proposed to ensure the uniform ultimate boundedness of MAS under FDIA. It should be pointed out that the attack model used in these studies assumes the attack persists on the agent after it occurs, which simplifies the design of the attack compensation mechanism. However, in reality, since attackers need to conserve energy for future attacks and conceal their presence, continuous attacks are not feasible. To better reflect this, the FDIA model based on a Bernoulli distribution is considered in [2, 4, 8], and a distributed impulsive

X.-G. Guo et al., *Anti-disturbance Intrusion- and Fault-Tolerant Control of Multi-agent Systems*, https://doi.org/10.1007/978-981-96-9614-7_5

controller is proposed to deal with the control signals attacked by FDIA, ensuring that the MAS achieves mean-square bounded synchronization [4, 8]. In these theoretical research, it is typically assumed that the system's state information is measurable. However, in practical applications, only output information is available, and the state maybe remains unmeasurable [9–11]. Additionally, most of these studies are based on undirected communication topologies and do not consider the effects of *disturbance*. Since the symmetry property of the *Laplacian matrix* does not hold in directed network topologies, and *disturbance* significantly impact MAS performance [12, 13], these results may not be applicable in cases involving directed topology and *disturbance*. Thus, designing an observer-based event-triggered control strategy for MASs under external *disturbance* and FDIA in directed network topologies is the core issue addressed in this chapter.

Furthermore, due to the interconnected nature of MASs, performance degradation or even instability can occur, whether a fault affects a single agent or multiple agents [14]. As a result, fault detection in MASs has gained increasing attention [15–21]. Since faults are usually detected through the cooperation mechanism of MAS based on the communication network, it is crucial to study the fault detection mechanism of MAS under *cyber attacks*. In [20], a co-design of fault detection and *complete consensus* of MASs under hidden DoS attack is explored, establishing a mixed H_-/H_∞ performance index to balance the fault detection and *complete consensus* performance. However, these results rely on fixed thresholds for fault detection, which can lead to false alarms due to non-zero initial observation errors, measurement noise, or *disturbance*. To address this limitation, a distributed fault detection and isolation strategy based on local observers is proposed for networked multi-robot systems in [15]. Meanwhile, the same information (local and received from neighboring agents) used by the local observer is also used to construct the residual and derive the adaptive threshold. However, when the *disturbance* is large, the adaptive threshold for fault detection may also increase, potentially rendering the fault detection mechanism ineffective. Therefore, developing a reliable fault detection mechanism for MASs under *disturbance* and FDIA is a more promising direction.

As mentioned earlier, the main challenges in current research are as follows.

(a) In a directed communication network, how to design an observer-based event-triggered control strategy that ensures *complete consensus* for nonlinear MASs while minimizing communication resource wastage in the presence of *disturbance* and FDIA.
(b) How to develop a feasible fault detection mechanism to reduce false alarms and mis-alarms when the nonlinear MAS is subjected to *disturbance* and FDIA.

To overcome these difficulties, and based on the control strategy in Chap. 4, this chapter studies the secure *complete consensus* problem and fault detection problem for nonlinear MAS under *disturbance* and FDIA. Moreover, the highlights of this chapter are as follows:

(1) Event-triggered state observer for system state estimation: Considering the impact of external *disturbance* and stochastic FDIA, a state observer is designed to eliminate the assumption that the agent's states are measurable, as in [2, 4, 8]. Moreover, it is worth mentioning that the state observation errors are first treated as *disturbance*, and then attenuated by H_∞ norm bounds, along with the external disturbances. Furthermore, unlike the assumption in Chap. 3 that the upper bound of the *disturbance* is known, this chapter removes this assumption.
(2) Novel event-triggered mechanism (ETM): Contrary to the memory ETM in Chaps. 3 and 4, the ETM in this chapter does not require the storage of past data. Additionally, unlike the distributed ETM in [3], which uses a constant threshold for the trigger condition, the proposed method employs a state-dependent threshold. Furthermore, an exponential term ℓe^{-t} as in (5.26) is introduced to help prevent *Zeno behavior* more effectively.
(3) Novel fault detection mechanism based on an adaptive threshold: To reduce the computational burden, the same information used in the state observer is also utilized to construct the residual. Unlike the fault detection logic with fixed thresholds in [16, 19–21], the adaptive threshold helps to minimize false alarms and mis-alarm rates.
(4) *Disturbance* compensation mechanism for observer and fault detection mechanism: Different from the method of using *disturbance* observer in [22] and Chap. 3, the proposed mechanism addresses the challenges posed by unmeasurable states and enhances the accuracy of the state observer. In the *complete consensus* analysis of MASs, an adaptive mechanism is adopted to estimate the upper bounds of the external disturbances, while for fault detection, these bounds are assumed to be known in advance, as in [21, 23, 24]. Besides, the issue of large adaptive thresholds caused by significant *disturbance*, which may lead to mis-alarms as in [15], is mitigated by introducing a *disturbance* compensation mechanism, thereby improving fault detection performance.

5.2 Problem Formulation

As shown in Fig. 5.1, the MAS configuration considered in this chapter consists of sensors, state observers, ETM, controllers, actuators, and the fault detection mechanism, in which the directed communication network among agents may be subject to FDIAs [4, 8]. Besides, when a fault happens in an agent, the fault detection mechanism is to detect the fault promptly.

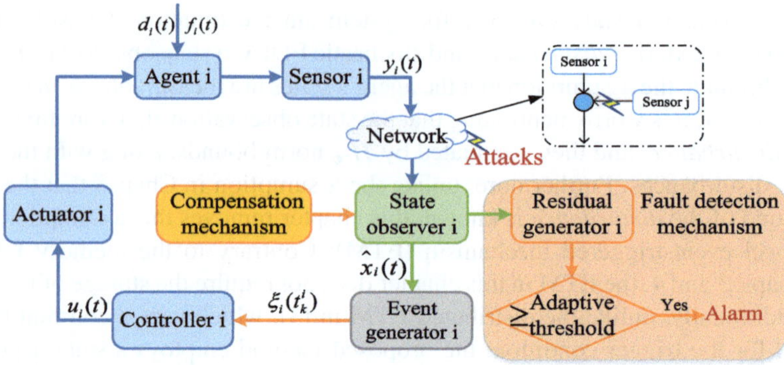

Fig. 5.1 Framework of the ith agent under FDIAs

5.2.1 System Model

Consider a multi-agent system with N followers, where the ith follower is described by

$$\begin{cases} \dot{x}_i(t) = Ax_i(t) + Bu_i(t) + Dd_i(t) + \phi(x_i(t)) \\ y_i(t) = Cx_i(t),\ i = 1, \cdots, N \end{cases} \tag{5.1}$$

where $x_i(t) \in \mathbb{R}^n$, $u_i(t) \in \mathbb{R}^m$, and $y_i(t) \in \mathbb{R}^p$ represent the state, input, and output of the ith agent, respectively. $d_i(t) \in \mathbb{R}$ is the external *disturbance* which is bounded to $\bar{d}_i$, i.e.,

$$\|d_i(t)\| \le \bar{d}_i. \tag{5.2}$$

Furthermore, $\phi(x_i) = [\phi_1(x_i), \cdots, \phi_n(x_i)]^T$ is a nonlinear function vector. Besides, A, B, C, and D are constant matrices.

The dynamics of the leader is described as

$$\begin{cases} \dot{x}_0(t) = Ax_0(t) + \phi(x_0(t)) \\ y_0(t) = Cx_0(t) \end{cases} \tag{5.3}$$

where $x_0(t) \in \mathbb{R}^n$ and $y_0(t) \in \mathbb{R}^p$ are the state and the output of the leader, and $\phi(x_0)$ denotes the nonlinear dynamics of the leader.

Assumption 5.1 (A, B) is stabilizable and (A, C) is detectable.

Assumption 5.2 ([25]) The directed graph $\bar{\mathscr{G}}$ of MASs (5.3) has a directed spanning tree, and the leader is the root node.

Assumption 5.3 For any $x_1(t), x_2(t) \in \mathbb{R}^n$, the nonlinear function $\phi(x_i)$ satisfies

$$\|\phi(x_2) - \phi(x_1)\| \leq \|\Lambda(x_2 - x_1)\| \tag{5.4}$$

where Λ is a symmetric positive definite matrix.

Assumption 5.4 The initial state of the agents $x(0) = \text{col}\{x_1(0), \cdots, x_N(0)\}$ is unknown but bounded as $\|x(0)\| \leq \Delta$, where Δ is a known positive constant.

Remark 5.1 Assumption 5.3, known as the Lipschitz condition, is used to describe a class of nonlinear systems, including complex circuits, robotic manipulators, and others [26]. Assumption 5.4 is a common assumption in fault detection based on adaptive threshold method [24, 27], and it can indicate that the robots start from a known bounded region [15] for example.

5.2.2 False Data Injection Attack Model

Since MASs operate in an open network environment, the information exchanged between agents is susceptible to FDIAs, which can compromise the integrity and accuracy of the transmitted information. For instance, in large-scale smart grid systems consisting of multiple grid sub-regions, continuous monitoring is crucial for ensuring the system's reliable operation. State estimation is typically used to monitor the system by analyzing meter readings and power models to estimate the state of the grid. However, FDIAs can corrupt the meter readings during transmission, leading to incorrect state estimation and potentially disrupting the normal functioning of the smart grid [28]. To make the FDIA model consistent with the reality, let independent stochastic variables $\gamma_{ij}(t)$ indicate whether the attacker launches an attack on the transmission of sensor information from agents j and i, i.e.,

$$\gamma_{ij}(t) = \begin{cases} 1, & \text{edge } (j, i) \text{ suffers from attacks} \\ 0, & \text{otherwise} \end{cases} \tag{5.5}$$

and the decision variable $\gamma_{ij}(t)$ obeys the Bernoulli distribution with the probabilities

$$\text{Prob}\{\gamma_{ij}(t) = 1\} = \chi_{ij}, \quad \text{Prob}\{\gamma_{ij}(t) = 0\} = 1 - \chi_{ij}$$

where $\chi_{ij} \in [0, 1)$ is a constant. Then, $\nu_i(t) \in \mathbb{R}^n$ is defined as the attack signal sent by the attacker. Therefore, the ith agent receives an attacked signal from its neighbor j as

$$\bar{y}_{ij}(t) = y_j(t) + \gamma_{ij}(t)\nu_i(t). \tag{5.6}$$

In practice, because attackers have limited energy and power and cannot sustain continuous attacks, the following restriction holds:

$$\|\nu(t)\| \le \bar{\nu} \tag{5.7}$$

where $\nu(t) = \mathrm{col}\{\nu_1(t), \cdots, \nu_N(t)\}$ and $\bar{\nu}$ is a known positive constant.

Remark 5.2 In some existing studies (e.g., [1, 5]), FDIAs are modeled similarly to faults and disturbances, even though they have fundamentally different characteristics. In these models, FDIAs affect the agent for an extended period. However, in practical applications, the occurrence of attacks is often random. Therefore, a more realistic approach is to model FDIAs using sequences that follow a Bernoulli distribution. Additionally, constraint (5.7) is a common limitation for FDIAs, indicating the fact that the attackers have limited energy or power [3, 4].

5.3 Observer-Based Event-Triggered Control Strategy

In this section, an observer-based event-triggered control strategy is designed, which makes MAS (5.1) and (5.3) achieve a prescribed H_∞ *complete consensus* performance when subjected to external *disturbance* and FDIAs, and *Zeno behavior* is avoided. In addition, a fault detection mechanism based on adaptive threshold is proposed by using the designed state observer.

Define the state estimation error, output estimation error, complete consensus error, and the local neighborhood error as follows, respectively

$$e_{x_i}(t) = x_i(t) - \hat{x}_i(t) \tag{5.8a}$$

$$e_{y_i}(t) = y_i(t) - \hat{y}_i(t) \tag{5.8b}$$

$$\delta_i(t) = x_i(t) - x_0(t) \tag{5.8c}$$

$$\xi_i(t) = \sum_{j \in \mathcal{N}_i} a_{ij}(\hat{x}_j(t) - \hat{x}_i(t)) + a_{i0}(x_0(t) - \hat{x}_i(t)) \tag{5.8d}$$

where $\hat{x}_i(t)$ and $\hat{y}_i(t)$ are the state and output of the observer, respectively, and will be designed later.

5.3.1 Design of Observer and Controller

The state observer for each follower is given by

$$\begin{cases} \dot{\hat{x}}_i(t) = A\hat{x}_i(t) + Bu_i(t) + Dg_i(t) + \phi(\hat{x}_i(t)) \\ \qquad + L\Big[\sum_{j\in\mathcal{N}_i} a_{ij}(e_{y_i} - \bar{y}_{ij}(t) + \hat{y}_j(t)) + a_{i0}(e_{y_i} - e_{y_0})\Big] \\ \hat{y}_i(t) = C\hat{x}_i(t),\ i \in \mathcal{V} \end{cases} \tag{5.9}$$

where L is the observer gain to be designed. The initial state is selected as $\hat{x}_i(0) = 0$, and the adaptive *disturbance* compensation $g_i(t)$ is determined by

$$g_i(t) = \frac{W_1 e_{y_i}(t)\hat{\bar{d}}_i^2(t)}{\|W_1 e_{y_i}(t)\| \left\|\hat{\bar{d}}_i(t)\right\| + \vartheta_{i,1}} \tag{5.10}$$

where $\hat{\bar{d}}_i(t)$ is the estimate of the *disturbance* bound $\bar{d}_i$ with

$$\dot{\hat{\bar{d}}}_i(t) = -\eta_i\vartheta_{i,1}\hat{\bar{d}}_i(t) + \eta_i \|W_1 e_{y_i}(t)\| \tag{5.11}$$

with $\eta_i > 0$, $\vartheta_{i,1} > 0$, and W_1 being a gain matrix to be designed. Note that the leader acts as a command generator, so it is assumed that the leader's state $x_0(t)$ is measurable [11], i.e., $\hat{x}_0(t) = x_0(t)$, hence $e_{y_0}(t) = 0$.

Remark 5.3 During the transmission of data to agent i, the output information from its neighboring agents can be compromised by FDIAs. Consequently, when modeling the output received by the observer (5.9), attack signals are incorporated, and this modeling method is commonly used in the literature on stochastic FDIAs [3, 4, 8]. Furthermore, although the attack signals are unknown, they still have definite values in practice, and as long as they satisfy the constraint (5.7), the designed observer (5.9) can fulfill the desired state estimation objectives. Moreover, the *disturbance* compensation mechanism $g_i(t)$ in (5.10) relies only on agent i's own output information and the observer's output signal. Since output information is typically more accessible than state information, this approach is more practical for implementation.

To alleviate the network load resulting from the heightened agent information transmission, an event-triggered strategy is utilized to minimize the frequency of controller activations and conserve communication resources. Let $\{t_k^i\}$ denote the event-triggered time sequence of the ith agent. Utilizing sampled data and state observer (5.9), an event-triggered control strategy is designed as

$$u_i(t) = K\xi_i(t_k^i),\ t \in [t_k^i, t_{k+1}^i) \tag{5.12}$$

where K is the controller gain matrix, and the event-trigger rule of the sequence $\{t_k^i\}$ will be designed later.

Subsequently, we demonstrate the validity of the designed observer. By combining (5.1) and (5.9), we can obtain

$$\begin{aligned}\dot{e}_{x_i}(t) = & Ae_{x_i}(t) + D(d_i(t) - g_i(t)) + \phi(x_i) - \phi(\hat{x}_i)\\ & - L\Big[\sum_{j\in\mathscr{N}_i} a_{ij}(e_{y_i} - e_{y_j} - \gamma_{ij}(t)\nu_i(t)) + a_{i0}e_{y_i}\Big].\end{aligned} \tag{5.13}$$

Denote

$$\begin{aligned} e_x(t) &= \mathrm{col}\{e_{x_1}(t), \cdots, e_{x_N}(t)\}\\ \tilde{\gamma}_i(t) &= \sum_{j\in\mathscr{N}_i} \gamma_{ij}(t)\\ \tilde{\Upsilon} &= \mathrm{diag}\{\tilde{\gamma}_1(t), \cdots, \tilde{\gamma}_N(t)\}.\end{aligned}$$

Moreover, let

$$\Upsilon \triangleq \mathbb{E}[\tilde{\Upsilon}(t)] = \mathrm{diag}\Big\{\sum_{j\in\mathscr{N}_i} \chi_{1j}, \cdots, \sum_{j\in\mathscr{N}_i} \chi_{Nj}\Big\}. \tag{5.14}$$

Then, (5.13) can be written in the compact form as follows

$$\begin{aligned}\dot{e}_x(t) = & (I_N \otimes A - \mathscr{L}_1 \otimes LC)e_x(t) + (I_N \otimes D)(d(t) - g(t))\\ & + \phi(x) - \phi(\hat{x}) + (\tilde{\Upsilon} \otimes L)\nu(t)\end{aligned} \tag{5.15}$$

where

$$\begin{aligned} & d(t) = \mathrm{col}\{d_1(t), \cdots, d_N(t)\}, \quad g(t) = \mathrm{col}\{g_1(t), \cdots, g_N(t)\}\\ & \phi(x) = \mathrm{col}\{\phi(x_1), \cdots, \phi(x_N)\}, \phi(\hat{x}) = \mathrm{col}\{\phi(\hat{x}_1), \cdots, \phi(\hat{x}_N)\}\\ & \nu(t) = \mathrm{col}\{\nu_1(t), \cdots, \nu_N(t)\}.\end{aligned}$$

The result of the observer design is now presented in Theorem 5.1.

Theorem 5.1 *Under Assumptions 5.2 and 5.3, for given positive scalars β, β_1, κ_1 and positive definite matrix R, the state observer (5.9) can ensure that state estimation error is uniformly ultimately bounded in the presence of external disturbance and FDIAs, if there exist symmetric positive definite matrix P_1 and matrix W_1 satisfying the following LMI conditions:*

$$\begin{bmatrix} \mathrm{He}(P_1^{-1}A) + \frac{1}{\beta_1}\Lambda^T\Lambda + \beta\kappa_1 I_n - \frac{\kappa_1\lambda_0}{\lambda_{\max}(\Theta)}C^T R^{-1} C & P_1^{-1}\\ * & -\frac{1}{\beta_1}I_n\end{bmatrix} < 0 \tag{5.16}$$

$$\begin{bmatrix} -\varepsilon I & P_1^{-1}D - C^T W_1^T \\ * & -\varepsilon I \end{bmatrix} < 0 \tag{5.17}$$

where $\varepsilon > 0$ is a small constant, and λ_0 is defined as in Lemma 2.2. Furthermore, the observer gain is given by $L = \kappa_1 P_1 C^T R^{-1}$.

Proof Let $e_{\bar{d}_i}(t) = \bar{d}_i - \hat{\bar{d}}_i(t)$ with $\bar{d}_i$ being the bound on $d_i(t)$ as in (5.2). Then, by using (5.11), one has

$$\dot{e}_{\bar{d}_i}(t) = \eta_i \vartheta_{i,1} \hat{\bar{d}}_i(t) - \eta_i \left\| W_1 e_{y_i}(t) \right\|. \tag{5.18}$$

Next, construct the Lyapunov function as

$$V_1(t) = e_x^T(t)(\Theta \otimes P_1^{-1})e_x(t) + \sum_{i=1}^{N} \frac{\theta_i}{\eta_i} e_{\bar{d}_i}^2(t) \tag{5.19}$$

where Θ is a diagonal matrix as in Lemma 2.2. Use Definition 2.3 to calculate the infinitesimal operator of $V_1(t)$ and take the mathematical expectation to get

$$\begin{aligned} \mathbb{E}\{\mathscr{I} V_1(t)\} = \mathbb{E}\{&e_x^T[\Theta \otimes (P_1^{-1}A + A^T P_1^{-1}) - \Theta \mathscr{L}_1 \otimes P_1^{-1} LC \\ &- \mathscr{L}_1{}^T \Theta \otimes C^T L^T P_1^{-1}]e_x + 2e_x^T(\Theta \otimes P_1^{-1})[\phi(x) - \phi(\hat{x})] \\ &+ 2e_x^T(\Theta \otimes P_1^{-1}D)d - 2e_x^T(\Theta \otimes P_1^{-1}D)g \\ &+ 2e_x^T(\Theta \tilde{\Upsilon} \otimes P_1^{-1}L)\nu(t) + 2\sum_{i=1}^{N} \frac{\theta_i}{\eta_i} e_{\bar{d}_i}(t)\dot{e}_{\bar{d}_i}(t)\}. \end{aligned} \tag{5.20}$$

Based on Lemma 2.5, Assumption 5.3 and (5.14), we have

$$\begin{aligned} &\mathbb{E}\{2e_x^T(\Theta \tilde{\Upsilon} \otimes P_1^{-1}L)\nu(t)\} \\ &= \mathbb{E}\{2\kappa_1 e_x^T(\Theta \otimes I_n)(\tilde{\Upsilon} \otimes C^T R^{-1})\nu(t)\} \\ &= \mathbb{E}\{2\kappa_1 e_x^T(\Theta \otimes I_n)(\Upsilon \otimes C^T R^{-1})\nu(t)\} \\ &\leq \mathbb{E}\{\beta \kappa_1 e_x^T(\Theta \otimes I_n)e_x + \beta^{-1}\kappa_1 \nu^T(t)(\Upsilon \otimes R^{-1}C)(\Theta \otimes I_n) \\ &\quad \times (\Upsilon \otimes C^T R^{-1})\nu(t)\} \\ &\leq \mathbb{E}\{\beta \kappa_1 e_x^T(\Theta \otimes I_n)e_x\} + \beta^{-1}\kappa_1 \bar{\nu}^2 \lambda_{\max}\big((\Upsilon \otimes R^{-1}C) \\ &\quad \times (\Theta \otimes I_n)(\Upsilon \otimes C^T R^{-1})\big) \end{aligned} \tag{5.21}$$

and

$$\mathbb{E}\{2e_x^T(\Theta \otimes P_1^{-1})[\phi(x) - \phi(\hat{x})]\} \leq \mathbb{E}\{e_x^T[\Theta \otimes (\beta_1 P_1^{-1}P_1^{-1} + \frac{1}{\beta_1}\Lambda^T \Lambda)]e_x\}. \tag{5.22}$$

Additionally, in light of (5.17), (5.18) and the adaptive laws (5.10) and (5.11), it yields that

$$
\begin{aligned}
&\mathbb{E}\{2e_x^T(\Theta \otimes P_1^{-1}D)d - 2e_x^T(\Theta \otimes P_1^{-1}D)g + 2\sum_{i=1}^{N}\frac{\theta_i}{\eta_i}e_{\bar{d}_i}(t)\dot{e}_{\bar{d}_i}(t)\}\\
&\leq \mathbb{E}\left\{2\sum_{i=1}^{N}\|e_{x_i}^T(t)\theta_i P_1^{-1}D\|\,\bar{d}_i - 2\sum_{i=1}^{N}e_{x_i}^T(t)\theta_i P_1^{-1}D\,\frac{W_1 e_{y_i}(t)\hat{\bar{d}}_i^2(t)}{\|W_1 e_{y_i}(t)\|\,\left\|\hat{\bar{d}}_i(t)\right\| + \vartheta_{i,1}}\right.\\
&\quad\left. - 2\sum_{i=1}^{N}\theta_i e_{\bar{d}_i}(t)\,\|W_1 e_{y_i}(t)\| + 2\sum_{i=1}^{N}\theta_i\vartheta_{i,1}e_{\bar{d}_i}(t)\hat{\bar{d}}_i(t)\right\}\\
&\leq \mathbb{E}\left\{2\sum_{i=1}^{N}\theta_i\frac{\|W_1 e_{y_i}(t)\|\,\left\|\hat{\bar{d}}_i(t)\right\|\vartheta_{i,1}}{\|W_1 e_{y_i}(t)\|\,\left\|\hat{\bar{d}}_i(t)\right\| + \vartheta_{i,1}} + 2\sum_{i=1}^{N}\theta_i\vartheta_{i,1}[-(e_{\bar{d}_i} - \frac{1}{2}\bar{d}_i)^2 + \frac{1}{4}\bar{d}_i^2]\right\}\\
&\leq 2\sum_{i=1}^{N}\theta_i\vartheta_{i,1}(1 + \frac{1}{4}\bar{d}_i^2).
\end{aligned}
\tag{5.23}
$$

Then, substituting (5.21)–(5.23) into (5.20), it is clear that

$$
\mathbb{E}\{\mathscr{I}V_1(t)\} \leq \mathbb{E}\{-e_x^T(\Theta \otimes \Xi_1)e_x\} + \Delta_1
$$

where

$$
\Xi_1 = -(P_1^{-1}A + A^T P_1^{-1} + \beta_1 P_1^{-1}P_1^{-1} + \frac{1}{\beta_1}\Lambda^T\Lambda + \beta\kappa_1 I_n - \frac{\kappa_1\lambda_0}{\lambda_{\max}(\Theta)}C^T R^{-1}C)
$$

$$
\Delta_1 = 2\sum_{i=1}^{N}\theta_i\vartheta_{i,1}(1 + \frac{1}{4}\bar{d}_i^2) + \beta^{-1}\kappa_1\bar{\upsilon}^2\lambda_{\max}\big((\Upsilon \otimes R^{-1}C)(\Theta \otimes I_n)(\Upsilon \otimes C^T R^{-1})\big).
$$

From (5.16), we can see that $\Xi_1 > 0$, thus, it can be obtain that

$$
\mathbb{E}\{\mathscr{I}V_1(t)\} \leq -\lambda_{\min}(\Theta)\lambda_{\min}(\Xi_1)\mathbb{E}\{\|e_x\|^2\} + \Delta_1.
$$

Applying the Lyapunov stability theory (Lemma 2.9), the state estimation error is uniformly ultimately bounded with bound

$$
\mathbb{E}\{\|e_x(t)\|\} \leq \sqrt{\frac{\Delta_1}{\lambda_{\min}(\Theta)\lambda_{\min}(\Xi_1)}} \triangleq \bar{\Delta}_1
$$

which completes the proof. □

Remark 5.4 In contrast to [9, 29, 30], we utilize the adaptive updating rule (5.11) to estimate the upper bound $\bar{d}_i$ of *disturbance* $d_i(t)$, and subsequently employ this bound to counteract the disturbance's effects. Consequently, prior knowledge of the disturbance's upper bound is unnecessary in this part. The size of the UUB bound Δ_1 is influenced by $\bar{d}_i$'s, meaning that the larger *disturbance* $d_i(t)$ result in larger estimation error $\bar{\Delta}_1$. Additionally, Δ_1 is also affected by FDIAs. More frequent FDIA attacks (i.e., larger values in Υ in (5.14)) lead to increased Δ_1 and consequently larger estimation error $\bar{\Delta}_1$. However, it is worth noting that the bound of the state estimation error $\bar{\Delta}_1$ can be minimized by selecting appropriate design parameters. Furthermore, for the sake of simplicity in solving the problem, the LMI (5.17) implements the equality constraint $P_1^{-1}D = C^T W_1^T$, where ε in (5.17) represents the degree of similarity between $P_1^{-1}D$ and $C^T W_1^T$. A smaller ε indicates a closer similarity, thus it is crucial to keep ε sufficiently small to minimize errors [9].

Remark 5.5 Apart from being an efficient observer design for MASs subject to FDIAs, the observer design outlined in (5.9) and the *disturbance* compensation method detailed in (5.10) and (5.11) represent a notable advancement compared to the widely-used Extended State Observer (ESO) approach, which has been used extensively in the active *disturbance* rejection control literature [31–33]. Specifically, our approach only requires the disturbances $d_i(t)$ to be bounded, instead of the derivative of $d_i(t)$ being bounded as in ESO. This advantage becomes apparent when considering that while $\sin \omega t$ remains bounded for all frequencies ω, its derivative $\omega \cos \omega t$ becomes unbounded if ω is unbounded.

5.3.2 H_∞ Complete Consensus Performance Analysis

In this subsection, we study the design of the controller (5.12) to ensure that the MASs (5.1) and (5.3) achieve *complete consensus* with a desired H_∞ performance index $\sigma > 0$ in the presence of external *disturbance* and FDIAs, and propose a method to update the event-triggered time series $\{t_k^i\}$.

To achieve the control objective, define measurement error $e_i(t) = \xi_i(t_k^i) - \xi_i(t)$ with $\xi_i(t)$ as in (5.8d). Then, according to (5.8c) and substituting (5.12) into (5.1), we have

$$\begin{aligned}\dot{\delta}(t) =&(I_N \otimes A - \mathscr{L}_1 \otimes BK)\delta(t) + (\mathscr{L}_1 \otimes BK)e_x(t) + (I_N \otimes BK)e(t)\\ &+ (I_N \otimes D)d(t) + \phi(x) - \tilde{\phi}(x_0)\end{aligned} \tag{5.24}$$

where

$$\begin{aligned}&\delta(t) = \mathrm{col}\{\delta_1(t), \cdots, \delta_N(t)\}, \quad e(t) = \mathrm{col}\{e_1(t), \cdots, e_N(t)\}\\ &\phi(x) - \tilde{\phi}(x_0) = \mathrm{col}\{\phi(x_1) - \phi(x_0), \cdots, \phi(x_N) - \phi(x_0)\}.\end{aligned}$$

Next, with the observer-based event-triggered control strategy (5.12), Theorem 5.2 can be obtained.

Theorem 5.2 *Suppose that Assumptions 5.2 and 5.3 hold. Consider the MAS (5.1)–(5.3), under the controller (5.12) driven by the following event-triggered time sequence:*

$$t_{k+1}^i = \inf\{t > t_k^i | z_i(t) > 0\} \tag{5.25}$$

where

$$z_i(t) = \frac{\kappa_2}{\beta_4} \left\| B^T P_2 e_i(t) \right\|^2 - \hbar_i \left\| B^T P_2 \xi_i(t) \right\|^2 - \ell e^{-t} \tag{5.26}$$

with κ_2, β_4, $\hbar_i$, and ℓ being positive constants. If there exist symmetric positive definite matrix P_2 and positive scalars β_2, β_3, and σ such that

$$\begin{bmatrix} \Pi_{11} & P_2^{-1}\Lambda^T & P_2^{-1} \\ * & -\beta_3 I_n & 0 \\ * & * & -I_n \end{bmatrix} < 0 \tag{5.27}$$

$$[\beta_2\kappa_2 + 4\bar{\hbar}((\tilde{\mathscr{N}} + \sqrt{N\tilde{\mathscr{N}}})^2 + 1)]\lambda_{\max}(P_2 B B^T P_2) < \sigma^2 \tag{5.28}$$

where

$$\Pi_{11} = \mathrm{He}(A P_2^{-1}) + \beta_3 I_n + \frac{1}{\sigma^2} D D^T - [\kappa_2 \lambda_{\min}(\mathscr{L}_1 + \mathscr{L}_1^T) - \frac{\kappa_2}{\beta_2} \lambda_{\max}(\mathscr{L}_1 \mathscr{L}_1^T) - \bar{\lambda}] B B^T \tag{5.29}$$

$$\bar{\hbar} = \max\{\hbar_i\} \tag{5.30}$$

$$\lambda = \kappa_2\beta_4 + 4\bar{\hbar}((\tilde{\mathscr{N}} + \sqrt{N\tilde{\mathscr{N}}})^2 + 1). \tag{5.31}$$

Furthermore, $\tilde{\mathscr{N}} = \max_{i\in\mathscr{V}}\{|\mathscr{N}_i|\}$ represents the maximum number of neighbors of the agents and $|\mathscr{N}_i|$ denotes the number of agents in the set $\mathscr{N}_i$. Then, the MAS is asymptotically stable with H_∞ complete consensus performance

$$\mathbb{E}\left\{\int_0^{t_f} \delta^T(t)\delta(t)dt\right\} \le \mathbb{E}\left\{\sigma^2 \int_0^{t_f} \varsigma^T(t)\varsigma(t)dt + V_2(0)\right\} \tag{5.32}$$

where $\varsigma(t) = \mathrm{col}\{e_x(t), d(t)\}$, and $V_2(t)$ is defined in (5.34) below. In addition, Zeno behavior is excluded since the inter-event time is strictly positive, i.e.,

$$t_{k+1}^i - t_k^i = T_k^i \ge \frac{1}{\|A\|} \ln\left(\frac{\|A\| \sqrt{\frac{\beta_4}{\kappa_2}\ell e^{-t}}}{\left\|B^T P_2\right\| \alpha_k^i} + 1\right) > 0 \tag{5.33}$$

where α_k^i will be given in (5.46) below. Finally, the controller gain K in (5.12) is given by $K = \kappa_2 B^T P_2$.

Proof Consider the following Lyapunov function

$$V_2(t) = \delta^T(t)(I_N \otimes P_2)\delta(t) + N\ell e^{-t} \tag{5.34}$$

where ℓ is a positive constant. Taking its infinitesimal operator and expectation along (5.24), it is true that

$$\begin{aligned}\mathbb{E}\{\mathscr{I}V_2(t)\} = &\ \mathbb{E}\{\delta^T[I_N \otimes (P_2A + A^TP_2) - \mathscr{L}_1 \otimes P_2BK - \mathscr{L}_1{}^T \otimes K^TB^TP_2]\delta \\ &+ 2\delta^T(\mathscr{L}_1 \otimes P_2BK)e_x + 2\delta^T(I_N \otimes P_2BK)e - N\ell e^{-t} \\ &+ 2\delta^T(I_N \otimes P_2D)d + 2\delta^T(I_N \otimes P_2)[\phi(x) - \tilde{\phi}(\hat{x})]\}.\end{aligned} \tag{5.35}$$

Using $K = \kappa_2 B^T P_2$, it follows from Lemma 2.7 that

$$\begin{aligned}&\mathbb{E}\{2\delta^T(\mathscr{L}_1 \otimes \kappa_2 P_2BB^TP_2)e_x\} \\ &\le \mathbb{E}\{\frac{1}{\beta_2}\kappa_2\lambda_{\max}(\mathscr{L}_1\mathscr{L}_1^T)\delta^T(I_N \otimes P_2BB^TP_2)\delta + \beta_2\kappa_2\lambda_{\max}(P_2BB^TP_2)e_x^Te_x\}\end{aligned} \tag{5.36}$$

$$\mathbb{E}\{2\delta^T(I_N \otimes P_2D)d\} \le \mathbb{E}\{\frac{1}{\sigma^2}\delta^T(I_N \otimes P_2DD^TP_2)\delta + \sigma^2 d^T(t)d(t)\} \tag{5.37}$$

and

$$\begin{aligned}&\mathbb{E}\{2\delta^T(I_N \otimes \kappa_2 P_2BB^TP_2)e\} \\ &\le \mathbb{E}\{\frac{\kappa_2}{\beta_4}e^T(I_N \otimes P_2BB^TP_2)e + \beta_4\kappa_2\delta^T(I_N \otimes P_2BB^TP_2)\delta\}.\end{aligned} \tag{5.38}$$

By utilizing the event-triggered condition (5.25), for any $t \in [t_k^i, t_{k+1}^i)$, one can deduce that

$$\begin{aligned}&\frac{\kappa_2}{\beta_4}e^T(t)(I_N \otimes P_2BB^TP_2)e(t) \\ &\le \bar{h}\sum_{i=1}^{N}\left\|B^TP_2\xi_i(t)\right\|^2 + N\ell e^{-t} \\ &\le 2\bar{h}\left(\sum_{i=1}^{N}\left(\left\|B^TP_2q_i^a(t)\right\|^2 + \left\|B^TP_2q_i^b(t)\right\|^2\right)\right) + N\ell e^{-t}\end{aligned} \tag{5.39}$$

where

$$q_i^a(t) = \sum_{j\in\mathcal{N}_i} a_{ij}(\hat{x}_j(t) - \hat{x}_i(t)), \quad q_i^b(t) = x_0(t) - \hat{x}_i(t).$$

Next, denote

$$q^a(t) = \mathrm{col}\{q_1^a(t), \cdots, q_N^a(t)\}$$
$$q^b(t) = \mathrm{col}\{q_1^b(t), \cdots, q_N^b(t)\}.$$

Then, we can conclude that the following properties are satisfied:

$$\begin{aligned}\left\|B^T P_2 q_i^a(t)\right\| &\le \sum_{j\in\mathcal{N}_i}\left(\left\|B^T P_2 q_i^b(t)\right\| + \left\|B^T P_2 q_j^b(t)\right\|\right)\\ &\le \tilde{\mathcal{N}}\left\|B^T P_2 q_i^b(t)\right\| + \sqrt{\tilde{\mathcal{N}}}\left\|(I_N\otimes B^T P_2)q^b(t)\right\|\end{aligned} \tag{5.40}$$

$$\left\|(I_N\otimes B^T P_2)q^b(t)\right\|^2 \le 2\left\|(I_N\otimes B^T P_2)\delta(t)\right\|^2 + 2\left\|(I_N\otimes B^T P_2)e_x(t)\right\|^2. \tag{5.41}$$

Thus, one has

$$\begin{aligned}\sum_{i=1}^{N}\left\|B^T P_2 q_i^a(t)\right\|^2 &\le (\tilde{\mathcal{N}} + \sqrt{N\tilde{\mathcal{N}}})^2\left\|(I_N\otimes B^T P_2)q^b(t)\right\|^2\\ &\le 2(\tilde{\mathcal{N}} + \sqrt{N\tilde{\mathcal{N}}})^2\Big[\left\|(I_N\otimes B^T P_2)\delta(t)\right\|^2\\ &\quad + \left\|(I_N\otimes B^T P_2)e_x(t)\right\|^2\Big].\end{aligned} \tag{5.42}$$

In order to establish the desired H_∞ *complete consensus* performance (5.32) for the MAS, the following function is introduced:

$$\mathcal{H}(t) = \mathbb{E}\{\mathcal{I} V_2(t) + \delta^T(t)\delta(t) - \sigma^2\varsigma^T(t)\varsigma(t)\}. \tag{5.43}$$

Then, combining with (5.35)–(5.43), one has

$$\begin{aligned}\mathcal{H}(t) \le \mathbb{E}\{&\delta^T[I_N\otimes(P_2A + A^TP_2 - \kappa_2\lambda_{\min}(\mathcal{L}_1 + \mathcal{L}_1^T)P_2BB^TP_2\\ &+ \frac{\kappa_2}{\beta_2}\lambda_{\max}(\mathcal{L}_1\mathcal{L}_1^T)P_2BB^TP_2 + \beta_3P_2P_2 + \frac{1}{\beta_3}\Lambda^T\Lambda\\ &+ \frac{1}{\sigma^2}P_2DD^TP_2 + \bar{\lambda}P_2BB^TP_2)]\delta + [\beta_2\kappa_2 + \sigma^2 d^T(t)d(t)\\ &+ 4\bar{h}((\tilde{\mathcal{N}} + \sqrt{N\tilde{\mathcal{N}}})^2 + 1)]\lambda_{\max}(P_2BB^TP_2)e_x^Te_x\\ &+ \delta^T(t)\delta(t) - \sigma^2e_x^Te_x - \sigma^2d^T(t)d(t)\}.\end{aligned}$$

Using (5.27) and applying Schur complement lemma (Lemma 2.8), and premultiplying and postmultiplying P_2 on both sides of it, it can be inferred that (5.27) and (5.28) can guarantee $\mathscr{H}(t) < 0$, integration of which yields

$$\mathbb{E}\left\{V_2(t_f) - V_2(0) + \int_0^{t_f} [\delta^T(t)\delta(t) - \sigma^2\varsigma^T(t)\varsigma(t)]dt\right\} < 0. \tag{5.44}$$

Therefore, the MAS can achieve the desired H_∞ performance (5.32) with the proposed event-triggered control strategy (5.12).

In what follows, we will prove that *Zeno behavior* is excluded by proving that the minimum interval between any two event-intervals is positive. For $t \in [t_k^i, t_{k+1}^i)$, one has

$$\frac{d\left\|e_i(t)\right\|}{dt} = \frac{d}{dt}(e_i^T e_i)^{\frac{1}{2}} \leq \left\|\dot{e}_i\right\|. \tag{5.45}$$

Similar to the proof process of Theorem 5.1, we know that the complete consensus error is also bounded. Assume that the bound is $\tilde{\Delta}$. So, one gets

$$\begin{aligned}
\|\dot{e}_i(t)\| &= \left\|-\dot{\xi}_i(t)\right\| \\
&= \left\|\sum_{j\in\mathscr{N}_i} a_{ij}(\dot{\hat{x}}_j(t) - \dot{\hat{x}}_i(t)) + a_{i0}(\dot{x}_0(t) - \dot{\hat{x}}_i(t))\right\| \\
&= \left\| A\xi_i(t) + BK\Big(\sum_{j\in\mathscr{N}_i} a_{ij}\left(\xi_j(t_k^i) - \xi_i(t_k^i)\right) - a_{i0}\xi_i(t_k^i)\Big)\right. \\
&\quad + \sum_{j\in\mathscr{N}_i} a_{ij}\left(\phi(\hat{x}_j) - \phi(\hat{x}_i)\right) + a_{i0}\left(\phi(x_0) - \phi(\hat{x}_i)\right) \\
&\quad + D\Big(\sum_{j\in\mathscr{N}_i} a_{ij}(g_j(t) - g_i(t)) - a_{i0}g_i(t)\Big) \\
&\quad + L\Big[\sum_{j\in\mathscr{N}_i} a_{ij}\Big(\sum_{m\in\mathscr{N}_j} a_{jm}(e_{y_j} - e_{y_m} + \gamma_{jm}\nu_j) + a_{i0}e_{y_j} \\
&\quad - \sum_{j\in\mathscr{N}_i} a_{ij}(e_{y_i} - e_{y_j} + \gamma_{ij}\nu_i) - a_{i0}e_{y_i}\Big) \\
&\quad \left. - a_{i0}\Big(\sum_{j\in\mathscr{N}_i} a_{ij}(e_{y_i} - e_{y_j} + \gamma_{ij}\nu_i) + a_{i0}e_{y_i}\Big)\Big]\right\| \\
&\leq \|A\|\,\|e_i\| + \alpha_k^i
\end{aligned}$$

where

$$\alpha_k^i = \max_{t\in[t_k^i,t_{k+1}^i)} \Big\{ \Big\| A\xi_i(t_k^i) + BK[\sum_{j\in\mathscr{N}_i} a_{ij}(\xi_j(t_k^i) - \xi_i(t_k^i)) - a_{i0}\xi_i(t_k^i)]$$

$$+ D[\sum_{j\in\mathscr{N}_i} a_{ij}(g_j(t) - g_i(t)) - a_{i0}g_i(t)]\Big\| + (2\tilde{\mathscr{N}} + a_{i0})\|\Lambda\|(\bar{\Delta}_1 + \tilde{\Delta})$$

$$+ \|LC\|(4\tilde{\mathscr{N}}^2 + 2a_{i0}\tilde{\mathscr{N}} + 2\tilde{\mathscr{N}} + a_{i0})\bar{\Delta}_1 + \|L\|(2\tilde{\mathscr{N}}^2 + a_{i0}\tilde{\mathscr{N}})\bar{\upsilon}\Big\} > 0 \tag{5.46}$$

from which one obtains

$$\|e_i(t)\| \le \frac{\alpha_k^i}{\|A\|}\left(e^{\|A\|(t-t_k^i)} - 1\right). \tag{5.47}$$

Thus, we have

$$\left\|B^T P_2 e_i(t)\right\| \le \frac{\left\|B^T P_2\right\| \alpha_k^i}{\|A\|}\left(e^{\|A\|(t-t_k^i)} - 1\right). \tag{5.48}$$

According to the event-triggered condition (5.25) and using (5.48), it yields that:

$$\sqrt{\frac{\beta_4}{\kappa_2}\ell e^{-t}} \le \frac{\left\|B^T P_2\right\| \alpha_k^i}{\|A\|}\left(e^{\|A\|(t_{k+1}^i-t_k^i)} - 1\right) \tag{5.49}$$

which implies that (5.33) holds. Therefore, it is demonstrated that the designed event-triggered condition prevents the occurrence of *Zeno behavior*. This concludes the proof. □

Remark 5.6 To facilitate the solution of the matrix parameters in Theorem 5.2, we can first solve P_2 using (5.27) by tuning the parameters β_2, β_3, β_4, σ, κ_2, and $\hbar_i$. Subsequently, we compute $\lambda_{\max}(P_2BB^TP_2)$ and substitute it into (5.28) to verify if the required condition is met. If it is not, we adjust the aforementioned parameters accordingly, thereby simplifying the process by eliminating the need to solve two inequalities concurrently. Furthermore, the control strategy introduced in this chapter, tailored for directed communication topologies, enjoys broader applicability compared to the undirected communication topology discussed in [10]. This is because an *undirected graph* can be regarded as a special case of a *directed graph* where agents engage in bidirectional communication. Moreover, the asymmetry of the *Laplacian matrix* $\mathscr{L}_1$ and $\mathscr{L}$, alongside external disturbances and stochastic FDIAs, renders the task of controller design and *complete consensus* analysis more complex than that addressed in [10].

Remark 5.7 In Theorem 5.2, the solution of LMIs is related to the setting of the H_∞ performance bound σ. Specifically, the large the H_∞ performance bound σ, the easier it is to solve the LMIs. However, a large H_∞ performance bound σ will cause the performance of the MAS to deteriorate. Therefore, the choice of the H_∞ performance bound σ is a tradeoff. It should be noted that, in practical applications, users can minimize the H_∞ performance bound σ to optimize the performance of the MAS.

Remark 5.8 In analyzing various parameter sets for the ETM (5.25), we discovered that reducing $\hbar_i$ and increasing $\frac{\kappa_2}{\beta_4}$ can lead to more points being triggered, thereby enhancing system performance. By adjusting these parameters, we can strike a balance between system performance and communication overhead. Unlike the centralized ETM [34, 35], the ETM (5.25) does not require global state information, minimizing the unnecessary consumption of communication resources. Furthermore, in contrast to the distributed ETM in [3], which employs a constant threshold, our ETM uses a state-dependent threshold in the trigger function (5.26). This approach further lowers the frequency of control input updates and reduces the communication burden. Additionally, by incorporating a time-dependent term ℓe^{-t} into the trigger function (5.26), we can easily prevent *Zeno behavior*.

Remark 5.9 As stated in [36, 37], if a variable is bounded and converges to zero or stays within a limited area close to the origin after the system is stable, it can be considered a *disturbance* during stability analysis. In Theorem 5.1, it is shown that the state estimation errors resulting from FDIAs are UUB, with the boundary of these errors being adjustable and able to be made arbitrarily small by tuning the design parameters. Therefore, it is reasonable to regard the state observation error as *disturbance* during the *complete consensus* analysis process.

5.3.3 Fault Detection Mechanism

When agents suffer from faults, it will lead to performance degradation or even instability of the MAS, so we propose a fault detection mechanism to detect faults promptly. Meanwhile, to avoid false alarms and mis-alarms of the fault detection mechanism, an adaptive threshold is derived, which reduces the conservatism of the mechanism. In addition, in line with the assumptions made in [21, 23, 24], we presume that $\bar{d}_i$ is known beforehand for the purpose of developing the fault detection mechanism. The model of the faulty agent can be characterized as follows:

$$\begin{cases} \dot{x}_i(t) = Ax_i(t) + Bu_i(t) + Dd_i(t) + Ff_i(t) + \phi(x_i) \\ y_i(t) = Cx_i(t),\ i = 1, \cdots, N \end{cases} \tag{5.50}$$

where $f_i(t)$ denote the fault signals. When the upper bound of *disturbance* $\bar{d}_i$ is known, the adaptive *disturbance* compensation $g_i(t)$ in (5.9) can be redesigned as

$$g_i(t) = \frac{W_2 e_{y_i}(t)\bar{d}_i^2}{\|W_2 e_{y_i}(t)\|\bar{d}_i + \vartheta_{i,2}} \tag{5.51}$$

where W_2 is the gain matrix whose design process is the same as W_1 in (5.10), and $\vartheta_{i,2}$ is the same positive constant as $\vartheta_{i,1}$ in (5.10). These two parameters need to be redesigned here, different from those in (5.10). Moreover, it follows from (5.9) and (5.50) that

$$\begin{aligned}\dot{e}_x(t) &= (I_N \otimes A - \mathscr{L}_1 \otimes LC)e_x(t) + (I_N \otimes D)(d(t) - g(t)) \\ &\quad + \phi(x) - \phi(\hat{x}) + (I_N \otimes F)f(t) + (\tilde{\Upsilon} \otimes L)\nu(t)\end{aligned} \tag{5.52}$$

where $f(t) = \text{col}\{f_1(t), \cdots, f_N(t)\}$.

In order to detect the fault, define the residual as

$$r_i(t) = \sum_{j \in \mathscr{N}_i} a_{ij}(e_{y_i} - e_{y_j}) + a_{i0}(e_{y_i} - e_{y_0}) \tag{5.53}$$

$$r(t) = \text{col}\{r_1(t), \cdots, r_N(t)\}. \tag{5.54}$$

Then, we can obtain

$$\mathbb{E}\{r(t)\} = \mathbb{E}\{(\mathscr{L}_1 \otimes C)e_x\}. \tag{5.55}$$

Furthermore, one has

$$r_i(t) = \Psi_i r(t) \tag{5.56}$$

where

$$\Psi_i = \{\mathbf{0}_p, \cdots, \underbrace{\mathbf{I}_p}_{i\text{th}}, \cdots, \mathbf{0}_p\} \in \mathbb{R}^{p \times Np}.$$

To this end, we can build an evaluation function utilizing the residual signal $r_i(t)$ and detect faults by comparing it to an adaptive threshold. Additionally, it is worth noting that defining $r_i(t)$ in (5.53) as residual helps to decrease computational load, since the same term as $r_i(t)$ is already incorporated in the observer (5.9).

The fault detection algorithm is presented in Theorem 5.3.

Theorem 5.3 *The following fault detection logic can be used to detect the occurrence of faults in MAS (5.1) and (5.3)*

$$\begin{cases} J_{r_i(t)} > J_{th_i}(t), & f_i(t) \neq 0 \text{ or } f_j(t) \neq 0,\ j \in \mathscr{N}_i \\ J_{r_i(t)} < J_{th_i}(t), & \text{No fault} \end{cases} \tag{5.57}$$

where $J_{r_i(t)} = \mathbb{E}\{\|r_i(t)\|^2\}$ *is an evaluation function,* $J_{th_i}(t)$ *is the adaptive threshold that is designed as*

$$J_{th_i}(t) = \frac{\lambda_{\max}((\mathscr{L}_1^T \otimes C^T)\Psi_i^T \Psi_i(\mathscr{L}_1 \otimes C))}{\lambda_{\min}(\Theta)\lambda_{\min}(P_3^{-1})} \Big\{ \lambda_{\max}(\Theta)\lambda_{\max}(P_3^{-1})\Delta^2 e^{-\frac{\lambda_{\min}(\Xi_2)}{\lambda_{\max}(P_3^{-1})}t} + \frac{\kappa_3}{\beta_5}\varpi\bar{\nu}^2 \frac{\lambda_{\max}(P_3^{-1})}{\lambda_{\min}(\Xi_2)}\left(1 - e^{-\frac{\lambda_{\min}(\Xi_2)}{\lambda_{\max}(P_3^{-1})}t}\right)\Big\} \tag{5.58}$$

where P_3 is a symmetric positive definite matrix that ensures

$$\begin{aligned} \Xi_2 = &-(P_3^{-1}A + A^T P_3^{-1} + \beta_6 P_3^{-1} P_3^{-1} + \frac{1}{\beta_6}\Lambda^T\Lambda + \beta_5\kappa_3 I_n \\ &- \frac{\kappa_3\lambda_0}{\lambda_{\max}(\Theta)} C^T R_1^{-1} C + \beta_7\lambda_{\max}(\Theta) P_3^{-1} P_3^{-1}) > 0 \end{aligned} \tag{5.59}$$

$$\varpi = \lambda_{\max}\big((\Upsilon \otimes R_1^{-1} C)(\Theta \otimes I_n)(\Upsilon \otimes C^T R_1^{-1})\big) \tag{5.60}$$

where Υ is as in (5.14), R_1 is a positive definite matrix, and β_5, β_6, β_7, κ_3 are positive constants. The observer gain L is redesigned as $L = \kappa_3 P_3 C^T R_1^{-1}$.

Proof Choose the Lyapunov function as

$$V_3(t) = e_x^T(t)(\Theta \otimes P_3^{-1})e_x(t). \tag{5.61}$$

By employing a proof process analogous to that of Theorem 5.1, we can obtain

$$\begin{aligned} &\mathbb{E}\{\mathscr{I} V_3(t)\} \\ &= \mathbb{E}\Big\{ e_x^T[\Theta \otimes (P_3^{-1}A + A^T P_3^{-1} + \beta_6 P_3^{-1}P_3^{-1} + \frac{1}{\beta_6}\Lambda^T\Lambda + \beta_5\kappa_3 I_n + \frac{\kappa_3}{\beta_5}\varpi\bar{\nu}^2 \\ &\quad - \frac{\kappa_3\lambda_0}{\lambda_{\max}(\Theta)} C^T R_1^{-1} C + \beta_7\lambda_{\max}(\Theta)P_3^{-1}P_3^{-1})]e_x + \frac{1}{\beta_7} f^T(I_N \otimes F^T F) f \Big\} \\ &\le -\frac{\lambda_{\min}(\Xi_2)}{\lambda_{\max}(P_3^{-1})}\mathbb{E}\{V_3(t)\} + \frac{\kappa_3}{\beta_5}\varpi\bar{\nu}^2 + \frac{1}{\beta_7} f^T(I_N \otimes F^T F) f. \end{aligned} \tag{5.62}$$

From (5.61), it is obvious that

$$\lambda_{\min}(\Theta)\lambda_{\min}(P_3^{-1})\mathbb{E}\{\|e_x\|^2\} \le \mathbb{E}\{V_3(t)\} \le \lambda_{\max}(\Theta)\lambda_{\max}(P_3^{-1})\mathbb{E}\{\|e_x\|^2\}.$$

Based on the above inequality and integrating (5.62) results in

$$\begin{aligned} &\lambda_{\min}(\Theta)\lambda_{\min}(P_3^{-1})\mathbb{E}\{\|e_x\|^2\} \\ &\le e^{-\frac{\lambda_{\min}(\Xi_2)}{\lambda_{\max}(P_3^{-1})}t}\lambda_{\max}(\Theta)\lambda_{\max}(P_3^{-1})\mathbb{E}\,\|e_x(0)\|^2 + \frac{\kappa_3}{\beta_5}\varpi\bar{\nu}^2\int_0^t e^{-\frac{\lambda_{\min}(\Xi_2)}{\lambda_{\max}(P_3^{-1})}(t-\tau)}d\tau \\ &\quad + \frac{1}{\beta_7}\int_0^t f^T(I_N \otimes F^T F) f e^{-\frac{\lambda_{\min}(\Xi_2)}{\lambda_{\max}(P_3^{-1})}(t-\tau)}d\tau. \end{aligned} \tag{5.63}$$

Using (5.55) and (5.56), we have

$$\begin{aligned}\mathbb{E}\{r_i^T(t)r_i(t)\} &\leq \mathbb{E}\{e_x^T(\mathscr{L}_1^T \otimes C^T)\Psi_i^T\Psi_i(\mathscr{L}_1 \otimes C)e_x\}\\ &\leq \lambda_{\max}((\mathscr{L}_1^T \otimes C^T)\Psi_i^T\Psi_i(\mathscr{L}_1 \otimes C))\mathbb{E}\{\|e_x\|^2\}. \end{aligned}\tag{5.64}$$

Then, combining (5.63) and (5.64) and considering the *initial condition* $\hat{x}_i(0) = 0$, it is obtained that

$$\begin{aligned}&\frac{\lambda_{\min}(\Theta)\lambda_{\min}(P_3^{-1})}{\lambda_{\max}((\mathscr{L}_1^T \otimes C^T)\Psi_i^T\Psi_i(\mathscr{L}_1 \otimes C))}\mathbb{E}\{r_i^T(t)r_i(t)\}\\ &\leq \lambda_{\max}(\Theta)\lambda_{\max}(P_3^{-1})\,\|x(0)\|^2\, e^{-\frac{\lambda_{\min}(\Xi_2)}{\lambda_{\max}(P_3^{-1})}t} + \frac{\kappa_3}{\beta_5}\varpi\bar{\nu}^2\frac{\lambda_{\max}(P_3^{-1})}{\lambda_{\min}(\Xi_2)}\left(1 - e^{-\frac{\lambda_{\min}(\Xi_2)}{\lambda_{\max}(P_3^{-1})}t}\right)\\ &+ \frac{1}{\beta_7}\int_0^t f^T(I_N \otimes F^T F) f e^{-\frac{\lambda_{\min}(\Xi_2)}{\lambda_{\max}(P_3^{-1})}(t-\tau)} d\tau. \end{aligned}\tag{5.65}$$

Thus, if $J_{r_i(t)} > J_{th_i}(t)$, which means that the ith agent or one of its neighbors is faulty. This completes the proof. □

Remark 5.10 Similar to Theorem 5.1, the matrix parameters P_3 and W_2 can be solved by the following LMIs:

$$\begin{bmatrix} \Omega_{11} & P_3^{-1} & P_3^{-1} \\ * & -\frac{1}{\beta_6}I_n & 0 \\ * & * & -\frac{1}{\beta_7\lambda_{\max}(\Theta)}I_n \end{bmatrix} < 0 \tag{5.66}$$

$$\begin{bmatrix} -\varepsilon I & P_3^{-1}D - C^T W_2^T \\ * & -\varepsilon I \end{bmatrix} < 0 \tag{5.67}$$

where

$$\Omega_{11} = P_3^{-1}A + A^T P_3^{-1} + \frac{1}{\beta_6}\Lambda^T\Lambda + \beta_5\kappa_3 I_n - \frac{\kappa_3\lambda_0}{\lambda_{\max}(\Theta)}C^T R_1^{-1}C$$

and $\beta_5, \beta_6, \beta_7, \kappa_3, \varepsilon$ are positive constants. Furthermore, the adaptive threshold (5.58) ultimately converges to a constant value that is associated with the attack signal. A larger attack signal results in a higher adaptive threshold, potentially compromising the accuracy and responsiveness of the fault detection mechanism, or even leading to false alarms. However, considering the stealthiness and energy constraints of FDIAs, they typically have smaller amplitudes compared to fault signals. By adjusting parameters, we can set the adaptive threshold to a sufficiently low level to maintain the effectiveness of the fault detection mechanism.

Remark 5.11 Note that, in contrast to the *complete consensus* analysis in Theorems 5.1 and 5.2, which does not necessitate knowing the upper bounds of disturbances, Theorem 5.3 necessitates these bounds and assumes they are pre-known for effective fault detection. This is a common assumption in fault detection literature [21, 23, 24]. When the upper bounds are known, the *disturbance* compensation (5.51) can be employed to counteract the disturbance's impact on fault detection, thereby preventing issues where the adaptive threshold incorporates a *disturbance* upper bound [15], which degrades the performance of fault detection. Moreover, the *disturbance* compensation (5.51) is also applicable in scenarios of *complete consensus* analysis where the *disturbance* upper bound is known.

5.4 Simulation Examples

In this section, simulation results are given to confirm the validity of our main findings. Let us consider the communication topology depicted in Fig. 5.2, and the parameter matrices specified in (5.1) are listed below.

$$A = \begin{bmatrix} -2.9 & 0.3 & 0.4 & 1.2 \\ -0.1 & -0.2 & 0.6 & 1.5 \\ 1.2 & 2.1 & -2.8 & 3.4 \\ 1 & -2 & -2.5 & -2.5 \end{bmatrix}, \quad B = \begin{bmatrix} 0 & 0 \\ 0 & 0 \\ -1 & 0.5 \\ -0.1 & 0.2 \end{bmatrix}$$

$$C = \begin{bmatrix} 1 & 0 & 0 & 0 \\ 0 & 1 & 0 & 0 \\ 0 & 0 & 1 & 1 \end{bmatrix}, \quad D = \begin{bmatrix} 0.1 & 0.1 & -0.1 & -0.1 \end{bmatrix}^T.$$

The nonlinear function $\phi(x_i)$ is considered as $\phi(x_i) = [0, 0, 0, 0.01\sin(x_{i1}(t))]^T$, and it follows from Assumption 5.3 that $\Lambda = \text{diag}\{0.01, \cdots, 0.01\}$. From Fig. 5.2, it is easy to obtain $\mathscr{L}_1$ as

$$\mathscr{L}_1 = \begin{bmatrix} 2 & -1 & 0 & 0 \\ 0 & 2 & -1 & -1 \\ -1 & 0 & 1 & 0 \\ 0 & 0 & -1 & 1 \end{bmatrix}.$$

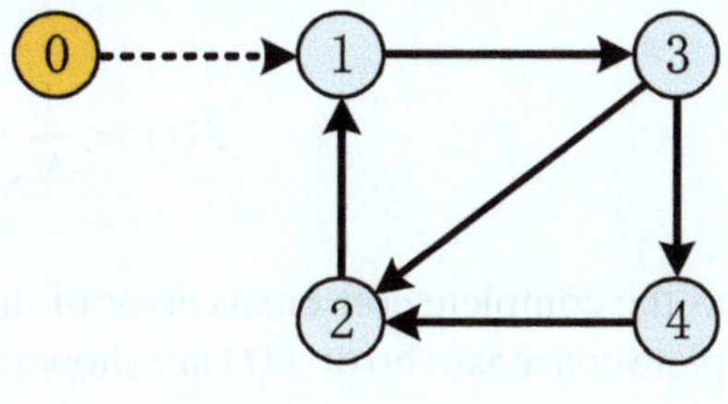

Fig. 5.2 Communication topology

Then, based on Lemma 2.2, we have $\Theta = \text{diag}\{4, 2.5, 7, 3.5\}$. Without loss of generality, we assume that the disturbances $d_1 = 1.2\cos(0.7t)$ and $d_4 = 0.5\cos(0.7t)$ acting on agent 1 and agent 4 all the time, respectively, and there are no disturbances to other agents. The probabilities of FDIAs occurring on each edge are chosen as $\chi_{12} = \chi_{23} = \chi_{43} = 0.02$, $\chi_{24} = \chi_{31} = 0.03$, and the attack characteristic is set to $\bar{\nu} = 0.1$. Moreover, take the initial states of each agent as

$$
\begin{aligned}
&x_0(0) = [-3, 2.5, 2, -2.8]^T, \quad x_1(0) = [-0.3, 0.4, -0.3, 0.3]^T \\
&x_2(0) = [0.4, -0.2, 0.3, 0.2]^T, \quad x_3(0) = [0.4, 0.1, 0.3, -0.1]^T \\
&x_4(0) = [0.6, -0.3, 0.3, -0.4]^T.
\end{aligned}
$$

First, we demonstrate the validity of the designed *complete consensus* control protocol (5.9), (5.10) and (5.12). Based on Theorems 5.1 and 5.2, select the parameters $\beta = 0.08$, $\beta_1 = 0.1$, $\kappa_1 = 200$, $R = I_3$, $\beta_2 = 60$, $\beta_3 = 0.85$, $\varepsilon = 0.00005$, $\sigma = 0.387$ and the triggering parameters $\kappa_2 = 0.0034$, $\beta_4 = 1$, $\hbar_i = 0.003$, $\ell = 0.00001$. Then, the observer gain L and the *disturbance* bound estimation gain matrix W_1, the controller gain K and the weighting matrix P_2 are obtained as follows after solving (5.16) and (5.17) and (5.27) and (5.28), respectively

$$
\begin{aligned}
L &= \begin{bmatrix} 5.6805 & 0.9824 & 2.3019 \\ 0.9824 & 14.1506 & -4.4867 \\ 0.5960 & 0.8026 & 8.3620 \\ 1.7059 & -5.2893 & 9.7204 \end{bmatrix}, \\
W_1 &= \begin{bmatrix} 4.5925 & 0.2255 & -2.7408 \end{bmatrix}, \\
K &= \begin{bmatrix} 0.0004 & -0.0001 & -0.0016 & 0.0002 \\ -0.0001 & 0.0005 & 0.0007 & 0.0004 \end{bmatrix}, \\
P_2 &= \begin{bmatrix} 0.4952 & 0.1179 & -0.1308 & 0.2451 \\ 0.1179 & 1.5879 & -0.0550 & 0.8306 \\ -0.1308 & -0.0550 & 0.4903 & -0.1585 \\ 0.2451 & 0.8306 & -0.1585 & 1.0038 \end{bmatrix}.
\end{aligned}
$$

Then, set the parameter in the adaptive law (5.11) for the 1st and 4th agent as $\eta_1 = 7$, $\vartheta_{1,1} = 0.017$, $\eta_4 = 0.7$, and $\vartheta_{4,1} = 0.05$. The *disturbance* signals $d_i(t)$ and the adaptive parameters $\hat{\bar{d}}_i(t)$ $(i = 1, 4)$ are given in Fig. 5.3, it can be seen that $\hat{\bar{d}}_i(t)$ can estimate the upper bounds of the *disturbance* well. Figure 5.4 depicts the moments when the information passed on different edges is attacked by FDIAs. Define

$$
\tilde{J}(t) = \frac{1}{N}\sqrt{\sum_{i=1}^{N} \|x_i(t) - x_0(t)\|^2}
$$

as the complete consensus error of the system, then the observer errors and the complete consensus error $\tilde{J}(t)$ are shown in Figs. 5.5 and 5.6, respectively. It is observable

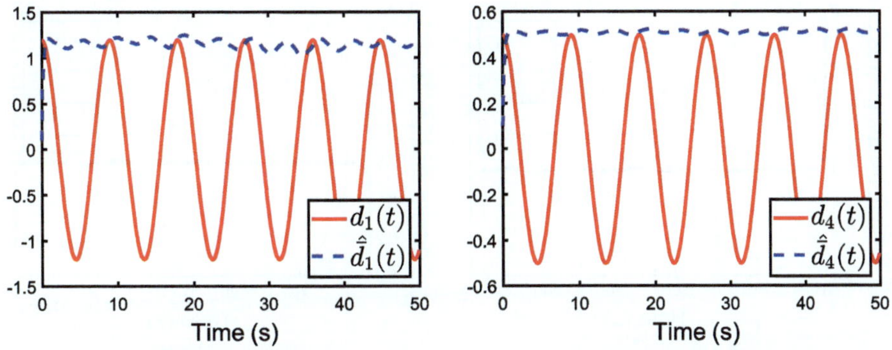

(a) $d_1(t)$ and $\hat{\hat{d}}_1(t)$ (b) $d_4(t)$ and $\hat{\hat{d}}_4(t)$

Fig. 5.3 The disturbance signals $d_i(t)$ and the adaptive parameters $\hat{\hat{d}}_i(t)$ $(i = 1, 4)$

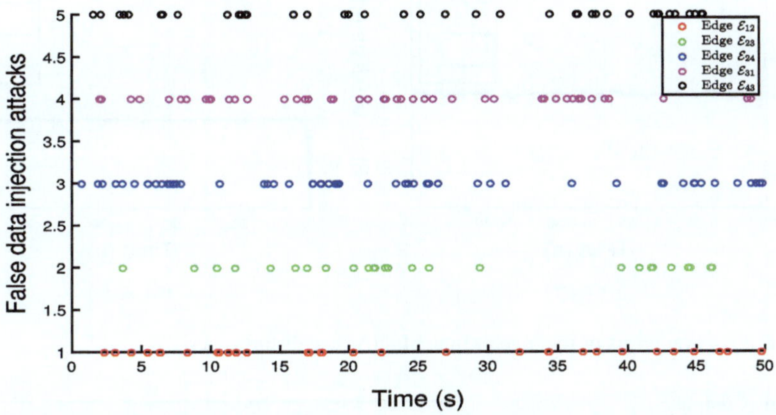

Fig. 5.4 The moment when the transmitted data on different edges is attacked by FDIAs

that the designed observer and H_∞ controller exhibit exceptional performance, even when confronted with FDIAs and external *disturbance*. Figure 5.7 shows the event-triggered instants and release intervals of different agents, clearly showing that the proposed ETM (5.25) is effective in conserving communication resources and preventing *Zeno behavior*. Figure 5.8 compares the performance of the observer (5.9) with or without *disturbance* compensation $g_i(t)$ in (5.10). It is obvious that incorporating *disturbance* compensation results in observed values that more closely align with the real values. Additionally, Table 5.1 summarizes the trigger counts for each agent under different control strategies. Overall, the proposed method yields fewer trigger points.

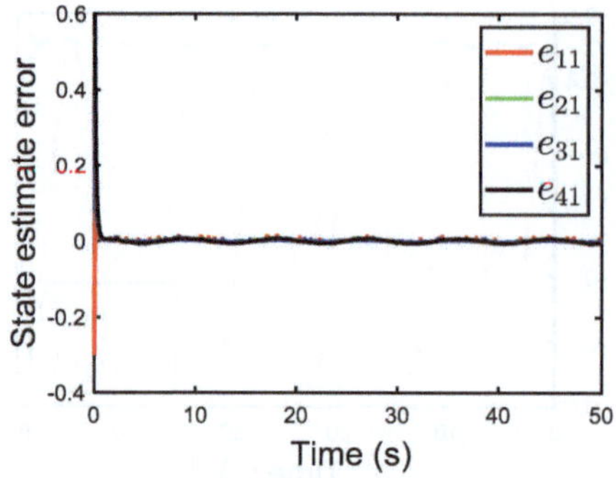

(a) $e_{i1}(t)$

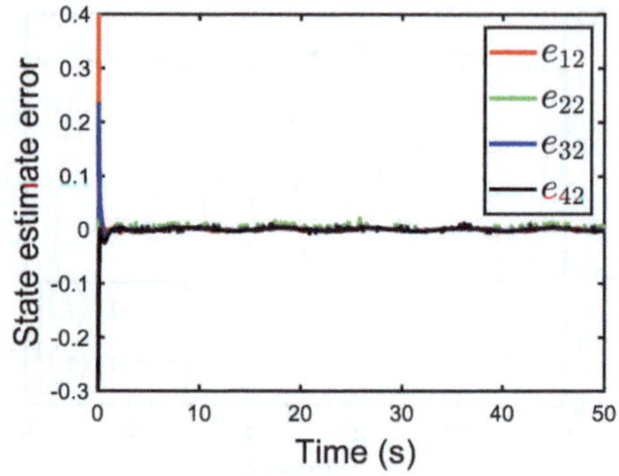

(b) $e_{i2}(t)$

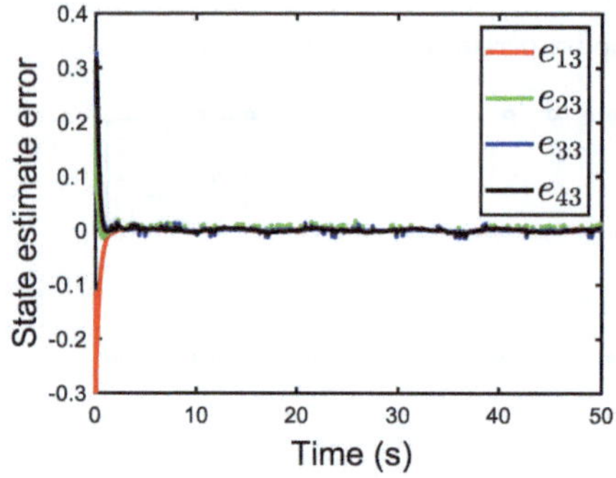

(c) $e_{i3}(t)$

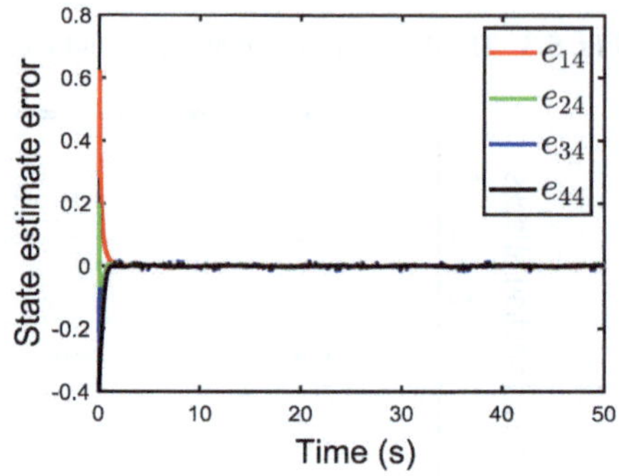

(d) $e_{i4}(t)$

Fig. 5.5 Observer errors of four agents under FDIAs and disturbances

Fig. 5.6 Complete consensus error $\tilde{J}(t)$ under FDIAs and disturbances

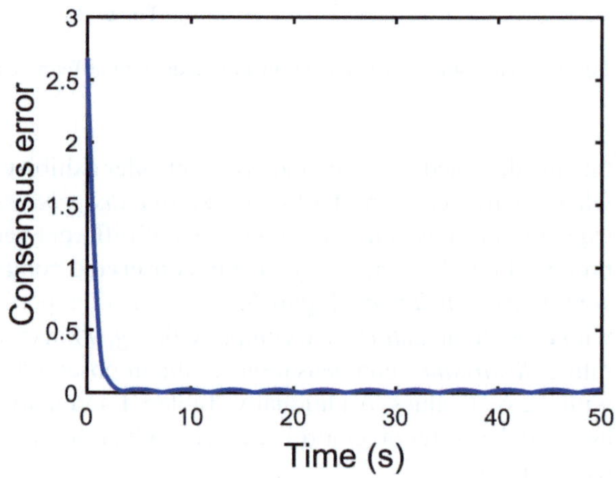

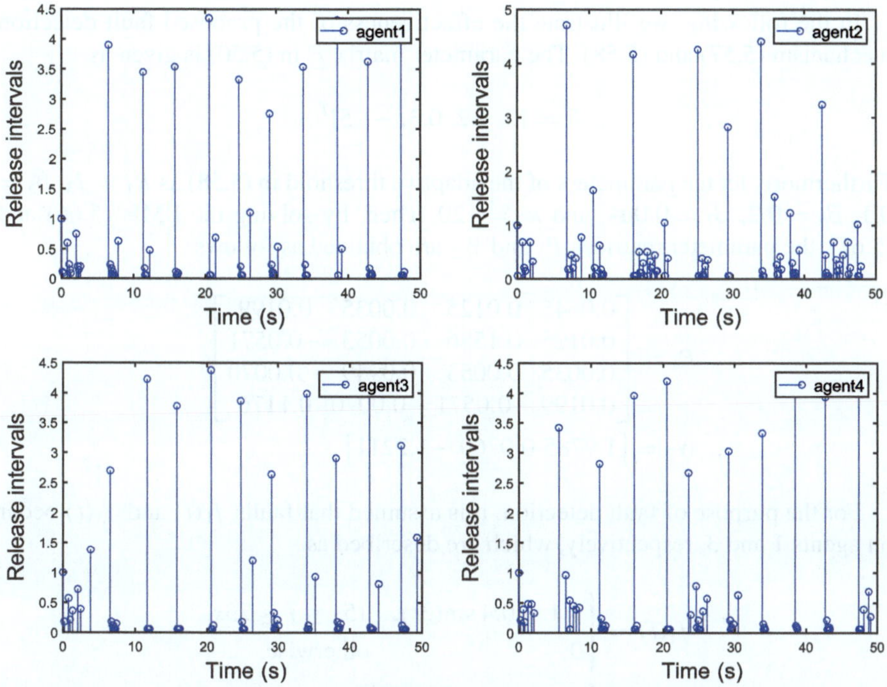

Fig. 5.7 Release instants and intervals of different agents

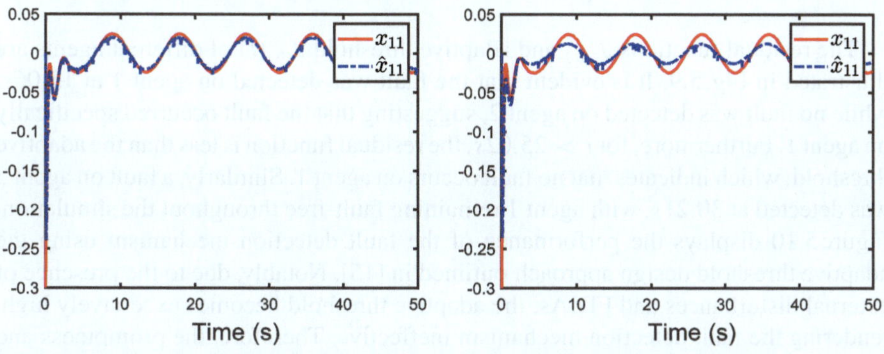

Fig. 5.8 The observer performance for the first element of the state $x_1(t)$ with (left) and without (right) disturbance compensation $g_i(t)$ in observer (5.9) under FDIAs and disturbances

Table 5.1 Trigger numbers of each agent under different control schemes

Control scheme	Agent 1	Agent 2	Agent 3	Agent 4
Event-triggered strategy (5.25)	76	100	87	87
Method in [3]	124	119	96	108
Method in [35]	122	122	122	122

In the following, we illustrate the effectiveness of the proposed fault detection mechanism (5.57) and (5.58). The parameter matrix F in (5.50) is given as

$$F = [3, -2, 0.3, -1.5]^T.$$

Furthermore, let the parameters of the adaptive threshold in (5.58) as $R_1 = I_3$, $\beta_5 = 0.1$, $\beta_6 = 0.2$, $\beta_7 = 0.004$, and $\kappa_3 = 120$. Then, by solving the LMIs (5.66) and (5.67), the parameter matrices P_3 and W_2 are obtained as follows

$$P_3 = \begin{bmatrix} 0.0645 & 0.0125 & 0.0035 & 0.0199 \\ 0.0125 & 0.1586 & 0.0053 & -0.0571 \\ 0.0035 & 0.0053 & 0.0949 & -0.0070 \\ 0.0199 & -0.0571 & -0.0070 & 0.1176 \end{bmatrix},$$
$$W_2 = \begin{bmatrix} 1.9785 & 0.0760 & -1.2211 \end{bmatrix}.$$

For the purpose of fault detection, it is assumed that faults $f_1(t)$ and $f_3(t)$ occur on agents 1 and 3, respectively, which are described as

$$f_1(t) = \begin{cases} 1.1 + 0.4\sin(3t), & 15s \le t \le 25s, \\ 0, & \text{otherwise}, \end{cases}$$

$$f_3(t) = \begin{cases} 0.9 + 0.2\cos(3t), & 30s \le t \le 40s, \\ 0, & \text{otherwise}. \end{cases}$$

The residual functions $J_{r_i(t)}$ and adaptive threshold $J_{th_i}(t)$ of different agents are illustrated in Fig. 5.9. It is evident that the fault was detected on agent 1 at $15.06s$, while no fault was detected on agent 2, suggesting that the fault occurred specifically on agent 1. Furthermore, for $t > 25.02s$, the residual function is less than the adaptive threshold, which indicates that no fault occurs on agent 1. Similarly, a fault on agent 3 was detected at $30.21s$, with agent 1 remaining fault-free throughout the simulation. Figure 5.10 displays the performance of the fault detection mechanism using the adaptive threshold design approach outlined in [15]. Notably, due to the presence of external disturbances and FDIAs, the adaptive threshold becomes excessively high, rendering the fault detection mechanism ineffective. Therefore, the promptness and efficacy of the proposed fault detection mechanism are confirmed.

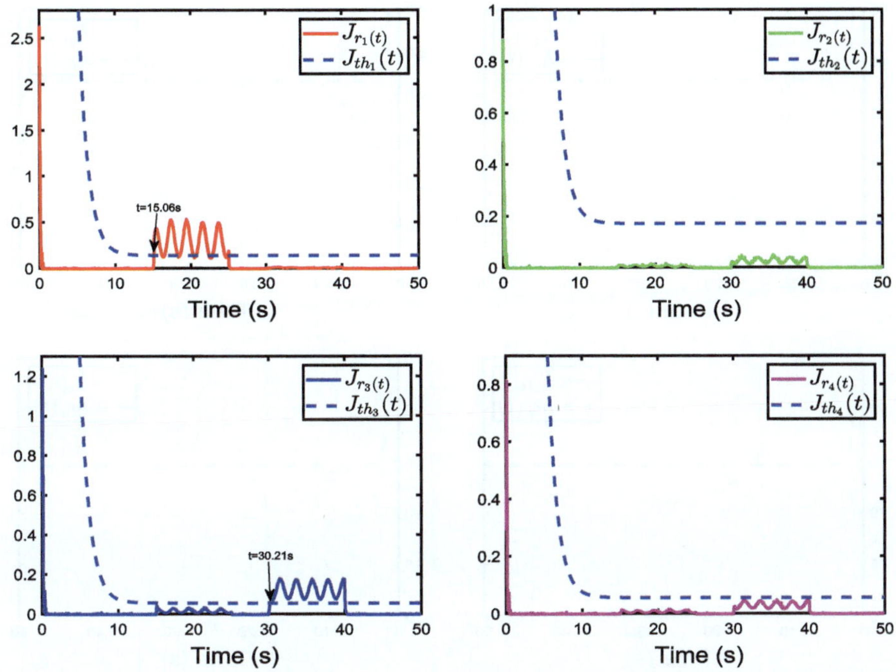

Fig. 5.9 The performance of the fault detection mechanism using the adaptive threshold design method proposed in Theorem 5.3

5.5 Conclusion

This chapter tackles the problem of achieving security *complete consensus* for a class of nonlinear MASs with a directed communication topology, while accounting for external *disturbance*, fault, and FDIA. A state observer is designed to estimate the system's state, and an adaptive compensation approach is introduced to mitigate the observer's accuracy disruption caused by disturbances. To alleviate the communication load, an event-triggered control strategy, grounded on this observer, is proposed, ensuring the MAS attains the desired H_∞ performance without exhibiting *Zeno behavior*. Furthermore, in the event of an agent fault, a fault detection mechanism, leveraging the state observer, is established to promptly identify the fault. A simulation example demonstrates the efficacy of the proposed methodologies. It's worth noting that this chapter solely considers a communication graph with a directed spanning tree and a static event-triggered mechanism (ETM). Hence, extending these findings to an adaptive ETM will be the focus of the next chapter. Additionally, exploring the extension of these results to scenarios involving multiple attacks and attack detection problems presents an intriguing research direction.

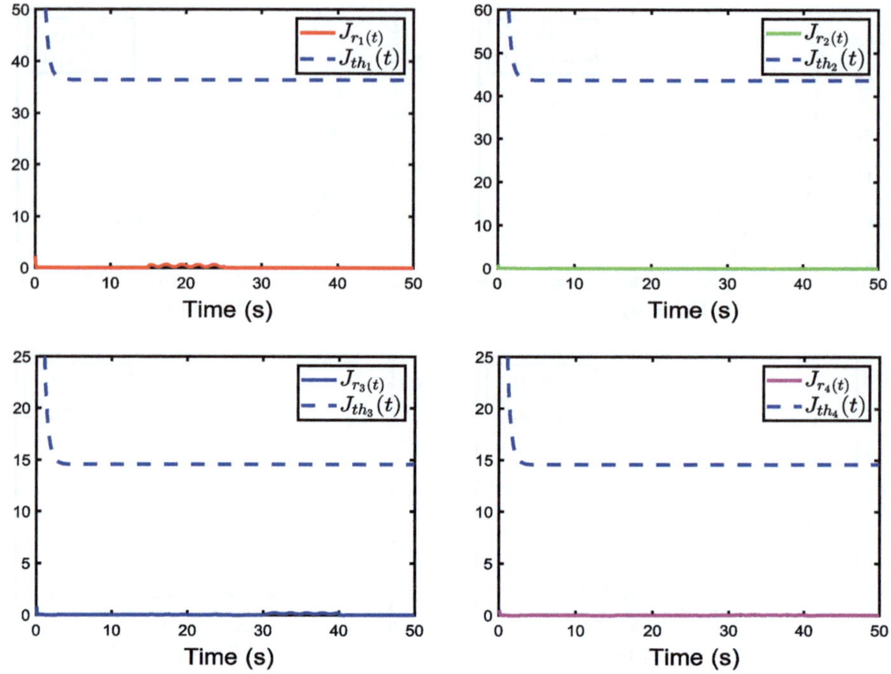

Fig. 5.10 The performance of the fault detection mechanism using the adaptive threshold design method proposed in [15]

References

1. L. Hu, Z. Wang, Q.L. Han, X. Liu, State estimation under false data injection attacks: security analysis and system protection. Automatica **87**, 176–183 (2018)
2. Y. Cui, Y. Liu, W. Zhang, F.E. Alsaadi, Sampled-based consensus for nonlinear multiagent systems with deception attacks: the decoupled method. IEEE Trans. Syst. Man Cybern. Syst. **51**(1), 561–573 (2021)
3. D. Ding, Z. Wang, D.W.C. Ho, G. Wei, Observer-based event-triggering consensus control for multiagent systems with lossy sensors and cyber-attacks. IEEE Trans. Cybern. **47**(8), 1936–1947 (2017)
4. W.L. He, X.Y. Gao, W.M. Zhong, F. Qian, Secure impulsive synchronization control of multi-agent systems under deception attacks. Inf. Sci. **459**, 354–368 (2018)
5. X. Huang, J.X. Dong, Reliable leader-to-follower formation control of multiagent systems under communication quantization and attacks. IEEE Trans. Syst. Man Cybern. Syst. **50**(1), 89–99 (2020)
6. X.M. Li, Q. Zhou, P. Li, H.Y. Li, R.Q. Lu, Event-triggered consensus control for multi-agent systems against false data-injection attacks. IEEE Trans. Cybern. **50**(5), 1856–1866 (2020)
7. M. Meng, G. Xiao, B. Li, Adaptive consensus for heterogeneous multi-agent systems under sensor and actuator attacks. Automatica **122**, Art. No. 109242 (2020)
8. L. Zhao, G.H. Yang, Cooperative adaptive fault-tolerant control for multi-agent systems with deception attacks. J. Franklin Inst. **357**(6), 3419–3433 (2020)

9. X.G. Guo, D.Y. Zhang, J.L. Wang, C.K. Ahn, Adaptive memory event-triggered observer-based control for nonlinear multi-agent systems under DoS attacks. IEEE/CAA J. Autom. Sinica **8**(10), 1644–1656 (2021)
10. Y. Xu, Z.G. Wu, Distributed adaptive event-triggered fault-tolerant synchronization for multi-agent systems. IEEE Trans. Ind. Electron. **68**(2), 1537–1547 (2021)
11. Y. Yang, D. Yue, C. Dou, Output-based event-triggered schemes on leader-following consensus of a class of multi-agent systems with lipschitz-type dynamics. Inf. Sci. **459**, 327–340 (2018)
12. X. Ruan, J. Feng, C. Xu, J. Wang, Observer-based dynamic event-triggered strategies for leader-following consensus of multi-agent systems with disturbances. IEEE Trans. Netw. Sci. Eng. **7**(4), 3148–3158 (2020)
13. C. Wang, L. Guo, C. Wen, X. Yu, J. Huang, Attitude coordination control for spacecraft with disturbances and event-triggered communication. IEEE Trans. Aerosp. Electron. Syst. **57**(1), 586–596 (2021)
14. X.G. Guo, B.Q. Wang, J.L. Wang, Z.G. Wu, L. Guo, Adaptive event-triggered PIO-based anti-disturbance fault-tolerant control for MASs with process and sensor faults. IEEE Trans. Netw. Sci. Eng. **11**(1), 77–88 (2024)
15. F. Arrichiello, A. Marino, F. Pierri, Observer-based decentralized fault detection and isolation strategy for networked multirobot systems. IEEE Trans. Control Syst. Technol. **23**(4), 1465–1476 (2015)
16. M. Davoodi, N. Meskin, K. Khorasani, Simultaneous fault detection and consensus control design for a network of multi-agent systems. Automatica **66**, 185–194 (2016)
17. X.G. Guo, M.E. Tian, Q. Li, C.K. Ahn, Y.H. Yang, Multiple-fault diagnosis for spacecraft attitude control systems using RBFNN-based observers. Aerosp. Sci. Technol. **106**, Art. No. 106195 (2020)
18. X.G. Guo, D.Y. Zhang, J.L. Wang, J.H. Park, L. Guo, Event-triggered observer-based $\mathscr{H}_\infty$ consensus control and fault detection of multiagent systems under stochastic false data injection attacks. IEEE Trans. Netw. Sci. Eng. **9**(2), 481–494 (2022)
19. Y. Wu, Z. Wang, Z. Huang, Distributed fault detection for nonlinear multi-agent systems under fixed-time observer. J. Franklin Inst. **356**(13), 7515–7532 (2019)
20. D. Zhang, Z. Ye, X. Dong, Co-design of fault detection and consensus control protocol for multi-agent systems under hidden DoS attack. IEEE Trans. Circuits Syst. Regul. Pap. **68**(5), 2158–2170 (2021)
21. Z.H. Zhang, G.H. Yang, Distributed fault detection and isolation for multiagent systems: an interval observer approach. IEEE Trans. Syst. Man Cybern. Syst. **50**(6), 2220–2230 (2020)
22. C.L. Liu, L. Shan, Y.Y. Chen, Y. Zhang, Average-consensus filter of first-order multi-agent systems with disturbances. IEEE Trans. Circuits Syst. Express Briefs **65**(11), 1763–1767 (2018)
23. V. Reppa, M.M. Polycarpou, C.G. Panayiotou, Adaptive approximation for multiple sensor fault detection and isolation of nonlinear uncertain systems. IEEE Trans. Neural Netw. Learn. Syst. **25**(1), 137–153 (2014)
24. X. Wang, Z. Fei, H. Yan, Y. Xu, Dynamic event-triggered fault detection via zonotopic residual evaluation and its application to vehicle lateral dynamics. IEEE Trans. Ind. Inf. **16**(11), 6952–6961 (2020)
25. X.G. Guo, P.M. Liu, J.L. Wang, C.K. Ahn, Event-triggered adaptive fault-tolerant pinning control for cluster consensus of heterogeneous nonlinear multi-agent systems under aperiodic DoS attacks. IEEE Trans. Netw. Sci. Eng. **8**(2), 1941–1956 (2021)
26. A.U. Rehman, M. Rehan, N. Iqbal, C.K. Ahn, LPV scheme for robust adaptive output feedback consensus of Lipschitz multiagents using Lipschitz nonlinear protocol. IEEE Trans. Syst. Man Cybern. Syst. **51**(11), 7040–7050 (2021)
27. C. Keliris, M.M. Polycarpou, T. Parisini, A robust nonlinear observer-based approach for distributed fault detection of input-output interconnected systems. Automatica **53**, 408–415 (2015)
28. Y. Liu, P. Ning, M.K. Reiter, False data injection attacks against state estimation in electric power grids. ACM Trans. Info. Syst. Sec. **14**(1), 1–33 (2011)

29. N. Sarrafan, J. Zarei, Bounded observer-based consensus algorithm for robust finite-time tracking control of multiple nonholonomic chained-form systems. IEEE Trans. Autom. Control **66**(10), 4933–4938 (2021)
30. Z. Zuo, Q.L. Han, B. Ning, An explicit estimate for the upper bound of the settling time in fixed-time leader-following consensus of high-order multivariable multiagent systems. IEEE Trans. Ind. Electron. **66**(8), 6250–6259 (2019)
31. M. Ran, Q. Wang, C. Dong, L.H. Xie. Active disturbance rejection control for uncertain time-delay nonlinear systems. Automatica **112**, Art. No. 108692 (2020)
32. H. Sun, R. Madonski, S. Li, Y. Zhang, W. Xue, Composite control design for systems with uncertainties and noise using combined extended state observer and kalman filter. IEEE Trans. Ind. Electron. **69**(4), 4119–4128 (2022)
33. Y. Zhu, J. Qiao, L. Guo, Adaptive sliding mode disturbance observer-based composite control with prescribed performance of space manipulators for target capturing. IEEE Trans. Ind. Electron. **66**(3), 1973–1983 (2019)
34. S. Chen, H. Jiang, Z. Yu, Observer-based event-triggered consensus of leader-following linear multi-agent systems with input saturation and switching topologies. Neurocomputing **364**, 138–151 (2019)
35. S.L. Du, T. Liu, D.W.C. Ho, Dynamic event-triggered control for leader-following consensus of multiagent systems. IEEE Trans. Syst. Man Cybern. **50**(9), 3243–3251 (2020)
36. X.G. Guo, X. Fan, C.K. Ahn, Adaptive event-triggered fault detection for interval type-2 T-S fuzzy systems with sensor saturation. IEEE Trans. Fuzzy Syst. **29**(8), 2310–2321 (2021)
37. M.M. Hossain, C. Peng, Observer-based event triggering $\mathscr{H}_\infty$ LFC for multi-area power systems under DoS attacks. Inf. Sci. **543**, 437–453 (2021)

Chapter 6
Anti-disturbance Control of MASs Under Multiple Disturbances and FDIAs

Abstract This chapter introduces a composite hierarchical anti-disturbance control (CHADC) strategy for nonlinear multi-agent systems (MASs) subjected to stochastic false data injection attacks (FDIAs) and multiple disturbances under a directed communication topology. To mitigate and counteract the adverse effects of two types of disturbances, a disturbance observer (DO) is designed to counteract the *disturbance* generated by exogenous system, and the H_∞ control method is applied to attenuate the bounded compensation errors and other coupled error variables caused by the norm bounded disturbances and FDIAs simultaneously. To ensure the *complete consensus* performance of MASs, an observer-based control strategy is designed, and a novel adaptive compensation technique is proposed to not only evaluate the upper bounds of the unknown but bounded disturbances but also improve the accuracy of the state observer. Furthermore, a novel ETM is developed, which eliminates the need for continuous communication between neighboring agents, reducing the controller update frequency and communication load, while avoiding *Zeno behavior*.

6.1 Introduction

In Chaps. 3–5, observer-based anti-disturbance control strategies are proposed for nonlinear multi-agent systems (MASs) with external disturbances. However, it is noteworthy that these chapters only take into account a single type of *disturbance*, i.e., norm bounded *disturbance*. In reality, MASs frequently encounter a variety of disturbances, making the aforementioned methods inadequate for addressing each disturbance based on its unique characteristics, and may even fail and result in the decline of system control accuracy or even instability. For systems with various sources of disturbances, composite hierarchical anti-disturbance control (CHADC) is an effective method [1], and has been applied in areas such as Markovian jump system [2], spacecraft control [3] and so on. However, few studies have addressed the *complete consensus* control problem of MASs under multiple disturbances, which motivated us to tackle this nontrivial control problem. Furthermore, due to the complex operational environment, MASs frequently confront the risk of *cyber attacks*.

X.-G. Guo et al., *Anti-disturbance Intrusion- and Fault-Tolerant Control of Multi-agent Systems*, https://doi.org/10.1007/978-981-96-9614-7_6

In the discovered attacks, false data injection attacks (FDIAs) destroy the information by tampering with transmitted data, which has strong concealment [4, 5]. Regrettably, despite the significant damage caused by the aforementioned attacks, many existing multi-agent anti-disturbance techniques are confined to scenarios with secure networks [6, 7]. Thus, to better reflect the reality, it is highly appealing to devise a secure *complete consensus* control strategy that can resist multiple *disturbance* when MASs are targeted by FDIAs.

On the other hand, as the expansion of the scale of MASs, minimizing the wastage of communication resources has become a pressing concern [8, 9]. To alleviate the burden on bandwidth and ensure system performance, event-triggered control has been incorporated into the anti-disturbance control framework for MASs (see [10–16], and the references therein). In [10, 13], an event-triggered mechanism (ETM) is proposed that eliminates the need for continuous communication between neighboring agents to update the triggering condition, thereby reducing the communication burden. However, the controller still needs continuous access to the disturbance observer (DO) information, which can lead to unnecessary resource usage. In [12], a self-triggering algorithm is introduced to avoid constant monitoring of the triggering condition, but the controller remains continuously updated. [15, 16] propose a scheme that benefits from updating the controller only at trigger times. However, a drawback of this ETM is that it necessitates continuous acquisition of neighbor agents' states, thereby failing to effectively reduce the communication load. While some of the results have been obtained that balance reducing controller updates with continuous monitoring of neighbor agents' states [14], they often rely on undirected graphs or assume the system state is measurable [11]. Notably, directed graphs pose greater challenges for research on MASs due to the asymmetry of the *Laplacian matrix*, compared to undirected graphs. In addition, acquiring system state information can be difficult or even impossible in some scenarios [17, 18], rendering the aforementioned results inapplicable. Therefore, the development of an observer-based event-triggered *complete consensus* protocol under a *directed graph*, which avoids both continuous neighbor information monitoring and constant controller updates, serves as a central motivation for this chapter.

Considering the aforementioned discussions, in this chapter, via the event-triggered control, the secure *complete consensus* and CHADC problems are investigated for nonlinear MASs subjected to multiple disturbances and stochastic FDIAs. The main contributions are summarized as follows:

(1) A CHADC method based on *disturbance* observer and H_∞ control: Unlike the secure network environment and single *disturbance* in Chaps. 3–5, FDIAs and multiple disturbances are considered in this chapter. To mitigate and counteract the adverse effects of these disturbances, a DO is designed to offset the *disturbance* produced by exogenous system, while the H_∞ control method is applied to attenuate bounded *disturbance*. Furthermore, rather than just obtaining an UUB complete consensus errors as in [19], the H_∞ control method can also attenuate the bounded compensation errors and other coupled error variables caused by the norm bounded disturbances and FDIAs to improve the system performance.

(2) An output feedback control strategy based on state observer: Different from [7, 11] which require the state to be measurable, an observer is designed to estimate the actual system state. Additionally, by incorporating a DO and an adaptive *disturbance* compensation technique based on the estimated upper bound of *disturbance*, the accuracy of the observer and the overall consensus performance of the MASs under multiple disturbances are significantly enhanced. Notably, this approach does not require *a prior* knowledge of the upper bounds of bounded *disturbance* signals, as was necessary in Chap. 4, nor does it assume boundedness of their derivatives, as was done in [20, 21].

(3) A novel ETM: In contrast to the results which only avoid the continuous communication of the controller in Chap. 5 or the continuous monitoring of the triggering condition in [15, 16], the proposed ETM does not require constant information transmission with neighbors and ensures the intermittent communication of the controller, thereby saving network resources and reducing the frequent activation of physical devices. Besides, *Zeno behavior* is excluded.

6.2 Problem Formulation

The structure of MASs discussed in this chapter is illustrated in Fig. 6.1. During the transmission of the control signal $u_i(t)$ and output signal $y_i(t)$ of each agent, they may be susceptible to FDIAs, leading to a decrease in data accuracy. In addition, in the presence of disturbances, the CHADC strategy is employed to maintain the performance of the MASs.

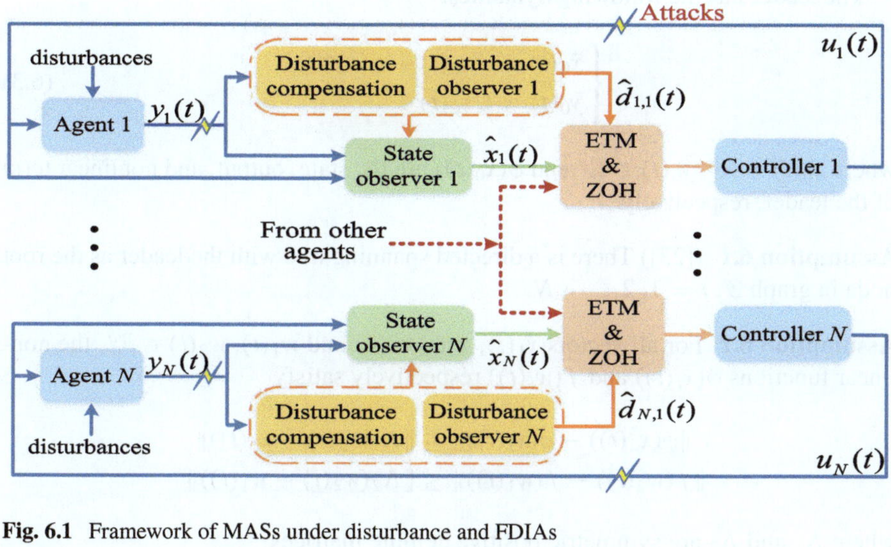

Fig. 6.1 Framework of MASs under disturbance and FDIAs

6.2.1 System Model

Considering a nonlinear MAS consisting of a leader and N followers under a *directed graph*, the dynamics of the ith follower is given by

$$\begin{cases} \dot{x}_i(t) = Ax_i(t) + B[u_i(t) + d_{i,1}(t)] + Dd_{i,2}(t) + \phi(x_i(t)) \\ y_i(t) = Cx_i(t),\ i = 1, \cdots, N \end{cases} \tag{6.1}$$

where $x_i(t) \in \mathbb{R}^n$ is the state vector; $u_i(t) \in \mathbb{R}^m$ is the control input signal; $y_i(t) \in \mathbb{R}^p$ is the measured output. A, B, C, and D are known system matrices. $\phi(x_i(t)) \in \mathbb{R}^n$ represents a nonlinear term. Suppose that the matrix pair (A, B, C) is stabilizable and detectable. Besides, two kinds of disturbances $d_{i,1}(t)$ and $d_{i,2}(t)$ are included in (6.1), where $d_{i,2}(t)$ denote a class of arbitrary bounded but unknown disturbances without a definite model, i.e.,

$$\|d_{i,2}(t)\| \leq \bar{d}_{i,2}$$

while $d_{i,1}(t) \in \mathbb{R}^m$ represent a class of unknown disturbances with a definite model generated by the following nonlinear exogenous system:

$$\begin{cases} \dot{w}_i(t) = Ww_i(t) + Ff(w_i(t)) \\ d_{i,1}(t) = Vw_i(t) \end{cases} \tag{6.2}$$

where $w_i(t) \in \mathbb{R}^q$ is the state vector of the nonlinear exogenous system; $f(w_i(t))$ is a continuous unknown nonlinear function; W, F, V are known constant matrices, and the pair (W, BV) is assumed to be observable [22].

The leader has the following dynamics:

$$\begin{cases} \dot{x}_0(t) = Ax_0(t) + \phi(x_0(t)) \\ y_0(t) = Cx_0(t) \end{cases} \tag{6.3}$$

where $x_0(t) \in \mathbb{R}^n$, $y_0(t) \in \mathbb{R}^p$ and $\phi(x_0(t))$ are the state, output, and nonlinear term of the leader, respectively.

Assumption 6.1 ([23]) There is a directed spanning tree with the leader as the root node in graph $\bar{\mathcal{G}}$. $i = 1, 2, \cdots, N$.

Assumption 6.2 For all vectors $x_1(t)$, $x_2(t) \in \mathbb{R}^n$ and $w_1(t)$, $w_2(t) \in \mathbb{R}^q$, the nonlinear functions $\phi(x_i(t))$ and $f(w_i(t))$ respectively satisfy

$$\begin{aligned} \|\phi(x_2(t)) - \phi(x_1(t))\| &\leq \|\Lambda_1(x_2(t) - x_1(t))\| \\ \|f(w_2(t)) - f(w_1(t))\| &\leq \|\Lambda_2(w_2(t) - w_1(t))\| \end{aligned}$$

where Λ_1 and Λ_2 are symmetric positive definite matrices.

Remark 6.1 Generally speaking, the *disturbance* model (6.2) is capable of representing a wide range of practical disturbances. Specifically, when $F \neq 0$, it captures a type of non-harmonic *disturbance* generated by a nonlinear exogenous system; if $F = 0$ and $W = 0$, it signifies an unknown constant *disturbance*. Moreover, by setting $F = 0$ and $W = \begin{bmatrix} 0 & \omega_0 \\ -\omega_0 & 0 \end{bmatrix}$ with $\omega_0 > 0$, the model represents a kind of harmonic *disturbance* with unknown phase and amplitude, and the frequency is known or unknown depending on whether the harmonic frequency ω_0 is known or not [22, 24].

6.2.2 *False Data Injection Attack Model*

Due to the intricate network architecture and heavy reliance on the network in MASs, they are more prone to FDIAs. In this chapter, we assume that attack signals infiltrate the follower's controller-actuator and sensor-controller channels, as depicted in Fig. 6.1. Considering the stochastic nature of the FDIAs, the actual control signal $\tilde{u}_i(t)$ received by the actuator is

$$\tilde{u}_i(t) = u_i(t) + \Gamma_i^u(t)\nu_i^u(t) \tag{6.4}$$

where $\nu_i^u(t)$ is the attack signal injected into the controller channel, the random variable $\Gamma_i^u(t)$ which obeys the Bernoulli distribution is the attacker's decision variable, and the probabilities of $\Gamma_i^u(t)$ are

$$\text{Prob}\{\Gamma_i^u(t) = 1\} = \chi_i^u, \quad \text{Prob}\{\Gamma_i^u(t) = 0\} = 1 - \chi_i^u$$

with $\chi_i^u \in [0, 1)$ being a constant.

Similarly, the actual output signal $\tilde{y}_i(t)$ can be written as

$$\tilde{y}_i(t) = y_i(t) + \Gamma_i^y(t)\nu_i^y(t) \tag{6.5}$$

where $\nu_i^y(t)$ is the attack signal injected into the sensor channel, and the probabilities of random decision variable $\Gamma_i^y(t)$ are:

$$\text{Prob}\{\Gamma_i^y(t) = 1\} = \chi_i^y, \quad \text{Prob}\{\Gamma_i^y(t) = 0\} = 1 - \chi_i^y$$

with $\chi_i^y \in [0, 1)$ being a constant.

Assumption 6.3 Define $\nu^u(t) = \text{col}\{\nu_1^u(t), \cdots, \nu_N^u(t)\}$, $\nu^y(t) = \text{col}\{\nu_1^y(t), \cdots, \nu_N^y(t)\}$, and complete consensus error $\delta_i(t) = x_i(t) - x_0(t)$, then

(1) The attack signal $\nu^u(t)$ satisfies $\|\nu^u(t)\|^2 \leq \kappa\|\delta(t)\|^2$, where $\delta(t) = \text{col}\{\delta_1(t), \cdots, \delta_N(t)\}$ and $\kappa > 0$.

(2) The attack signal $\nu^y(t)$ is bounded, i.e., $\|\nu^y(t)\| \leq \bar{\nu}$, where $\bar{\nu} > 0$.

Therefore, under the action of FDIAs, the model of MAS (6.1) can be represented as

$$\begin{cases} \dot{x}_i(t) = Ax_i(t) + B[\tilde{u}_i(t) + d_{i,1}(t)] + Dd_{i,2}(t) + \phi(x_i(t)) \\ y_i(t) = Cx_i(t), \ i = 1, \cdots, N. \end{cases} \tag{6.6}$$

Remark 6.2 This chapter considers FDIAs that simultaneously target both the sensor-controller and controller-actuator channels, which is more general and challenging compared to previous studies that focus on attacking a single channel [25, 26]. Moreover, from the attacker's perspective, the energy of the FDIAs is often limited, so Assumption 6.3 is reasonable and common [5, 27, 28].

6.3 Event-Triggered Anti-disturbance Control Strategy

In this section, we propose a CHADC method to guarantee the *complete consensus* of the MASs under multiple disturbances and FDIAs. Meanwhile, a novel ETM is introduced, which not only decreases the frequency of controller updates but also eliminates the need for constant monitoring of neighboring agents' states. Furthermore, this ETM ensures that no agent exhibits *Zeno behavior*.

6.3.1 Observer and Controller Design

As the state of MAS is unmeasurable, a state observer for each agent is proposed as

$$\begin{cases} \dot{\hat{x}}_i(t) = A\hat{x}_i(t) + B[\tilde{u}_i(t) + \hat{d}_{i,1}(t)] + Dg_i(t) + \phi(\hat{x}_i(t)) \\ \qquad\quad - L_1(\tilde{y}_i(t) - \hat{y}_i(t)) \\ \hat{y}_i(t) = C\hat{x}_i(t), \ i = 1, \cdots, N \end{cases} \tag{6.7}$$

where $\hat{x}_i(t)$, $\hat{y}_i(t)$, and $\hat{d}_{i,1}(t)$ are the estimated values of the state $x_i(t)$, output $y_i(t)$ and *disturbance* $d_{i,1}(t)$. The observer gain matrix L_1 is to be devised and the *disturbance* compensation term $g_i(t)$ is designed as

$$g_i(t) = \frac{H(\tilde{y}_i(t) - \hat{y}_i(t))\hat{\bar{d}}_{i,2}^2(t)}{\left\| H(\tilde{y}_i(t) - \hat{y}_i(t)) \right\| \left\| \hat{\bar{d}}_{i,2}(t) \right\| + \vartheta_i} \tag{6.8}$$

where $\vartheta_i > 0$, H is a gain matrix, and the adaptive parameter $\hat{\bar{d}}_{i,2}(t)$ is given by

$$\dot{\hat{\bar{d}}}_{i,2}(t) = -\frac{1}{\eta_i}\vartheta_i\hat{\bar{d}}_{i,2}(t) + \frac{1}{\eta_i}\left\| H(\tilde{y}_i(t) - \hat{y}_i(t)) \right\| \tag{6.9}$$

with η_i being a positive constant. Besides, to resist *disturbance* $d_{i,1}$, the following DO is considered:

$$\begin{cases} \dot{\hat{w}}_i(t) = W\hat{w}_i(t) + Ff(\hat{w}_i(t)) - L_2(\tilde{y}_i(t) - \hat{y}_i(t)) \\ \hat{d}_{i,1}(t) = V\hat{w}_i(t) \end{cases} \tag{6.10}$$

where $\hat{w}_i(t)$ and L_2 are the state and gain matrix of the DO, respectively.

Remark 6.3 As shown in Fig. 6.1, to improve the accuracy of the observer under multiple disturbances, the DO (6.10) and adaptive compensation mechanism $g_i(t)$ in (6.8) are introduced into the distributed state observer (6.7), in which the DO estimates the modeling *disturbance* $d_{i,1}(t)$ and subsequently compensates for it. For the norm bounded *disturbance* $d_{i,2}(t)$, the adaptive compensation mechanism (6.8) mitigates its impact using the estimated upper bound of the disturbance *disturbance*. Notably, this approach solely relies on the output information $\tilde{y}_i(t)$ in (6.5) which is more accessible than state information, and it avoids reliance on *disturbance* models by estimating the disturbance's upper bound. This makes it more practical for real-world applications. Subsequently, we will develop a CHADC mechanism that combines DO and H_∞ control method to ensure the *complete consensus* of MASs.

In MASs, the energy consumption of physical institutions is often huge, and a lot of network resources will be occupied in the process of information transmission. Therefore, to achieve *complete consensus* and counteract *disturbance* within limited network bandwidth, a control law driven by an ETM is developed as follows when $t \in [t_k^i, t_{k+1}^i)$:

$$u_i(t) = -\hat{d}_{i,1}(t_k^i) + cK\Big[\sum\nolimits_{j\in\mathcal{N}_i} a_{ij}(\hat{x}_j(t_k^i) - \hat{x}_i(t_k^i)) + a_{i0}(x_0(t_k^i) - \hat{x}_i(t_k^i))\Big] \tag{6.11}$$

where $c > 0$ represents the coupling strength, K is a feedback matrix to be designed, t_k^i is the moment when the kth event of agent i occurs, and its update rule (i.e., event-triggered condition) will be specified in (6.28). Furthermore, it is necessary to point out that the leader's state is measurable because it can be considered as a command generator [29].

Remark 6.4 In some existing event-triggered anti-disturbance results based on DO for MASs (see [10, 13]), continuous information transmission between neighboring agents is avoided, but they still necessitate constant access to the DO's state, resulting in frequent controller updates. In contrast, the controller formulated in (6.11) is updated only when the trigger condition is met, improving the agents' lifespan. Meanwhile, the output feedback control strategy does not require the assumption that the state is directly accessible, as in [7, 11], which makes this method more challenging.

Denoting $e_{x_i}(t) = x_i(t) - \hat{x}_i(t)$ as the state estimation error, and *disturbance* error is represented by $e_{w_i}(t) = w_i(t) - \hat{w}_i(t)$. Then, let $e_i(t) = \left[e_{x_i}^T(t)\ e_{w_i}^T(t)\right]^T$, according to (6.2), (6.6), (6.7) and (6.10), one can derive the following augmented error system:

$$\dot{e}_i(t) = (\bar{A} + \bar{L}\bar{C})e_i(t) + \bar{D}(d_{i,2}(t) - g_i(t)) + \bar{F}\varphi_i(t) + \bar{L}\Gamma_i^y(t)\nu_i^y(t) \tag{6.12}$$

with

$$\bar{A} = \begin{bmatrix} A & BV \\ 0 & W \end{bmatrix}, \quad \bar{L} = \begin{bmatrix} L_1 \\ L_2 \end{bmatrix}, \quad \bar{C} = \begin{bmatrix} C & 0 \end{bmatrix}, \quad \bar{D} = \begin{bmatrix} D \\ 0 \end{bmatrix}$$

$$\bar{F} = \begin{bmatrix} I & 0 \\ 0 & F \end{bmatrix}, \quad \varphi_i(t) = \begin{bmatrix} \phi(x_i(t)) - \phi(\hat{x}_i(t)) \\ f(w_i(t)) - f(\hat{w}_i(t)) \end{bmatrix}.$$

Next, define $e(t) = \text{col}\{e_1(t), \cdots, e_N(t)\}$, it follows from (6.12) that

$$\begin{aligned} \dot{e}(t) = {} & [I_N \otimes (\bar{A} + \bar{L}\bar{C})]e(t) + (I_N \otimes \bar{D})(d_2(t) - g(t)) \\ & + [\Gamma^y(t) \otimes \bar{L}]\nu^y(t) + (I_N \otimes \bar{F})\varphi(t) \end{aligned}$$

where

$$\begin{aligned} & g(t) = \text{col}\{g_1(t), \cdots, g_N(t)\}, \quad \Gamma^y(t) = \text{diag}\{\Gamma_1^y(t), \cdots, \Gamma_N^y(t)\} \\ & d_2(t) = \text{col}\{d_{1,2}(t), \cdots, d_{N,2}(t)\}, \quad \varphi(t) = \text{col}\{\varphi_1(t), \cdots, \varphi_N(t)\}. \end{aligned}$$

Subsequently, Theorem 6.1 outlines the methods for designing the parameters of the state observer and DO.

Theorem 6.1 *Consider the MAS (6.1) subjected to multiple disturbances and FDIAs. Under Assumptions 6.1–6.3, for given positive parameters β_1, β_2, and ϵ, if there exist matrices Q and H and a symmetric positive definite matrix P_1 such that*

$$\begin{bmatrix} \text{He}(P_1\bar{A} + Q\bar{C}) + \beta_1\Lambda^T\Lambda & P_1\bar{F} & Q \\ * & -\beta_1 I_n & 0 \\ * & * & -\beta_2 I_n \end{bmatrix} < 0 \tag{6.13}$$

$$\begin{bmatrix} -\epsilon I & P_1\bar{D} - \bar{C}^T H^T \\ * & -\epsilon I \end{bmatrix} < 0 \tag{6.14}$$

hold with $\Lambda = \text{diag}\{\Lambda_1, \Lambda_2\}$, and the observer gain $\bar{L}$ is designed as $\bar{L} = P_1^{-1}Q$. Then, if we choose the state observer and disturbance observer as in (6.7) and (6.10), the state estimation error $e_{x_i}(t)$ and disturbance estimation error $e_{w_i}(t)$ are UUB.

Proof Choose a Lyapunov function as follows

$$V_1(t) = e^T(t)(I_N \otimes P_1)e(t) + \sum_{i=1}^{N} \eta_i e^2_{\bar{d}_{i,2}}(t) \tag{6.15}$$

where $\eta_i > 0$ and $e_{\bar{d}_{i,2}}(t) = \bar{d}_{i,2} - \hat{\bar{d}}_{i,2}(t)$. Next, by calculating $\Im V_1(t)$ and taking mathematical expectations, one has

$$\begin{aligned}\mathbb{E}\{\mathscr{I} V_1(t)\} = {} & \mathbb{E}\{e^T(t)[I_N \otimes (P_1(\bar{A} + \bar{L}\bar{C}) + (\bar{A} + \bar{L}\bar{C})^T P_1)]e(t) \\ & + 2e^T(t)(I_N \otimes P_1\bar{D})[d_2(t) - g(t)] + 2\sum_{i=1}^{N} \eta_i e_{\bar{d}_{i,2}}(t)\dot{e}_{\bar{d}_{i,2}}(t)\} \\ & + 2e^T(t)(I_N \otimes P_1\bar{F})\varphi(t) + 2e^T(t)[\Gamma^y(t) \otimes P_1\bar{L}]\nu^y(t). \end{aligned} \tag{6.16}$$

From *Young's inequality* in Lemma 2.7 and Assumption 6.2, it yields

$$\mathbb{E}\{2e^T(t)(I_N \otimes P_1\bar{F})\varphi(t)\} \le \mathbb{E}\{e^T(t)[I_N \otimes (\frac{1}{\beta_1} P_1\bar{F}\bar{F}^T P_1 + \beta_1\Lambda^T\Lambda)]e(t)\}. \tag{6.17}$$

Denote $\mathbb{E}\{\Gamma^y(t)\} = \bar{\Gamma}^y$ with $\bar{\Gamma}^y = \mathrm{diag}\{\chi_1^y, \cdots, \chi_N^y\}$. Based on Lemma 2.5, we have

$$\begin{aligned} & \mathbb{E}\{2e^T(t)(\Gamma^y(t) \otimes P_1\bar{L})\nu^y(t)\} \\ & \le \mathbb{E}\{\frac{1}{\beta_2} e^T(t)(I_N \otimes P_1\bar{L}\bar{L}^T P_1)e(t) + \beta_2\lambda_{\max}(\bar{\Gamma}^{y^T}\bar{\Gamma}^y)\bar{\nu}^2\}. \end{aligned} \tag{6.18}$$

Notice that if $\epsilon > 0$ is selected small enough, $P_1\bar{D} = \bar{C}^T H^T$ can be achieved from LMI (6.14) [6]. In addition, by the definition of $e_{\bar{d}_{i,2}}(t)$ in (6.15), it is known that

$$\dot{e}_{\bar{d}_{i,2}}(t) = \frac{1}{\eta_i}\vartheta_i\hat{\bar{d}}_{i,2}(t) - \frac{1}{\eta_i}\left\| H(\tilde{y}_i(t) - \hat{y}_i(t))\right\| .$$

Then, combined with (6.8), we obtain

$$
\begin{aligned}
&2\mathbb{E}\left\{e^T(t)(I_N \otimes P_1\bar{D})[d_2(t)-g(t)] + \sum_{i=1}^{N}\eta_i e_{\bar{d}_{i,2}}(t)\dot{e}_{\bar{d}_{i,2}}(t)\right\} \\
&\leq 2\mathbb{E}\left\{\sum_{i=1}^{N}\left\|e_i^T(t)P_1\bar{D}\right\|\bar{d}_{i,2} - \sum_{i=1}^{N}e_i^T(t)P_1\bar{D}\,\frac{H(\tilde{y}_i(t)-\hat{y}_i(t))\hat{\bar{d}}_{i,2}^2(t)}{\left\|H(\tilde{y}_i(t)-\hat{y}_i(t))\right\|\left\|\hat{\bar{d}}_{i,2}(t)\right\|+\vartheta_i}\right. \\
&\quad\left. + \sum_{i=1}^{N}\vartheta_i e_{\bar{d}_{i,2}}(t)\hat{\bar{d}}_{i,2}(t) - \sum_{i=1}^{N}e_{\bar{d}_{i,2}}(t)\left\|H(\tilde{y}_i(t)-\hat{y}_i(t))\right\|\right\} \\
&\leq 2\mathbb{E}\left\{\sum_{i=1}^{N}\left\|e_i^T(t)P_1\bar{D}+\nu_i^{y^T}(t)\chi_i^{y^T}H^T\right\|\bar{d}_{i,2} + \sum_{i=1}^{N}\varpi_i\bar{d}_{i,2}\right. \\
&\quad - \sum_{i=1}^{N}\frac{[e_i^T(t)P_1\bar{D}+\nu_i^{y^T}(t)\chi_i^{y^T}H^T]H[\bar{C}e_i(t)+\chi_i^y\nu_i^y(t)]\hat{\bar{d}}_{i,2}^2(t)}{\left\|H[\bar{C}e_i(t)+\chi_i^y\nu_i^y(t)]\right\|\left\|\hat{\bar{d}}_{i,2}(t)\right\|+\vartheta_i} \\
&\quad + \sum_{i=1}^{N}\vartheta_i e_{\bar{d}_{i,2}}(t)\hat{\bar{d}}_{i,2}(t) - \sum_{i=1}^{N}e_{\bar{d}_{i,2}}(t)\left\|H[\bar{C}e_i(t)+\chi_i^y\nu_i^y(t)]\right\| \\
&\quad\left. + \sum_{i=1}^{N}\nu_i^{y^T}(t)\chi_i^{y^T}H^T\frac{H[\bar{C}e_i(t)+\chi_i^y\nu_i^y(t)]\hat{\bar{d}}_{i,2}^2(t)}{\left\|H[\bar{C}e_i(t)+\chi_i^y\nu_i^y(t)]\right\|\left\|\hat{\bar{d}}_{i,2}(t)\right\|+\vartheta_i}\right\} \\
&\leq 2\mathbb{E}\left\{\sum_{i=1}^{N}\left\|e_i^T(t)P_1\bar{D}+\nu_i^{y^T}(t)\chi_i^{y^T}H^T\right\|\left\|\hat{\bar{d}}_{i,2}(t)\right\| + \sum_{i=1}^{N}\varpi_i\bar{d}_{i,2}\right. \\
&\quad - \sum_{i=1}^{N}\frac{[e_i^T(t)P_1\bar{D}+\nu_i^{y^T}(t)\chi_i^{y^T}H^T][H\bar{C}e_i(t)+H\chi_i^y\nu_i^y(t)]}{\left\|H[\bar{C}e_i(t)+\chi_i^y\nu_i^y(t)]\right\|\left\|\hat{\bar{d}}_{i,2}(t)\right\|+\vartheta_i}\hat{\bar{d}}_{i,2}^2(t) \\
&\quad\left. + \sum_{i=1}^{N}\vartheta_i e_{\bar{d}_{i,2}}(t)(\bar{d}_{i,2}-e_{\bar{d}_{i,2}}(t)) + \sum_{i=1}^{N}\left\|\hat{\bar{d}}_{i,2}(t)\right\|\left\|\nu_i^{y^T}(t)\chi_i^{y^T}H^T\right\|\right\} \\
&\leq \mathbb{E}\left\{2\sum_{i=1}^{N}\vartheta_i + 2\sum_{i=1}^{N}\varpi_i\bar{d}_{i,2} - 2\sum_{i=1}^{N}\vartheta_i e_{\bar{d}_{i,2}}^2(t) + \beta_3\sum_{i=1}^{N}\vartheta_i e_{\bar{d}_{i,2}}^2(t) + \sum_{i=1}^{N}\vartheta_i\|e_{\bar{d}_{i,2}}(t)\|\right. \\
&\quad\left. + \frac{1}{\beta_3}\sum_{i=1}^{N}\vartheta_i\bar{d}_{i,2}^2(t) + 2\sum_{i=1}^{N}\left\|\nu_i^{y^T}(t)\chi_i^{y^T}H^T\right\|\bar{d}_{i,2} + \sum_{i=1}^{N}\frac{1}{\vartheta_i}\left\|\nu_i^{y^T}(t)\chi_i^{y^T}H^T\right\|^2\right\} \\
&\leq \mathbb{E}\left\{2\sum_{i=1}^{N}\vartheta_i + 6\sum_{i=1}^{N}\left\|\nu_i^{y^T}(t)\chi_i^{y^T}H^T\right\|\bar{d}_{i,2} + \frac{1}{\beta_3}\sum_{i=1}^{N}\vartheta_i\bar{d}_{i,2}^2\right. \\
&\quad\left. -(1-\beta_3)\sum_{i=1}^{N}\vartheta_i e_{\bar{d}_{i,2}}^2(t) + \sum_{i=1}^{N}\frac{1}{\vartheta_i}\left\|\nu_i^{y^T}(t)\chi_i^{y^T}H^T\right\|^2\right\}
\end{aligned}
\tag{6.19}
$$

where $0 < \beta_3 < 1$, and

$$\varpi_i = \left\|e_i^T(t)P_1\bar{D}\right\| + \left\|\nu_i^{y^T}(t)\chi_i^{y^T}H^T\right\| - \left\|e_i^T(t)P_1\bar{D} + \nu_i^{y^T}(t)\chi_i^{y^T}H^T\right\| \le 2\left\|\nu_i^{y^T}(t)\chi_i^{y^T}H^T\right\|.$$

Furthermore, let $Q = P_1\bar{L}$, and substituting (6.17)–(6.19) into (6.16), we can obtain

$$\begin{aligned}
&\mathbb{E}\{\mathscr{I}V_1(t)\}\\
&\le \mathbb{E}\{-e^T(t)(I_N \otimes \Xi_1)e(t) - (1-\beta_3)\sum_{i=1}^{N}\vartheta_i e_{\bar{d}_{i,2}}^2(t) + \Delta\}\\
&\le \mathbb{E}\left\{-\frac{\lambda_{\min}(\Xi_1)}{\lambda_{\max}(P_1)}e^T(t)(I_N \otimes P_1)e(t) - \frac{(1-\beta_3)\underline{\vartheta}}{\bar{\eta}}\sum_{i=1}^{N}\vartheta_i e_{\bar{d}_{i,2}}^2(t) + \Delta\right\}
\end{aligned} \tag{6.20}$$

where

$$\begin{aligned}
\Delta = {}& \beta_2\lambda_{\max}(\bar{\Gamma}^{y^T}\bar{\Gamma}^y)\bar{\nu}^2 + 2\sum_{i=1}^{N}\vartheta_i + 6\sum_{i=1}^{N}\left\|\nu_i^{y^T}(t)\chi_i^{y^T}H^T\right\|\bar{d}_{i,2}\\
&+ \frac{1}{\beta_3}\sum_{i=1}^{N}\vartheta_i\bar{d}_{i,2}^2 + \sum_{i=1}^{N}\frac{1}{\vartheta_i}\left\|\nu_i^{y^T}(t)\chi_i^{y^T}H^T\right\|^2 > 0
\end{aligned} \tag{6.21}$$

$$\Xi_1 = -(P_1\bar{A} + \bar{A}^TP_1 + Q\bar{C} + \bar{C}^TQ^T + \frac{1}{\beta_1}P_1\bar{F}\bar{F}^TP_1 + \beta_1\Lambda^T\Lambda + \frac{1}{\beta_2}QQ^T). \tag{6.22}$$

Then, by applying Schur complement lemma (Lemma 2.8) to (6.13), it holds that $\Xi_1 > 0$, and (6.20) can be rewritten as

$$\mathbb{E}\{\mathscr{I}V_1(t)\} \le \mathbb{E}\{-\alpha V_1(t) + \Delta\} \tag{6.23}$$

where

$$\alpha = \min\left\{\frac{\lambda_{\min}(\Xi_1)}{\lambda_{\max}(P_1)}, \frac{(1-\beta_3)\underline{\vartheta}}{\bar{\eta}}\right\} \tag{6.24}$$

$$\bar{\eta} = \max\{\eta_1, \cdots, \eta_N\} \tag{6.25}$$

$$\underline{\vartheta} = \min\{\vartheta_1, \cdots, \vartheta_N\} \tag{6.26}$$

which implies that $V_1(t)$ is bounded. Therefore, the estimation errors $e_{x_i}(t)$, $e_{w_i}(t)$, and $e_{\bar{d}_{i,2}}(t)$ are UUB. The proof is completed. □

Remark 6.5 From the state observer (6.7) and *disturbance* observer (6.10), it is evident that attack signals $\nu_i^y(t)$ and $\nu_i^u(t)$ are included. Although their values are unknown to us, they have specific values in practice, and once Assumption 6.3 is satisfied, the designed observers can guarantee that the estimation errors $e_{x_i}(t)$ and $e_{w_i}(t)$ remain UUB. Additionally, compared to methods that rely on knowing the disturbance's upper bound, as in [6, 30], our approach does not necessitate this assumption. Furthermore, unlike the *disturbance* rejection control method based on extended state observer in [20], which requires that the *disturbance* and its derivatives are bounded and satisfies $\lim_{t\to\infty} \dot{d}(t) = 0$, the proposed disturbance compensation method (6.8) only needs to ensure that the *disturbance* is bounded. Moreover, based on (6.23), we know that the estimation errors $e_{x_i}(t)$, $e_{w_i}(t)$, and $e_{\bar{d}_{i,2}}(t)$ are affected by the energy and frequency of FDIAs and the upper bound of *disturbance* $d_{i,2}(t)$. Therefore, it is crucial to adjust the parameters to minimize these estimation errors, thereby enhancing accuracy.

6.3.2 Complete Consensus Performance Analysis

In this subsection, sufficient conditions for MAS (6.1) and (6.3) subjected to multiple disturbances and FDIAs are obtained to achieve H_∞ complete consensus performance. First, the state of followers under FDIAs can be obtained by substituting (6.11) into (6.6) as follows:

$$\begin{aligned}\dot{x}_i(t) = {} & Ax_i(t) + B\Big[-\hat{d}_{i,1}(t_k^i) + cK\Big(\sum\nolimits_{j\in\mathcal{N}_i} a_{ij}(\hat{x}_j(t_k^i) - \hat{x}_i(t_k^i)) + a_{i0}(x_0(t_k^i) - \hat{x}_i(t_k^i))\Big) \\ & + \Gamma_i^u(t)\nu_i^u(t) + d_{i,1}(t)\Big] + Dd_{i,2}(t) + \phi(x_i(t)). \end{aligned} \tag{6.27}$$

Before moving forward, for each agent, denote $\varepsilon_i(t) = \hat{x}_i(t_k^i) - \hat{x}_i(t)$, $\varepsilon_{i0}(t) = x_0(t_k^i) - x_0(t)$, and $\tilde{\varepsilon}_i(t) = \hat{d}_{i,1}(t_k^i) - \hat{d}_{i,1}(t)$ as measurement errors, and the time sequence $\{t_k^i\}$ is generated by

$$t_{k+1}^i = \inf\{t > t_k^i | r_i(t) > 0\} \tag{6.28}$$

and

$$\begin{aligned} r_i(t) = {} & \theta_i\Big(\rho_1\left\|B^T P_2\varepsilon_i(t)\right\|^2 + \rho_2 a_{i0}\left\|B^T P_2\varepsilon_{i0}(t)\right\|^2 + \beta_5\left\|\tilde{\varepsilon}_i(t)\right\|^2\Big) \\ & - \hbar_{i1}\theta_i\Bigg(\left\|B^T P_2\sum_{j\in\mathcal{N}_i} a_{ij}(\hat{x}_j(t_k^i) - \hat{x}_i(t_k^i))\right\|^2 + \left\|B^T P_2 a_{i0}(x_0(t_k^i) - \hat{x}_i(t_k^i))\right\|^2\Bigg) \\ & - \hbar_{i2}\theta_i\left\|\hat{\bar{d}}_{i,2}(t)\right\|^2 - \upsilon_i\ell_i e^{-\upsilon_i t} \end{aligned} \tag{6.29}$$

where

$$\rho_1 = \kappa_1 c + \hbar_1(1 + 1/\kappa_2)(\theta_{\max}/\theta_{\min})\lambda_{\max}(\mathscr{L}^T\mathscr{L}) + \hbar_1(1 + 1/\kappa_3)(1 + 1/\kappa_4)$$
$$\rho_2 = \kappa_1 c + \hbar_1(1 + 1/\kappa_3)(1 + \kappa_4), \quad \hbar_1 = \max\{\hbar_{i1}\}, \quad \max\{\hbar_{i2}\} \leq \frac{2\sigma^2}{3}.$$

Furthermore, $\hbar_{i1}$, $\hbar_{i2}$, β_5, υ_i, ℓ_i are positive constants. In addition, the remaining parameters will be given in Theorem 6.2.

Remark 6.6 Unlike the ETM presented in [14] that uses state-independent thresholds, our ETM design process incorporates a dynamically adjustable threshold based on system performance. This approach allows for a more optimal balance between resource conservation and control performance enhancement. Meanwhile, it is noteworthy that the introduction of estimated value $\hat{\hat{d}}_{i,2}(t)$ and exponential term $\upsilon_i\ell_i e^{-\upsilon_i t}$ as in Chap. 5 can facilitate the H_∞ complete consensus analysis and the exclusion of *Zeno behavior*.

Recall that $\delta_i(t) = x_i(t) - x_0(t)$. Then, let $\delta(t) = \{\delta_1(t), \cdots, \delta_N(t)\}$, and combining (6.3) and (6.27) yields

$$\begin{aligned}\dot{\delta}(t) = &(I_N \otimes A - c\mathscr{L}_1 \otimes BK)\delta(t) + (c\mathscr{L}_1 \otimes BK)e_x(t) + \phi(x)\\ &- (c\mathscr{L}_1 \otimes BK)\varepsilon(t) + (c\mathscr{A}_0 \otimes BK)\varepsilon_0(t) - (I_N \otimes B)\tilde{\varepsilon}(t)\\ &+ (I_N \otimes BV)e_w(t) + [\Gamma^u(t) \otimes B]\nu^u(t) + (I_N \otimes D)d_2(t) - \tilde{\phi}(x_0) \end{aligned} \tag{6.30}$$

where

$$\begin{aligned}&e_x(t) = \mathrm{col}\{e_{x_1}(t), \cdots, e_{x_N}(t)\}, \quad \varepsilon_0(t) = \mathrm{col}\{\varepsilon_{10}(t), \cdots, \varepsilon_{N0}(t)\}\\ &\varepsilon(t) = \mathrm{col}\{\varepsilon_1(t), \cdots, \varepsilon_N(t)\}, \quad e_w(t) = \mathrm{col}\{e_{w_1}(t), \cdots, e_{w_N}(t)\}\\ &\tilde{\varepsilon}(t) = \mathrm{col}\{\tilde{\varepsilon}_1(t), \cdots, \tilde{\varepsilon}_N(t)\}, \quad \Gamma^u(t) = \mathrm{diag}\{\Gamma^u_1(t), \cdots, \Gamma^u_N(t)\}\\ &\phi(x) - \tilde{\phi}(x_0) = \mathrm{col}\{\phi(x_1) - \phi(x_0), \cdots, \phi(x_N) - \phi(x_0)\}.\end{aligned}$$

Theorem 6.2 *Assume that Assumptions 6.1–6.3 hold. For given positive scalars* $\kappa_{1,2,3}$, $\beta_i (i = 4, \cdots, 7)$, *and* σ, *if there exists matrix* $P_2 > 0$ *such that*

$$\begin{bmatrix} \Phi_{11} & P_2^{-1} & P_2^{-1}\Lambda_1^T & P_2^{-1} \\ * & -\frac{1}{\beta_6\kappa\lambda^2_{\max}(\tilde{\Gamma}^u)}I_n & 0 & 0 \\ * & * & -\frac{1}{\beta_7}I_n & 0 \\ * & * & * & -I_n \end{bmatrix} < 0 \tag{6.31}$$

$$\left[2\hbar_1\Big((1 + \kappa_2)(\mathscr{N} + \sqrt{N\tilde{\mathscr{N}}})^2 + 1 + \kappa_3\Big) + \beta_4 c\right]\lambda_{\max}(P_2BB^TP_2) < \sigma^2 \tag{6.32}$$

where

$$\begin{aligned}\Phi_{11} = & AP_2^{-1}+P_2^{-1}A^T + \Bigg(\frac{c\lambda_{\max}(\mathscr{L}_1^T\mathscr{L}_1)\lambda_{\max}(\Theta)}{\beta_4\lambda_{\min}(\Theta)} + \frac{1}{\beta_5}+\frac{1}{\beta_6}+\frac{c\lambda_{\max}(\mathscr{L}_1^T\mathscr{L}_1)\lambda_{\max}(\Theta)}{\kappa_1\lambda_{\min}(\Theta)} \\ & + \frac{c}{\kappa_1} + 2\hbar_1\Big((1+\kappa_2)(\tilde{\mathscr{N}}+\sqrt{N\tilde{\mathscr{N}}})^2+1+\kappa_3\Big) - \frac{c\lambda_0}{\lambda_{\max}(\Theta)}\Bigg)BB^T \\ & + \frac{1}{\sigma^2}BVV^TB^T + \frac{4}{\sigma^2}DD^T + \frac{1}{\beta_7}I \end{aligned} \tag{6.33}$$

and the scalar κ is defined in Assumption 6.3. Then, in the presence of multiple disturbances and FDIAs, under the complete consensus control protocol (6.11) and ETM (6.28), the H_∞ complete consensus performance can be guaranteed, i.e.,

$$\mathbb{E}\left\{\int_0^{t_f}\delta^T(t)(\Theta\otimes I)\delta(t)dt\right\} \le \mathbb{E}\left\{\sigma^2\int_0^{t_f}\zeta^T(t)(I\otimes\Theta)\zeta(t)dt + V_2(0)\right\} \tag{6.34}$$

with σ being the error attenuation level, $\zeta(t) = \mathrm{col}\{e_x(t), e_w(t), e_{\bar{d}_2}(t), \hat{\bar{d}}_2(t)\}$, and the controller gain matrix is $K = B^TP_2$.

Proof Constructing a Lyapunov function candidate as

$$V_2(t) = \delta^T(t)(\Theta\otimes P_2)\delta(t) + \sum_{i=1}^N \ell_i e^{-\upsilon_i t} \tag{6.35}$$

in which $P_2, \ell_i, \upsilon_i > 0$ such that $V_2(t) > 0$. Based on the definition of $\mathscr{I}V(t)$, it follows from (6.30) and (6.35) by taking mathematical expectations that

$$\begin{aligned}\mathbb{E}\{\mathscr{I}V_2(t)\} = \mathbb{E}\Big\{ & \delta^T(t)[\Theta\otimes(P_2A+A^TP_2) - c\Theta\mathscr{L}_1\otimes P_2BK - c\mathscr{L}_1{}^T\Theta\otimes K^TB^TP_2]\delta(t) \\ & + 2\delta^T(t)(c\Theta\mathscr{L}_1\otimes P_2BK)e_x(t) - 2\delta^T(t)(c\Theta\mathscr{L}_1\otimes P_2BK)\varepsilon(t) \\ & + 2\delta^T(t)(c\Theta\mathscr{A}_0\otimes P_2BK)\varepsilon_0(t) - 2\delta^T(t)(\Theta\otimes P_2B)\tilde{\varepsilon}(t) \\ & + 2\delta^T(t)(\Theta\otimes P_2BV)e_w(t) + 2\delta^T(t)(\Theta\Gamma^u(t)\otimes P_2B)\nu^u(t) \\ & + 2\delta^T(t)(\Theta\otimes P_2D)d_2(t) + 2\delta^T(t)(\Theta\otimes P_2)[\phi(x)-\tilde{\phi}(x_0)] \\ & - \sum_{i=1}^N \ell_i\upsilon_i e^{-\upsilon_i t}\Big\}. \end{aligned} \tag{6.36}$$

Using Lemma 2.5, we can obtain

$$\begin{aligned}
&\mathbb{E}\{2\delta^T(t)(c\Theta\mathscr{L}_1 \otimes P_2BK)e_x(t)\} \\
&\leq \mathbb{E}\Big\{\frac{c\lambda_{\max}(\mathscr{L}_1^T\mathscr{L}_1)\lambda_{\max}(\Theta)}{\beta_4\lambda_{\min}(\Theta)}\delta^T(t)(\Theta \otimes P_2BB^TP_2)\delta(t) \\
&\quad + \beta_4 ce_x^T(t)(\Theta \otimes P_2BB^TP_2)e_x(t)\Big\}
\end{aligned} \tag{6.37}$$

and

$$\begin{aligned}
&\mathbb{E}\{-2\delta^T(t)(\Theta \otimes P_2B)\tilde{\varepsilon}(t)\} \\
&\leq \mathbb{E}\{\frac{1}{\beta_5}\delta^T(t)(\Theta \otimes P_2BB^TP_2)\delta(t) + \beta_5\tilde{\varepsilon}^T(t)(\Theta \otimes I)\tilde{\varepsilon}(t)\}.
\end{aligned} \tag{6.38}$$

Subsequently, let

$$\mathbb{E}\{\Gamma^u(t)\} = \tilde{\Gamma}^u \tag{6.39}$$

where $\tilde{\Gamma}^u = \mathrm{diag}\{\chi_1^u, \cdots, \chi_N^u\}$. Based on the *Young's inequality* in Lemma 2.7 and Assumption 6.2, by following the similar steps in (6.37) and (6.38), one has

$$\begin{aligned}
&\mathbb{E}\{\mathscr{I}V_2(t)\} \\
&\leq \mathbb{E}\Big\{\delta^T(t)[\Theta \otimes (P_2A + A^TP_2) - c\Theta\mathscr{L}_1 \otimes P_2BK - c\mathscr{L}_1{}^T\Theta \otimes K^TB^TP_2]\delta(t) \\
&\quad + \delta^T(t)\Big[\Theta \otimes \Big(\frac{c\lambda_{\max}(\mathscr{L}_1^T\mathscr{L}_1)\lambda_{\max}(\Theta)}{\beta_4\lambda_{\min}(\Theta)}P_2BB^TP_2 + \frac{1}{\beta_5}P_2BB^TP_2 \\
&\quad + \frac{c\lambda_{\max}(\mathscr{L}_1^T\mathscr{L}_1)\lambda_{\max}(\Theta)}{\kappa_1\lambda_{\min}(\Theta)}P_2BB^TP_2 + \frac{c}{\kappa_1}P_2BB^TP_2 + \frac{1}{\sigma^2}P_2BVV^TB^TP_2 \\
&\quad + \frac{1}{\beta_6}P_2BB^TP_2 + \frac{4}{\sigma^2}P_2DD^TP_2 + \frac{1}{\beta_7}P_2P_2 + \beta_7\Lambda_1^T\Lambda_1\Big)\Big]\delta(t) \\
&\quad + \beta_4 ce_x^T(t)(\Theta \otimes P_2BB^TP_2)e_x(t) + \kappa_1 c\varepsilon^T(t)(\Theta \otimes P_2BB^TP_2)\varepsilon(t) \\
&\quad + \kappa_1 c\sum_{i=1}^N \theta_i a_{i0}\varepsilon_{i0}^T(t)P_2BB^TP_2\varepsilon_{i0}(t) + \beta_5\tilde{\varepsilon}^T(t)(\Theta \otimes I)\tilde{\varepsilon}(t) \\
&\quad + \sigma^2 e_w^T(t)(\Theta \otimes I)e_w(t) + \beta_6\lambda_{\max}^2(\tilde{\Gamma}^u)\nu^{u^T}(t)(\Theta \otimes I)\nu^u(t) \\
&\quad + \frac{\sigma^2}{4}d_2^T(t)(\Theta \otimes I)d_2(t) - \sum_{i=1}^N \ell_i\upsilon_i e^{-\upsilon_i t}\Big\}.
\end{aligned} \tag{6.40}$$

Further, in each event interval $[t_k^i, t_{k+1}^i)$, with the event-triggered condition (6.28), one gets

$$\begin{aligned}
&\rho_1 \sum_{i=1}^{N} \theta_i \left\| B^T P_2 \varepsilon_i(t) \right\|^2 + \rho_2 \sum_{i=1}^{N} \theta_i a_{i0} \left\| B^T P_2 \varepsilon_{i0}(t) \right\|^2 + \beta_5 \sum_{i=1}^{N} \theta_i \|\tilde{\varepsilon}_i(t)\|^2 \\
&\le \hbar_1 \sum_{i=1}^{N} \theta_i \left(\left\| B^T P_2 \sum_{j \in \mathcal{N}_i} a_{ij} (\hat{x}_j(t_k^i) - \hat{x}_i(t_k^i)) \right\|^2 + \left\| B^T P_2 a_{i0} (x_0(t_k^i) - \hat{x}_i(t_k^i)) \right\|^2 \right) \\
&\quad + \frac{2\sigma^2}{3} \sum_{i=1}^{N} \theta_i \left\| \hat{\tilde{d}}_{i,2}(t) \right\|^2 + \sum_{i=1}^{N} \upsilon_i \ell_i e^{-\upsilon_i t}
\end{aligned} \tag{6.41}$$

where

$$\begin{aligned}
&\theta_i \left\| B^T P_2 \sum_{j \in \mathcal{N}_i} a_{ij} (\hat{x}_j(t_k^i) - \hat{x}_i(t_k^i)) \right\|^2 + \theta_i \left\| B^T P_2 a_{i0} (x_0(t_k^i) - \hat{x}_i(t_k^i)) \right\|^2 \\
&\le \theta_i \left(\left\| B^T P_2 \sum_{j \in \mathcal{N}_i} a_{ij} (\hat{x}_j(t) - \hat{x}_i(t)) \right\| + \left\| B^T P_2 \sum_{j \in \mathcal{N}_i} a_{ij} (\varepsilon_j(t) - \varepsilon_i(t)) \right\| \right)^2 \\
&\quad + \theta_i \left(\left\| B^T P_2 a_{i0} (x_0(t) - \hat{x}_i(t)) \right\| + \left\| B^T P_2 a_{i0} (\varepsilon_{i0}(t) - \varepsilon_i(t)) \right\| \right)^2 \\
&\le (1 + \kappa_2) \theta_i \left\| B^T P_2 \sum_{j \in \mathcal{N}_i} a_{ij} (\hat{x}_j(t) - \hat{x}_i(t)) \right\|^2 \\
&\quad + (1 + \frac{1}{\kappa_2}) \theta_i \left\| B^T P_2 \sum_{j \in \mathcal{N}_i} a_{ij} (\varepsilon_j(t) - \varepsilon_i(t)) \right\|^2 \\
&\quad + (1 + \kappa_3) \theta_i \left\| B^T P_2 a_{i0} (x_0(t) - \hat{x}_i(t)) \right\|^2 \\
&\quad + (1 + \frac{1}{\kappa_3}) \theta_i \left\| B^T P_2 a_{i0} (\varepsilon_{i0}(t) - \varepsilon_i(t)) \right\|^2
\end{aligned} \tag{6.42}$$

can be obtained according to the additive property of norm and Lemma 2.7. To simplify the analysis, define

$$\begin{aligned}
p_i(t) &= x_0(t) - \hat{x}_i(t) \\
q_i(t) &= \sum_{j \in \mathcal{N}_i} a_{ij} (\hat{x}_j(t) - \hat{x}_i(t)).
\end{aligned}$$

Next, we accumulate the items in (6.42) from agent 1 to agent N, respectively. By the relationship $p_i(t) = e_{x_i}(t) - \delta_i(t)$, one obtains:

$$\sum_{i=1}^{N}\theta_i\left\|B^T P_2 a_{i0} p_i(t)\right\|^2 \leq \left\|(\sqrt{\Theta}\otimes B^T P_2)p(t)\right\|^2$$
$$\leq 2\left\|(\sqrt{\Theta}\otimes B^T P_2)\delta(t)\right\|^2 + 2\left\|(\sqrt{\Theta}\otimes B^T P_2)e_x(t)\right\|^2 \tag{6.43}$$

where $p(t) = \mathrm{col}\{p_1(t), \cdots, p_N(t)\}$, and one has

$$\sum_{i=1}^{N}\theta_i\left\|B^T P_2 q_i(t)\right\|^2 \leq (\tilde{\mathcal{N}} + \sqrt{N\tilde{\mathcal{N}}})^2\left\|(\sqrt{\Theta}\otimes B^T P_2)p(t)\right\|^2 \tag{6.44}$$

since

$$\left\|B^T P_2 q_i(t)\right\| \leq \sum_{j\in\mathcal{N}_i} a_{ij}\big(\left\|B^T P_2 p_i(t)\right\| + \left\|B^T P_2 p_j(t)\right\|\big)$$
$$\leq \tilde{\mathcal{N}}\left\|B^T P_2 p_i(t)\right\| + \sqrt{\tilde{\mathcal{N}}}\left\|(I_N\otimes B^T P_2)p(t)\right\|. \tag{6.45}$$

Moreover, notice that

$$\sum_{i=1}^{N}\theta_i\left\|B^T P_2\sum_{j\in\mathcal{N}_i} a_{ij}(\varepsilon_j(t)-\varepsilon_i(t))\right\|^2$$
$$= \left\|(\sqrt{\Theta}\mathcal{L}\otimes B^T P_2)\varepsilon(t)\right\|^2$$
$$\leq \frac{\lambda_{\max}(\Theta)}{\lambda_{\min}(\Theta)}\lambda_{\max}(\mathcal{L}^T\mathcal{L})\varepsilon^T(t)(\Theta\otimes P_2 B B^T P_2)\varepsilon(t) \tag{6.46}$$

and

$$\sum_{i=1}^{N}\theta_i\left\|B^T P_2 a_{i0}(\varepsilon_{i0}(t)-\varepsilon_i(t))\right\|^2$$
$$\leq \sum_{i=1}^{N}\theta_i\left(a_{i0}\left\|B^T P_2\varepsilon_{i0}(t)\right\| + a_{i0}\left\|B^T P_2\varepsilon_i(t)\right\|\right)^2$$
$$\leq (1+\kappa_4)\sum_{i=1}^{N}\theta_i a_{i0}\left\|B^T P_2\varepsilon_{i0}(t)\right\|^2 + (1+\frac{1}{\kappa_4})\sum_{i=1}^{N}\theta_i\left\|B^T P_2\varepsilon_i(t)\right\|^2. \tag{6.47}$$

Then, it can be inferred from (6.41)–(6.47) that

$$
\begin{aligned}
&\rho_1 \sum_{i=1}^{N} \theta_i \left\| B^T P_2 \varepsilon_i(t) \right\|^2 + \rho_2 \sum_{i=1}^{N} \theta_i a_{i0} \left\| B^T P_2 \varepsilon_{i0}(t) \right\|^2 + \beta_5 \sum_{i=1}^{N} \theta_i \|\tilde{\varepsilon}_i(t)\|^2 \\
&\leq 2\hbar_1(1+\kappa_2)(\tilde{\mathcal{N}} + \sqrt{N\tilde{\mathcal{N}}})^2 \Big(\left\|(\sqrt{\Theta} \otimes B^T P_2)\delta(t)\right\|^2 + \left\|(\sqrt{\Theta} \otimes B^T P_2) e_x(t)\right\|^2 \Big) \\
&\quad + \hbar_1(1+\frac{1}{\kappa_2}) \frac{\lambda_{\max}(\Theta)}{\lambda_{\min}(\Theta)} \lambda_{\max}(\mathscr{L}^T \mathscr{L}) \varepsilon^T(t)(\Theta \otimes P_2 B B^T P_2)\varepsilon(t) \\
&\quad + 2\hbar_1(1+\kappa_3)\Big(\left\|(\sqrt{\Theta} \otimes B^T P_2)\delta(t)\right\|^2 + \left\|(\sqrt{\Theta} \otimes B^T P_2) e_x(t)\right\|^2 \Big) \\
&\quad + \hbar_1(1+\frac{1}{\kappa_3})\Big((1+\kappa_4) \sum_{i=1}^{N} \theta_i a_{i0} \left\| B^T P_2 \varepsilon_{i0}(t) \right\|^2 + \sum_{i=1}^{N} \upsilon_i \ell_i e^{-\upsilon_i t} \\
&\quad + (1+\frac{1}{\kappa_4}) \sum_{i=1}^{N} \theta_i \left\| B^T P_2 \varepsilon_i(t) \right\|^2 \Big) + \frac{2\sigma^2}{3} \sum_{i=1}^{N} \theta_i \left\| \hat{\bar{d}}_{i,2}(t) \right\|^2.
\end{aligned} \tag{6.48}
$$

Recall that ρ_1 and ρ_2 are defined in (6.29). By combining similar terms on both sides of inequality (6.48), it yields

$$
\begin{aligned}
&\kappa_1 c \varepsilon^T(t)(\Theta \otimes P_2 B B^T P_2)\varepsilon(t) + \kappa_1 c \sum_{i=1}^{N} \theta_i a_{i0} \varepsilon_{i0}^T(t) P_2 B B^T P_2 \varepsilon_{i0}(t) + \beta_5 \tilde{\varepsilon}^T(t)(\Theta \otimes I)\tilde{\varepsilon}(t) \\
&\leq 2\hbar_1 \Big[(1+\kappa_2)(\tilde{\mathcal{N}} + \sqrt{N\tilde{\mathcal{N}}})^2 + 1 + \kappa_3\Big] \Big(\delta^T(t)(\Theta \otimes P_2 B B^T P_2)\delta(t) \\
&\quad + e_x^T(t)(\Theta \otimes P_2 B B^T P_2) e_x(t)\Big) + \frac{2\sigma^2}{3} \hat{\bar{d}}_2^T(t)(\Theta \otimes I)\hat{\bar{d}}_2(t) + \sum_{i=1}^{N} \upsilon_i \ell_i e^{-\upsilon_i t}
\end{aligned} \tag{6.49}
$$

where $\hat{\bar{d}}_2(t) = \text{col}\{\hat{\bar{d}}_{1,2}(t), \cdots, \hat{\bar{d}}_{N,2}(t)\}$. Additionally, it is clear that

$$
\begin{aligned}
&\frac{\sigma^2}{4} d_2^T(t)(\Theta \otimes I) d_2(t) + \frac{2\sigma^2}{3} \hat{\bar{d}}_2^T(t)(\Theta \otimes I)\hat{\bar{d}}_2(t) \\
&\leq \frac{\sigma^2}{4} [(e_{\bar{d}_2}(t) + \hat{\bar{d}}_2(t))^T (\Theta \otimes I)(e_{\bar{d}_2}(t) + \hat{\bar{d}}_2(t))] + \frac{2\sigma^2}{3} \hat{\bar{d}}_2^T(t)(\Theta \otimes I)\hat{\bar{d}}_2(t) \\
&\leq \frac{\sigma^2}{4} [e_{\bar{d}_2}^T(t)(\Theta \otimes I) e_{\bar{d}_2}(t) + 3 e_{\bar{d}_2}^T(t)(\Theta \otimes I) e_{\bar{d}_2}(t) + \hat{\bar{d}}_2^T(t)(\Theta \otimes I)\hat{\bar{d}}_2(t)] \\
&\quad + \frac{2\sigma^2}{3} \hat{\bar{d}}_2^T(t)(\Theta \otimes I)\hat{\bar{d}}_2(t) + \frac{1}{3} \hat{\bar{d}}_2^T(t)(\Theta \otimes I)\hat{\bar{d}}_2(t) \\
&\leq \sigma^2 e_{\bar{d}_2}^T(t)(\Theta \otimes I) e_{\bar{d}_2}(t) + \sigma^2 \hat{\bar{d}}_2^T(t)(\Theta \otimes I)\hat{\bar{d}}_2(t).
\end{aligned} \tag{6.50}
$$

In what follows, to analyze the H_∞ stability of MASs under multiple disturbances and FDIAs, consider a function as

$$\aleph(t) = \mathbb{E}\{\mathscr{I} V_2(t) + \delta^T(t)(\Theta \otimes I)\delta(t) - \sigma^2\zeta^T(t)(I \otimes \Theta)\zeta(t)\}. \tag{6.51}$$

Then, according to Lemma 2.2 and Assumption 6.3, by substituting (6.40), (6.49), and (6.50) into (6.51), it attains

$$\begin{aligned}
\aleph(t) \le \mathbb{E}\Bigg\{\delta^T(t)\Bigg[\Theta \otimes \Bigg(P_2A + A^TP_2 + \Bigg(\frac{c\lambda_{\max}(\mathscr{L}_1^T\mathscr{L}_1)\lambda_{\max}(\Theta)}{\beta_4\lambda_{\min}(\Theta)} + \frac{1}{\beta_5} + \frac{1}{\beta_6} + \frac{c}{\kappa_1} \\
&+ \frac{c\lambda_{\max}(\mathscr{L}_1^T\mathscr{L}_1)\lambda_{\max}(\Theta)}{\kappa_1\lambda_{\min}(\Theta)} + \frac{1}{\sigma^2}P_2BVV^TB^TP_2 + \beta_6\kappa\lambda_{\max}^2(\tilde{\Gamma}^u)I \\
&+ 2\hbar_1\Big((1+\kappa_2)(\tilde{\mathscr{N}} + \sqrt{N\tilde{\mathscr{N}}})^2 + 1 + \kappa_3\Big)\Bigg)P_2BB^TP_2 + \frac{4}{\sigma^2}P_2DD^TP_2 \\
&+ \frac{1}{\beta_7}P_2P_2 + \beta_7\Lambda_1^T\Lambda_1\Bigg) - c\lambda_0 \otimes P_2BB^TP_2\Bigg]\delta(t) - \sigma^2\zeta^T(t)(I \otimes \Theta)\zeta(t) \\
&+ \Big[2\hbar_1\Big((1+\kappa_2)(\tilde{\mathscr{N}} + \sqrt{N\tilde{\mathscr{N}}})^2 + 1 + \kappa_3\Big) + \beta_4 c\Big] \\
&\times \lambda_{\max}(P_2BB^TP_2)e_x^T(t)(\Theta \otimes I)e_x(t) + \sigma^2 e_w^T(t)(\Theta \otimes I)e_w(t) \\
&+ \sigma^2 e_{\bar{d}_2}^T(t)(\Theta \otimes I)e_{\bar{d}_2}(t) + \sigma^2\hat{\bar{d}}_2^T(t)(\Theta \otimes I)\hat{\bar{d}}_2(t) \\
&- \sum_{i=1}^N \ell_i v_i e^{-v_i t} + \sum_{i=1}^N v_i \ell_i e^{-v_i t} + \delta^T(t)(\Theta \otimes I)\delta(t)\Bigg\}.
\end{aligned}$$

From (6.31) and (6.32), one can conclude that $\aleph(t) < 0$ by using Lemma 2.8. Integrating both sides of $\aleph(t) < 0$, the following inequality can be obtained:

$$\begin{aligned}
&\mathbb{E}\left\{V_2(t_f) - V_2(0) + \int_0^{t_f} \delta^T(t)(\Theta \otimes I)\delta(t)dt\right\} \\
\le&\mathbb{E}\left\{\sigma^2\int_0^{t_f} \zeta^T(t)(I \otimes \Theta)\zeta(t)dt\right\}
\end{aligned} \tag{6.52}$$

which indicates that (6.34) holds. This completes the proof. □

Remark 6.7 The process of solving the matrix P_2 in Theorem 6.2 is outlined in Algorithm 6.1. Since (6.32) is a constant constraint, we can easily satisfy it by adjusting parameters. Besides, note that the solution of matrix P_2 depends on the *Laplacian matrix* of the system, so it is not fully distributed. As a potential area for future exploration, this will be a consideration for us. When selecting the parameters for the ETM (6.28), the tradeoff between better system performance and control actions and communication resource consumption need to be considered. For instance, smaller values of $\hbar_{i1}$, $\hbar_{i2}$, and ℓ_i will result in more frequent triggering, more controller updates, and a greater communication burden, but they also yield better system performance, and vice versa.

Algorithm 6.1

Step 1: Select the appropriate parameters $\kappa_{1,2,3}$, $\beta_i (i = 4, \cdots, 7)$, c, $\hbar_1$ and σ;
Step 2: **If** LMI (6.31) is feasible **then**
  go to **Step 3**;
 else
  go to **Step 1**;
Step 3: Calculate $\lambda_{\max}(P_2 B B^T P_2)$ using the matrix P_2 solved by LMI (6.31), and substitute it into (6.32);
Step 4: **If** (6.32) is satisfied **then**
  output P_2;
 else
  go to **Step 1**.

Remark 6.8 According to Theorem 6.1 and the definition of $e_{\bar{d}_{i,2}}(t)$ provided in (6.15), we know that $e_x(t)$, $e_w(t)$, $e_{\bar{d}_2}(t)$, and $\hat{\bar{d}}_2(t)$ are UUB. Then, in the *complete consensus* analysis, we can regard them as disturbances and mitigate their impact using the H_∞ control strategy [22, 31]. Hence, in comparison to a single anti-disturbance mechanism [13, 32], the composite anti-disturbance mechanism based on DO and H_∞ control strategy is more effective in enhancing the disturbance rejection capabilities of MASs facing multiple disturbances. Moreover, it is noteworthy that the observer-based event-triggered CHADC method is developed based on the directed interaction topology, which offers broader applicability than *undirected graph*.

6.3.3 Exclusion of Zeno Behavior

In this part, we will show that the proposed ETM is Zeno-free, and the main result is given in Theorem 6.3.

Theorem 6.3 *Under the complete consensus control protocol (6.11) driven by the ETM (6.28), due to the minimum time between events is positive, Zeno behavior is ruled out in MASs, i.e.,*

$$t_{k+1}^i - t_k^i \geq \frac{1}{\|\mathrm{M}\|} \ln \left(\frac{\|\mathrm{M}\| \sqrt{\upsilon_i \ell_i e^{-\upsilon_i t}}}{\|\Omega_i\| \alpha_k^i} + 1 \right) > 0 \tag{6.53}$$

where

$$\Omega_i = \text{diag}\{\sqrt{\theta_i \rho_1} B^T P_2, \sqrt{\theta_i \rho_2 a_{i0}} B^T P_2, \sqrt{\theta_i \beta_5} V\}$$

as well as M *and* α_k^i *are defined in (6.55).*

Proof Define $\bar{\varepsilon}_i(t) = \hat{w}_i(t_k^i) - \hat{w}_i(t)$, we then have $\tilde{\varepsilon}_i(t) = V\bar{\varepsilon}_i(t)$. Thus, the event-triggered function (6.29) can be rewritten as

$$\begin{aligned} r_i(t) = \bar{e}_i^T(t)\Omega_i^T \Omega_i \bar{e}_i(t) - \hbar_{i1}\theta_i \Bigg(\Bigg\| B^T P_2 \sum_{j \in \mathcal{N}_i} a_{ij}(\hat{x}_j(t_k^i) - \hat{x}_i(t_k^i)) \Bigg\|^2 \\ + \left\| B^T P_2 a_{i0}(x_0(t_k^i) - \hat{x}_i(t_k^i)) \right\|^2 \Bigg) - \hbar_{i2}\theta_i \left\| \hat{\bar{d}}_{i,2}(t) \right\|^2 - \upsilon_i \ell_i e^{-\upsilon_i t} \end{aligned} \tag{6.54}$$

where $\bar{e}_i(t) = \text{col}\{\varepsilon_i(t), \varepsilon_{i0}(t), \bar{\varepsilon}_i(t)\}$. For each agent, $\forall t \in [t_k^i, t_{k+1}^i)$, the derivative of $\|\bar{e}_i(t)\|$ satisfies

$$\|\dot{\bar{e}}_i(t)\| = \left\| - \begin{bmatrix} \dot{\hat{x}}_i(t) \\ \dot{x}_0(t) \\ \dot{\hat{w}}_i(t) \end{bmatrix} \right\| \leq \|\text{M}\| \, \|\bar{e}_i(t)\| + \alpha_k^i \tag{6.55}$$

in which $\text{M} = \text{diag}\{A, A, W\}$, and

$$0 < \alpha_k^i = \|\text{M}_1\| \, \bar{\nu} + \left\| \begin{bmatrix} -A\hat{x}_i(t_k^i) - B[\tilde{u}_i(t) + \hat{d}_{i,1}(t)] - Dg_i(t) \\ -\phi(\hat{x}_i(t)) + L_1 C e_{x_i}(t) \\ -Ax_0(t_k^i) - \phi(x_0(t)) \\ -W\hat{w}_i(t_k^i) - Ff(\hat{w}_i(t)) + L_2 C e_{x_i}(t) \end{bmatrix} \right\|$$

with $\text{M}_1 = \text{diag}\{L_1, 0, L_2\}$. It follows from (6.55) that

$$\|\bar{e}_i(t)\| \leq \frac{\alpha_k^i}{\|\text{M}\|} \left(e^{\|\text{M}\|(t - t_k^i)} - 1 \right). \tag{6.56}$$

Obviously, (6.56) can be converted into

$$\|\Omega_i \bar{e}_i(t)\| \leq \frac{\|\Omega_i\| \, \alpha_k^i}{\|\text{M}\|} \left(e^{\|\text{M}\|(t - t_k^i)} - 1 \right). \tag{6.57}$$

Then, by the ETM (6.28) with event-triggered function (6.54), one can achieve that

$$\sqrt{\upsilon_i \ell_i e^{-\upsilon_i t}} \leq \frac{\|\Omega_i\| \, \alpha_k^i}{\|\text{M}\|} \left(e^{\|\text{M}\|(t_{k+1}^i - t_k^i)} - 1 \right). \tag{6.58}$$

By solving (6.58), it shows that $t_{k+1}^i - t_k^i > 0$ is valid. Hence, Theorem 6.3 holds. □

6.4 Simulation Examples

This section offers numerical examples to validate our results. We consider a MAS comprising one leader and four followers, with its network structure depicted in Fig. 5.2. The system matrices of (6.1) and (6.2) are chosen as

$$A = \begin{bmatrix} -2.9 & 0.3 & 0.4 & 1.2 \\ -0.1 & -0.2 & 0.6 & 1.5 \\ 1.2 & 2.1 & -2.8 & 3.4 \\ 1 & -2 & -2.5 & -2.5 \end{bmatrix}, \quad B = \begin{bmatrix} 0 & 0 \\ 0 & 0 \\ -1 & 0.5 \\ -0.1 & 0.2 \end{bmatrix}$$

$$C = \begin{bmatrix} 1 & 0 & 0 & 0 \\ 0 & 1 & 0 & 0 \\ 0 & 0 & 1 & 1 \end{bmatrix}, \quad D = \begin{bmatrix} 0.1 & 0.1 & -0.1 & -0.1 \end{bmatrix}^T$$

$$W = \begin{bmatrix} 0 & 0.1 \\ -0.1 & 0 \end{bmatrix}, \quad F = \begin{bmatrix} -0.27 & -0.2 \\ 0.28 & 0 \end{bmatrix}, \quad V = \begin{bmatrix} 0.4 & 0.5 \\ 0.5 & 0.2 \end{bmatrix}.$$

The functions $\phi(x_i(t))$ and $f(w_i(t))$ are given as $\phi(x_i) = [0, 0, 0, 0.01\sin(x_{i1}(t))]^T$ and $f(w_i(t)) = [0.01\sin(w_{i1}(t)), 0]^T$, respectively. Then, according to Assumption 6.2, one has $\Lambda_1 = \text{diag}\{0.01, 0.01, 0.01, 0.01\}$ and $\Lambda_2 = \text{diag}\{0.01, 0.01\}$. In addition, it can be obtained from Fig. 5.2 and Lemma 2.2 that $\lambda_0 = 1.7859$ and $\Theta = \text{diag}\{4, 2.5, 7, 3.5\}$. Next, select the appropriate values for the positive scalars β_1, β_2, and ϵ, by using the LMI Toolbox in Matlab to solve (6.13) and (6.14) in Theorem 6.1, we get the observer gain L_1, L_2 and gain matrix H as

$$L_1 = \begin{bmatrix} -0.5251 & -0.3762 & 0.1873 \\ -0.3761 & -1.4251 & 0.9100 \\ 0.0862 & 0.2073 & -0.8614 \\ 0.1011 & 0.7028 & -0.9898 \end{bmatrix}$$

$$L_2 = \begin{bmatrix} -0.2387 & -0.7350 & -0.0121 \\ -0.3048 & -0.7429 & 0.4430 \end{bmatrix}$$

$$H = \begin{bmatrix} 0.0188 & -0.0051 & -0.0115 \end{bmatrix}.$$

Without loss of generality, it is assumed that FDIAs will occur on all agents, the probabilities of FDIAs are given as $\chi_i^u = \chi_i^y = 0.02$ $(i = 1, 2)$ and $\chi_i^u = \chi_i^y = 0.03$ $(i = 3, 4)$, and the attack energy limit parameters are set as $\kappa = 0.1$ and $\bar{\nu} = 0.1$. Then, choose $\kappa_1 = 2, \kappa_2 = \kappa_3 = \beta_4 = \beta_6 = \beta_7 = 10, \beta_5 = 0.5, \sigma = 1.41, c = 0.08$ and $\hbar_1 = 0.005$, based on Theorem 6.2, one can get the matrix P_2 and the gain matrix K are

$$P_2 = \begin{bmatrix} 0.3795 & 0.1339 & 0.0007 & 0.3258 \\ 0.1339 & 1.4411 & 0.1882 & 0.7283 \\ 0.0007 & 0.1882 & 0.3893 & 0.0623 \\ 0.3258 & 0.7283 & 0.0623 & 1.0152 \end{bmatrix}$$

$$K = \begin{bmatrix} -0.0333 & -0.2610 & -0.3955 & -0.1638 \\ 0.0655 & 0.2398 & 0.2071 & 0.2342 \end{bmatrix}.$$

In the simulation, the disturbances $d_{i,2}(t)$ are simulated by

$$d_{1,2}(t) = 0.2e^{-0.1t}\sin(5t), \quad d_{2,2}(t) = 0.3e^{-0.1t}\sin(5t)$$
$$d_{3,2}(t) = \frac{\sin(2t)}{3+t}, \quad d_{4,2}(t) = e^{-0.2t}\sin(5t).$$

Furthermore, the initial conditions of external *disturbance* (6.2) and *disturbance* observer (6.10) are chosen as

$$w_i(0) = i * [0.1 \ \ 0.2]^T, \quad \hat{w}_i(0) = i * [-0.1 \ \ 0.1]^T$$

where $i = 1, \cdots, 4$. In addition, the initial state values of each agent are shown in Table 6.1.

To save communication resources and obtain good complete consensus performance, let $\kappa_4 = 10$, and one has $\rho_1 = 0.2722$ and $\rho_2 = 0.1661$. Then, by selecting the appropriate values for other parameters, the simulation results are shown in Figs. 6.2, 6.3, 6.4, 6.5, 6.6 and 6.7. The attack moments of the FDIAs on each agent are plotted in Fig. 6.2. Figure 6.3 displays the *disturbance* signals $d_{i,1}(t) = [d_{i,1}^1(t), d_{i,1}^2(t)]^T$ and their observed values $\hat{d}_{i,1}(t) = [\hat{d}_{i,1}^1(t), \hat{d}_{i,1}^2(t)]^T$. The disturbances $d_{i,2}(t)$ and the adaptive parameters $\hat{\hat{d}}_{i,2}(t)$ are depicted in Fig. 6.4. From Figs. 6.3 and 6.4, it is evident that the designed DO (6.10) and adaptive law (6.9) provide accurate estimation performance even under FDIAs, effectively contributing to disturbance rejection. Figure 6.5 represents the evaluated errors of the state observer (6.7) with respect to $x_i(t)$, demonstrating that the introduction of DO and $g_i(t)$ ensures the observer's accuracy. Let $\tilde{J}(t) = \frac{1}{N}\sqrt{\sum_{i=1}^{N} \|x_i(t) - x_0(t)\|^2}$ be the complete consensus error, then we plot the complete consensus error of the MASs in Fig. 6.6, from which we can find

Table 6.1 Initial state values of each agent

Initial values	Agent 0	Agent 1	Agent 2	Agent 3	Agent 4
$x_{i1}/\hat{x}_{i1}$	–0.3/-0.3	–0.3/–0.1	0.4/0.1	0.25/0.3	0.6/0.8
$x_{i2}/\hat{x}_{i2}$	0.5/0.5	0.4/0.2	–0.2/0.2	0.1/0.4	–0.2/-0.6
$x_{i3}/\hat{x}_{i3}$	0.2/0.2	–0.3/0	0.3/0.3	0.15/0.1	0.3/0.1
$x_{i4}/\hat{x}_{i4}$	–0.8/–0.8	0.3/0.3	0.2/–0.1	–0.1/0.2	–0.4/-0.2

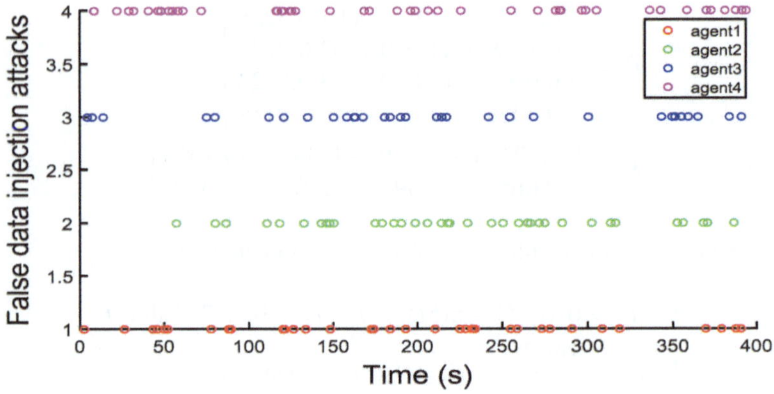

Fig. 6.2 The attack moments of the FDIAs on each agent

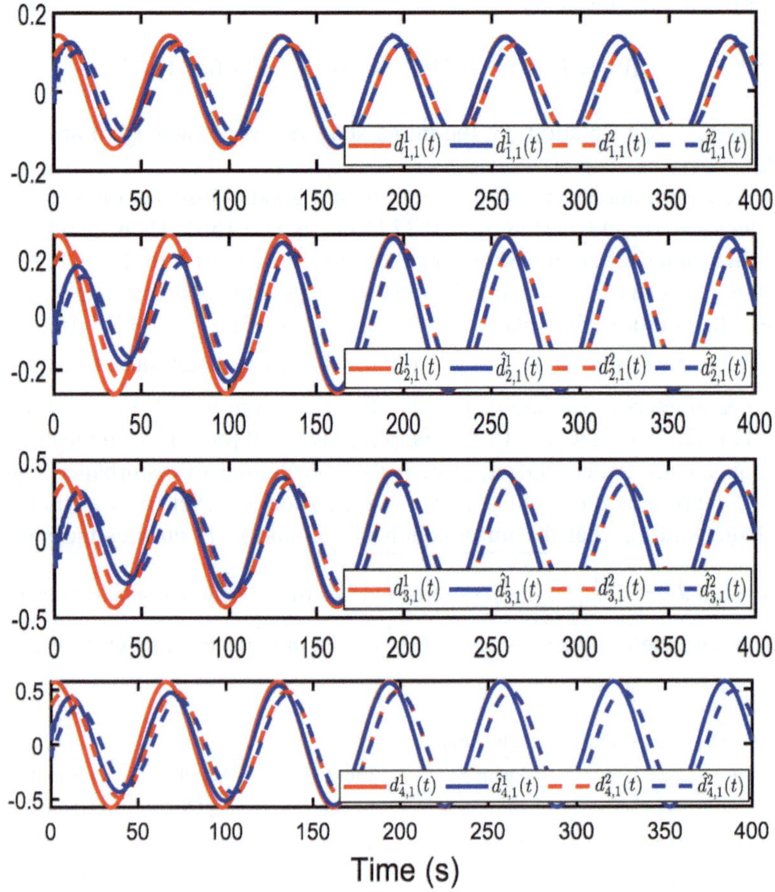

Fig. 6.3 The disturbances $d_{i,1}(t)$ and their estimated values $\hat{d}_{i,1}(t)(i = 1, \cdots, 4)$

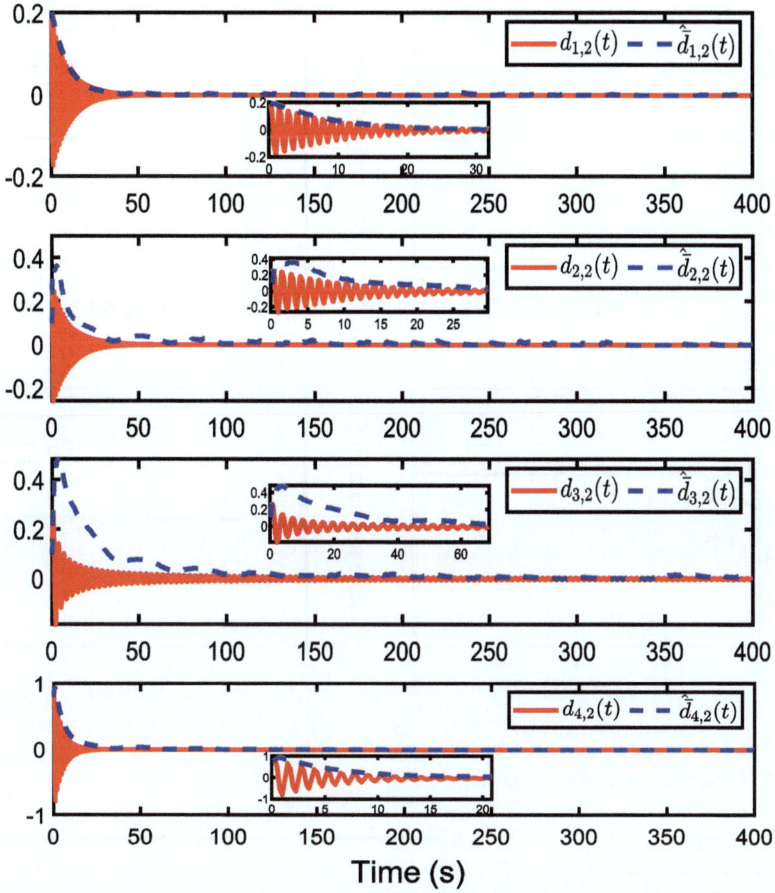

Fig. 6.4 The disturbances $d_{i,2}(t)$ and the adaptive parameters $\hat{\bar{d}}_{i,2}(t)(i = 1, \cdots, 4)$

that in the presence of multiple disturbances and FDIAs, the proposed CHADC protocol (6.11) based on ETM (6.28) shows a good ability to ensure *complete consensus* of the MAS and reject disturbances. The triggering instants and intervals of different agents are presented in Fig. 6.7, which indicates that communication resources are saved and the absence of *Zeno behavior*.

On the other hand, to highlight the benefits of the designed control strategy, Fig. 6.8 shows the observer errors and complete consensus errors using the method in [29]. It is evident that our method performs well. Furthermore, the number of trigger points of each agent under different methods is displayed in Table 6.2. Considering these results, we can summarize that our proposed method not only achieves good control performance but also conserves network resources and alleviates the communication load.

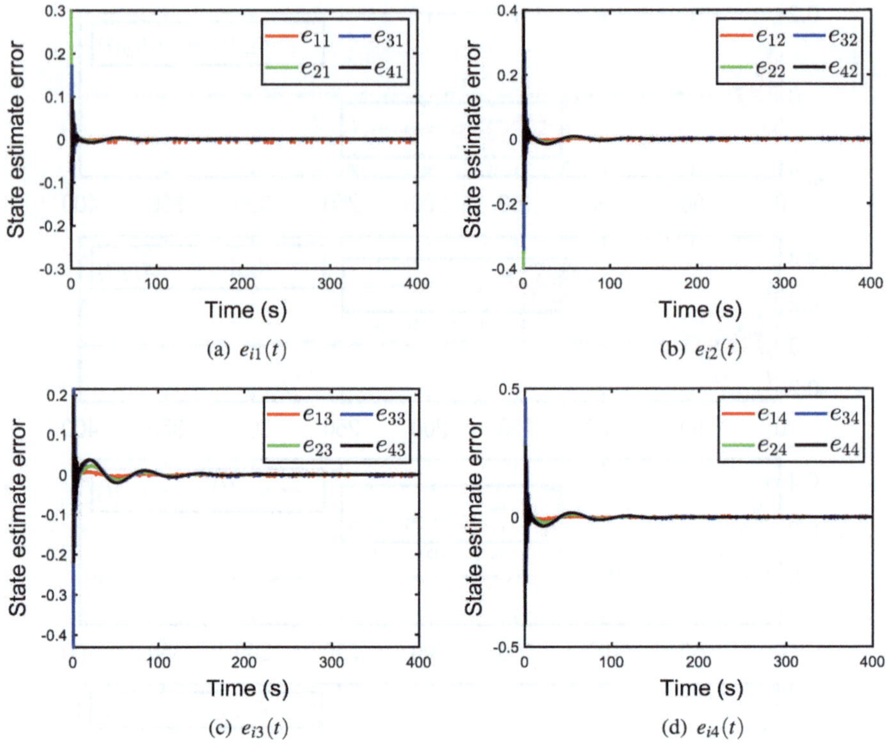

(a) $e_{i1}(t)$ (b) $e_{i2}(t)$

(c) $e_{i3}(t)$ (d) $e_{i4}(t)$

Fig. 6.5 Observer errors of four agents

Fig. 6.6 Complete consensus error $\tilde{J}(t)$ under control scheme (6.11)

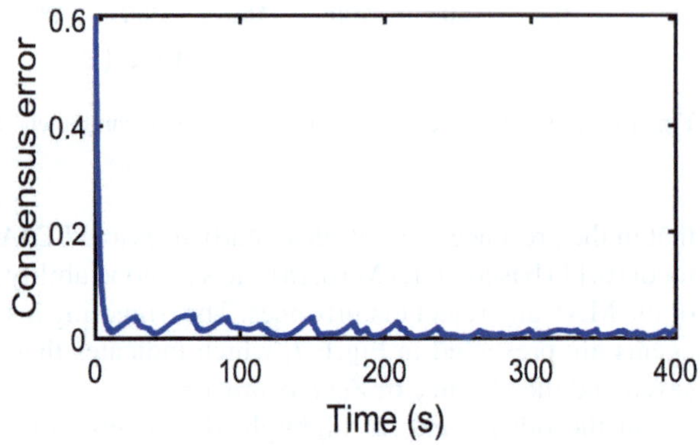

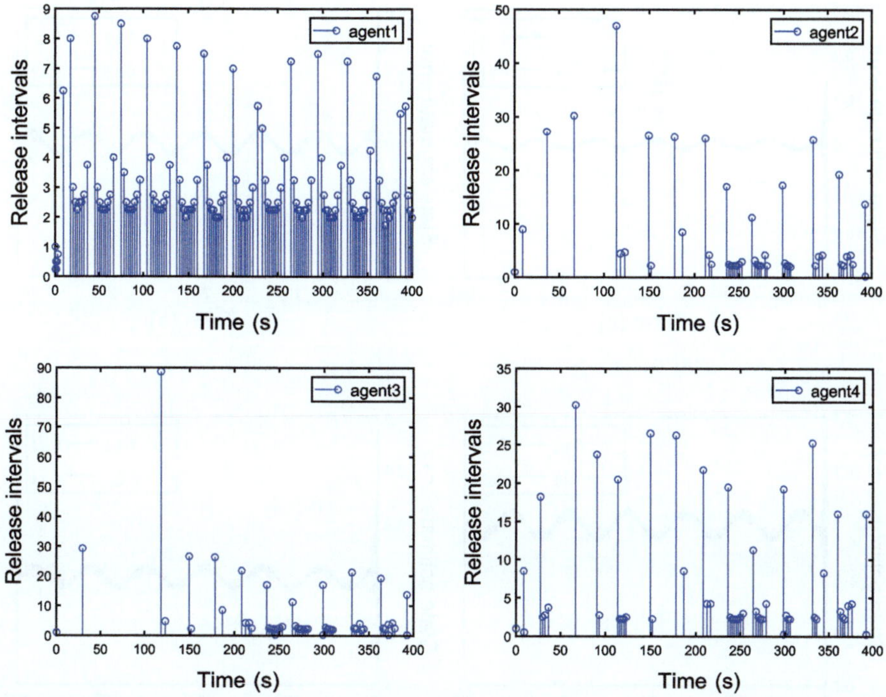

Fig. 6.7 The occurrence time and interval of events on each agent

6.5 Conclusion

This chapter develops an event-triggered control strategy to solve the secure *complete consensus* problem for MASs under multiple disturbances and FDIAs. A CHADC strategy based on DO and H_∞ control is designed to ensure the complete consensus performance of MASs in the presence of multiple disturbances and FDIAs. To reduce communication load, a novel ETM is introduced, which avoids *Zeno behavior*. Notably, the proposed ETM eliminates the need for continuous monitoring of neighboring agents' information, and the controller does not require continuous updates. The effectiveness of the method is validated through simulation results. In the following chapter, the focus will shift to fault-tolerant control for MASs.

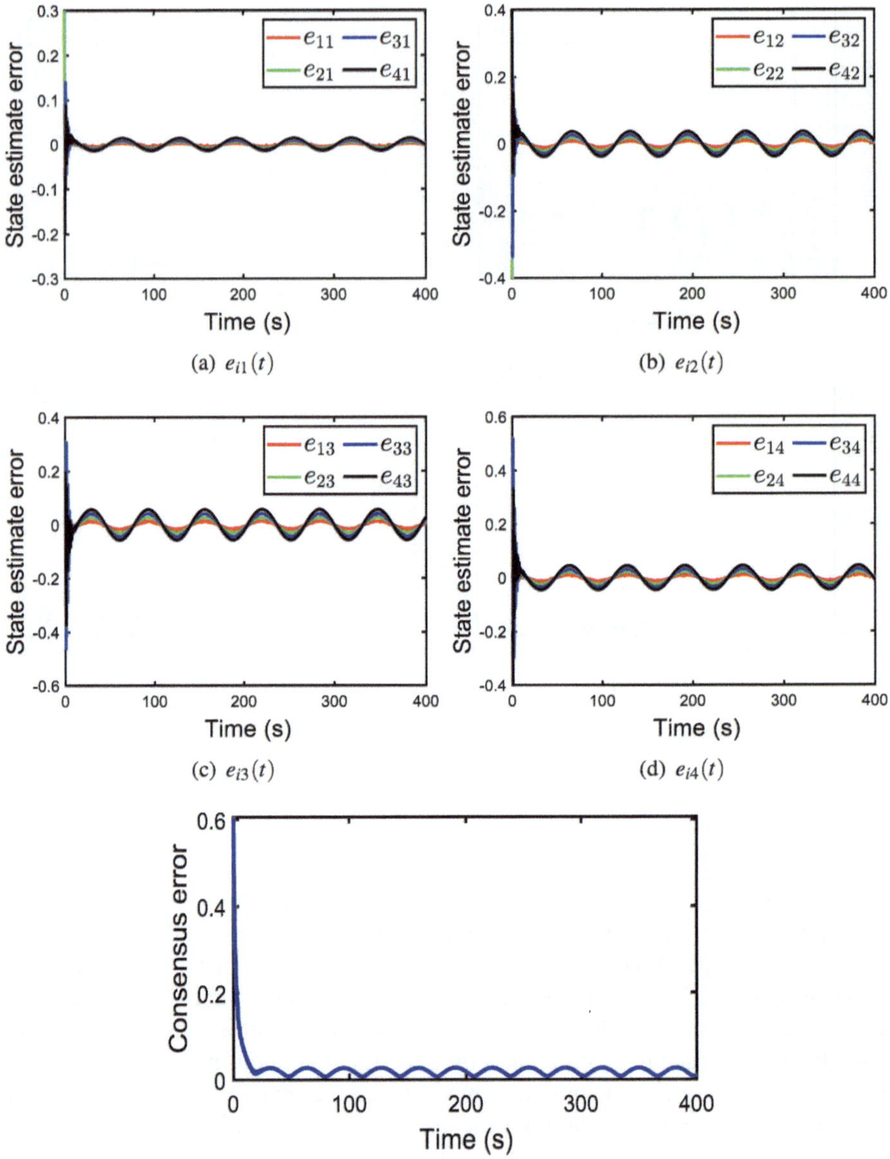

(a) $e_{i1}(t)$

(b) $e_{i2}(t)$

(c) $e_{i3}(t)$

(d) $e_{i4}(t)$

(e) Complete consensus error $\tilde{J}(t)$

Fig. 6.8 Observer errors and complete consensus error under the method in [29]

Table 6.2 The number of trigger points of each agent under different methods

Methods		Agents			
		1	2	3	4
Control scheme (6.11)	CON	133	46	51	54
	NEI	133	46	51	54
Method in [29]	CON	148	84	135	92
	NEI	1600	1600	1600	1600

CON denotes the number of controller updates, and NEI represents the number of communications between each follower and its neighbors

References

1. L. Guo, S. Cao, Anti-disturbance control theory for systems with multiple disturbances: a survey. ISA Trans. **53**(4), 846–849 (2014)
2. X. Yao, J.H. Park, L. Wu, L. Guo, Disturbance-observer-based composite hierarchical antidisturbance control for singular Markovian jump systems. IEEE Trans. Autom. Control **64**(7), 2875–2882 (2019)
3. Y. Zhu, L. Guo, J. Qiao, W. Li, An enhanced anti-disturbance attitude control law for flexible spacecrafts subject to multiple disturbances. Control. Eng. Pract. **84**, 274–283 (2019)
4. W. Qi, Y. Hou, G. Zong, C.K. Ahn, Finite-time event-triggered control for Semi-Markovian switching cyber-physical systems with FDI attacks and applications. IEEE Trans. Circuits Syst. I, Reg. Papers **68**(6), 2665–2674 (2021)
5. L. Zhao, G.H. Yang, Cooperative adaptive fault-tolerant control for multi-agent systems with deception attacks. J. Franklin Inst. **357**(6), 3419–3433 (2020)
6. X.G. Guo, D.Y. Zhang, J.L. Wang, C.K. Ahn, Adaptive memory event-triggered observer-based control for nonlinear multi-agent systems under DoS attacks. IEEE/CAA J. Autom. Sinica **8**(10), 1644–1656 (2021)
7. H. Modares, B. Kiumarsi, F.L. Lewis, F. Ferrese, A. Davoudi, Resilient and robust synchronization of multiagent systems under attacks on sensors and actuators. IEEE Trans. Cybern. **50**(3), 1240–1250 (2020)
8. C. Deng, C. Wen, J. Huang, X.M. Zhang, Y. Zou, Distributed observer-based cooperative control approach for uncertain nonlinear MASs under event-triggered communication. IEEE Trans. Autom. Control **67**(5), 2669–2676 (2022)
9. T.K. Tasooji, H.J. Marquez, Cooperative localization in mobile robots using event-triggered mechanism: theory and experiments. IEEE Trans. Autom. Sci. Eng. **19**(4), 3246–3258 (2022)
10. B. Cheng, Z. Li, Consensus disturbance rejection with event-triggered communications. J. Franklin Inst. **356**(2), 956–974 (2019)
11. C. Ma, H. Qiao, Distributed asynchronous event-triggered consensus of nonlinear multi-agent systems with disturbances: an extended dissipative approach. Neurocomputing **243**, 103–114 (2017)
12. Y.Y. Qian, L. Liu, G. Feng, Distributed event-triggered adaptive control for consensus of linear multi-agent systems with external disturbances. IEEE Trans. Cybern. **50**(5), 2197–2208 (2020)
13. L. Rong, X. Liu, G.P. Jiang, S. Xu, Event-driven multiagent consensus disturbance rejection with input uncertainties via adaptive protocols. IEEE Trans. Syst. Man Cybern. Syst. **52**(5), 2911–2919 (2022)
14. J. Sun, J. Yang, S. Li, X. Wang, G. Li, Event-triggered output consensus disturbance rejection algorithm for multi-agent systems with time-varying disturbances. J. Franklin Inst. **357**(17), 12870–12885 (2020)
15. K. Sun, H. Yu, X. Xia, Distributed control of nonlinear stochastic multi-agent systems with external disturbance and time-delay via event-triggered strategy. Neurocomputing **452**, 275–283 (2021)

16. Z.G. Wu, Y. Xu, Y.J. Pan, H. Su, Y. Tang, Event-triggered control for consensus problem in multi-agent systems with quantized relative state measurements and external disturbance. IEEE Trans. Circuits Syst. I, Reg. Papers **65**(7), 2232–2242 (2018)
17. L.Y. Hao, Y. Yu, T.S. Li, H. Li, Quantized output-feedback control for unmanned marine vehicles with thruster faults via sliding-mode technique. IEEE Trans. Cybern. **52**(9), 9363–9376 (2022)
18. Y. Shi, C. Liu, Y. Wang, Asymptotically stable filter for MVU estimation of states and homologous unknown inputs in heterogeneous multiagent systems. IEEE Trans. Autom. Sci. Eng. **19**(2), 884–894 (2022)
19. X. Huang, J.X. Dong, Reliable leader-to-follower formation control of multiagent systems under communication quantization and attacks. IEEE Trans. Syst. Man Cybern. Syst. **50**(1), 89–99 (2020)
20. H. Sun, R. Madonski, S. Li, Y. Zhang, W. Xue, Composite control design for systems with uncertainties and noise using combined extended state observer and kalman filter. IEEE Trans. Ind. Electron. **69**(4), 4119–4128 (2022)
21. X. Wang, S. Li, J. Lam, Distributed active anti-disturbance output consensus algorithms for higher-order multi-agent systems with mismatched disturbances. Automatica **74**, 30–37 (2016)
22. X. Yao, L. Guo, Composite anti-disturbance control for Markovian jump nonlinear systems via disturbance observer. Automatica **49**(8), 2538–2545 (2013)
23. T. Ménard, S.A. Ajwad, E. Moulay, P. Coirault, M. Defoort, Leader-following consensus for multi-agent systems with nonlinear dynamics subject to additive bounded disturbances and asynchronously sampled outputs. Automatica **121**, Art. No. 109176 (2020)
24. Y. Yuan, Z. Wang, L. Guo, Event-triggered strategy design for discrete-time nonlinear quadratic games with disturbance compensations: the noncooperative case. IEEE Trans. Syst. Man Cybern. Syst. **48**(11), 1885–1896 (2018)
25. Y. Cui, Y. Liu, W. Zhang, F.E. Alsaadi, Sampled-based consensus for nonlinear multiagent systems with deception attacks: the decoupled method. IEEE Trans. Syst. Man Cybern. Syst. **51**(1), 561–573 (2021)
26. G. Wen, X. Zhai, Z. Peng, A. Rahmani, Fault-tolerant secure consensus tracking of delayed nonlinear multi-agent systems with deception attacks and uncertain parameters via impulsive control. Commun. Nonlinear Sci. Numer. Simulat. **82**, Art. No. 105043 (2020)
27. X.M. Li, Q. Zhou, P. Li, H.Y. Li, R.Q. Lu, Event-triggered consensus control for multi-agent systems against false data-injection attacks. IEEE Trans. Cybern. **50**(5), 1856–1866 (2020)
28. S. Yuan, C. Yu, J. Sun, Adaptive event-triggered consensus control of linear multi-agent systems with cyber attacks. Neurocomputing **442**, 1–9 (2021)
29. Y. Yang, D. Yue, C. Dou, Output-based event-triggered schemes on leader-following consensus of a class of multi-agent systems with lipschitz-type dynamics. Inf. Sci. **459**, 327–340 (2018)
30. X. Wu, K. Xu, M. Lei, X. He, Disturbance-compensation-based continuous sliding mode control for overhead cranes with disturbances. IEEE Trans. Autom. Sci. Eng. **17**(4), 2182–2189 (2020)
31. X.G. Guo, X. Fan, C.K. Ahn, Adaptive event-triggered fault detection for interval type-2 T-S fuzzy systems with sensor saturation. IEEE Trans. Fuzzy Syst. **29**(8), 2310–2321 (2021)
32. L. Ma, Y.L. Wang, Q.L. Han, H_∞ cluster formation control of networked multiagent systems with stochastic sampling. IEEE Trans. Cybern. **51**(12), 5761–5772 (2021)

Chapter 7
Fault-Tolerant Sliding-Mode Control for MASs with ANASs and IQCs

Abstract In this chapter, by integrating pseudo-PID sliding-mode control (SMC) method with adaptive control techniques, two innovative fully distributed adaptive *fault-tolerant* control schemes are proposed to address the leader-following *complete consensus* issue in nonlinear multi-agent systems (MASs) under integral quadratic constraints (IQCs) and actuator malfunctions, both with and without asymmetric nonlinear actuator *saturations* (ANASs). For scenarios devoid of *saturations*, the proposed control scheme boasts a straightforward design and minimal computational demands, albeit relying on rudimentary system model details. To address this limitation, in scenarios involving *saturations*, the control scheme is revamped with the incorporation of an innovative anti-windup mechanism and fuzzy logic systems (FLSs), addressing computational complexity reduction. These controllers operate on local interaction data rather than global network data, guaranteeing the practical achievement of *complete consensus* in leader-follower systems within a *finite time* frame.

7.1 Introduction

In the previous chapters, the *complete consensus* control strategy of multi-agent systems (MASs) with *disturbance* and cyber attack is introduced. However, actuator faults in the physical systems of agents can lead to system performance degradation and even system instability [1]. In a networked environment, a single agent's failure can cascade, affecting the overall system's performance or causing a total collapse [2]. To minimize the detrimental effects of faults on system performance, it is crucial to investigate *fault-tolerant* control strategies for MASs [3–12]. Various control methods have been proposed to address the diverse characteristics of MASs, including fixed or time-varying topologies and linear or nonlinear dynamics. These methods include sliding-mode control (SMC) [9], event-triggered mechanisms (ETM) [7, 11], adaptive control method[3, 5, 6, 8, 10, 12], and fuzzy control method [4]. However, many of these approaches, such as [3, 5–7, 9–12], assume knowledge of the actuator efficiency's bounds, while others, such as [4, 8], focus solely on the *complete*

X.-G. Guo et al., *Anti-disturbance Intrusion- and Fault-Tolerant Control of Multi-agent Systems*, https://doi.org/10.1007/978-981-96-9614-7_7

consensus issue under actuator bias faults. There is a need for more efficient methodologies to address the challenge of unknown actuator efficiency factors and combined multiplicative and bias faults.

On the other hand, the practical implementation of control systems can be impeded by actuator *saturations* limitations, leading to diminished performance or system instability [13, 14]. Consequently, a thorough study of actuator *saturations* in MASs is essential [13, 15–18]. Extensive research has been conducted on MASs with symmetric linear [13, 16–18] and asymmetric linear [15] actuator *saturations*. However, these studies typically presume a linear relationship between the control input and the unsaturated area of the *saturations* model. In reality, actuators often exhibit nonlinear behavior. Thus, [19] introduces a broader model known as asymmetric nonlinear actuator *saturations* (ANAS) to capture the *saturations* traits of individual agents. Yet, this model is hampered by the strict monotonicity assumption for the unsaturated portion of nonlinear functions, restricting its broader utility. Moreover, the presence of actuator faults, when combined with integral quadratic constraints (IQCs) and ANASs, further complicates the challenges addressed in this research. It is of significant theoretical importance to develop innovative approaches to address the leader-follower consensus problem in MASs, considering the interplay between actuator faults, IQCs, and ANASs.

Motivated by these insights, two innovative fully distributed adaptive *fault-tolerant* control schemes are proposed to guarantee the practical leader-following *complete consensus* in *finite time* for nonlinear MASs. These systems are influenced by the interplay of actuator faults, IQCs, and ANASs. The following challenging issues are considered here.

(1) The upper and lower bounds of the actuator efficiency factors are not available.
(2) Actuator faults, IQCs, and ANASs coexist. Specifically, IQCs incorporate the absolute values of actuator fault information, and their interplay with actuator faults and ANASs significantly increases the complexity of the problem at hand.
(3) The acceleration information of the leader agent is inaccessible.
(4) The communication topology is assumed to be a *directed graph* with local neighbor data.

To address these challenges, the proposed approach is threefold: initially, a pseudo-PID sliding-mode surface is introduced. Following this, an adaptive strategy is implemented for actuator fault tolerance. Subsequently, an innovative anti-windup compensator is introduced to counteract the negative effects of ANASs. This chapter features the following highlights.

(1) Fully distributed pseudo-PID sliding-mode control strategy: In contrast to [6, 9] and Chaps. 3–6, which rely on both local and global network information, the designed fully distributed adaptive *fault-tolerant* control strategies require only local neighbor information, facilitated by Lemma 7.1. Moreover, the pseudo-PID sliding-mode surface is utilized to address scenarios where the leader's acceleration data is unavailable.

(2) For the ANAS case, a novel anti-windup compensator is employed to bypass the constraint that the control input is linearly related to the unsaturated region of the *saturations* model as seen in [19]. Lemma 7.2 theoretically confirms that this compensator can mitigate the negative effects induced by ANASs. Additionally, the proposed control strategy is independent of the number of fuzzy membership rules, thereby reducing computational complexity.
(3) Distinguishing from [6], the key distinctions are as follows: (a) The graph model in [6] is an *undirected graph*, while this chapter deals with a directed graph that includes a directed spanning tree. (b) Unlike [6], which requires global topology data, this chapter relies solely on local neighbor information. (c) Unlike [6], this chapter does not necessitate prior knowledge of the bounds for actuator efficiency factors. (d) The requirement for nonlinear functions to adhere to the differential mean value property is eliminated here, and there is no need for the nonlinear dynamics of the leader and followers to be identical. (e) The problem tackled in this chapter is more complex due to the interplay among actuator faults, IQCs, and ANASs. Consequently, the methodologies developed here are not a straightforward extension of those in [6].

7.2 Problem Formulation

7.2.1 System Model

Consider the $N+1$ MAS with a direct network; each agent in the network is a second-order integrator. The leader agent can be described as

$$\begin{aligned} \dot{x}_0(t) &= v_0(t) \\ \dot{v}_0(t) &= a_0(t) \end{aligned} \tag{7.1}$$

while the ith agent is governed by the following dynamic model:

$$\begin{aligned} \dot{x}_i(t) &= v_i(t) \\ \dot{v}_i(t) &= Sat(u_i^F(t)) + f_i(x_i, v_i, t) + \sum_{k=1}^{p_i} B_{ik} w_{ik}(t) \\ \xi_{ik}(t) &= H_{ik} v_i(t) + G_{ik} u_i^F(t), \quad i \in \mathscr{V}_N \end{aligned} \tag{7.2}$$

where $x_i(t) \in \mathbb{R}$ and $v_i(t) \in \mathbb{R}$ are the agent position and velocity state, respectively; $a_0(t)$ denotes leader acceleration; $u_i^F(t)$ represents the ith control input signal with a fault; $f_i(x_i, v_i, t)$ are nonlinear unmodeled dynamic uncertainties, which may contain parameter uncertainties, external *disturbances*, and so on [13, 20]; B_{ik}, H_{ik}, and G_{ik}, $k \in \mathscr{V}_{p_i}$ are the known real constant matrices of corresponding dimensions; the input and output behavior of external *disturbances* is described by IQCs, and $w_{ik}(t)$ and $\xi_{ik}(t)$ denote the IQC input and output, respectively.

To achieve the control objective, the following assumption in this chapter are needed.

Assumption 7.1 ([8, 21, 22]) The weighted *directed graph* $\bar{\mathscr{G}}$ of the leader-following system composed of (7.1) and (7.2) has at least one directed spanning tree, in which the root is labeled 0.

Based on (7.1) and (7.2), the relative position and velocity errors can be defined as

$$\begin{aligned} e_{x,i}(t) &= \sum_{j\in N_i} a_{ij}(x_i(t) - x_j(t)) \\ e_{v,i}(t) &= \sum_{j\in N_i} a_{ij}(v_i(t) - v_j(t)), \quad i \in \mathscr{V}_N. \end{aligned} \tag{7.3}$$

Then, the following lemma playing a key role in the derivation of our main results is presented.

Lemma 7.1 *Suppose that Assumption 7.1 holds. Then, we can obtain the following formula relationship, i.e., when*

$$\lim_{t\to\infty} e_{x,i}(t) = 0, \quad \lim_{t\to\infty} e_{v,i}(t) = 0, \quad i \in \mathscr{V}_N \tag{7.4}$$

then the follower agents will follow the leader agent, i.e.,

$$\begin{aligned} &\lim_{t\to\infty} |x_i(t) - x_0(t)| = 0 \\ &\lim_{t\to\infty} |v_i(t) - v_0(t)| = 0, \quad i \in \mathscr{V}_N. \end{aligned} \tag{7.5}$$

Proof Let $x(t) = [x_1(t), x_2(t), \cdots, x_N(t)]^T$ and $v(t) = [v_1(t), v_2(t), \cdots, v_N(t)]^T$. By using the properties of the *Laplacian matrix* $\mathscr{L}$, the following results can be derived:

$$\begin{aligned} e_x(t) &= \mathscr{L}_1 x(t) + \mathscr{L}_2 x_0(t) \\ &= \mathscr{L}_1 x(t) - \mathscr{L}_1 \mathbf{1} x_0(t) \\ &= \mathscr{L}_1 (x(t) - \mathbf{1} x_0(t)) \\ e_v(t) &= \mathscr{L}_1 v(t) + \mathscr{L}_2 v_0(t) \\ &= \mathscr{L}_1 v(t) - \mathscr{L}_1 \mathbf{1} v_0(t) \\ &= \mathscr{L}_1 (v(t) - \mathbf{1} v_0(t)). \end{aligned} \tag{7.6}$$

Based on Assumption 7.1, it follows from (7.4) and (7.6) that (7.5) holds since matrix $\mathscr{L}_1$ is nonsingular. The proof is completed. □

Remark 7.1 A variety of successful strategies have been implemented to address the leader-following *complete consensus* issue in MASs, as seen in [6, 9]. Typically, these approaches rely on both local neighbor data and global topology structure. However, it's noteworthy that Lemma 7.1 reveals an interesting finding: global topology information is not essential. Furthermore, when the tracking errors in position $e_x(t)$ and velocity $e_v(t)$ are zero, Lemma 7.1 guarantees leader-following *complete consensus*. Conversely, when these errors approach zero, Lemma 7.1 ensures a bounded or practical *complete consensus*, rather than an *complete consensus*.

7.2.2 IQC Constraints

The IQC model is known to be useful for characterizing the properties of external *disturbances* and uncertainties [6, 23]. In this chapter, the IQC model is used to describe the external *disturbances* occurring in the follower agent system (7.2). The prospective behavior of the input $w_{ik}(t)$ and the output $\xi_{ik}(t)$ in IQCs in (7.2) is represented by

$$w_{ik}(t) = \psi_{ik}(t, \xi_{ik}(\cdot)|_{t_0}^{t}), \quad i \in \mathscr{V}_N \tag{7.7}$$

where $\psi_{ik}(\cdot)$ denotes nonlinear time-varying dynamic constraints.

Definition 7.1 ([24]) For the MAS (7.2), given any locally square integrable input $\xi_{ik}(t)$ and any corresponding system solution, if assumption (7.7) holds, and there exist a sequence $\{t_a\}_{a=1}^{\infty}$ and a positive constant δ_{ik} $(i \in \mathscr{V}, k = 1, 2, \cdots, p_i)$ for a given $t_0 \geq 0$ in the interval $(0, t_s)$, that satisfy the following relationship:

$$\int_{t_0}^{t_a} \|w_{ik}(t)\|^2 \, dt \leq \int_{t_0}^{t_a} \|\xi_{ik}(t)\|^2 \, dt + \delta_{ik}, \forall t_a \tag{7.8}$$

where $t_a \to t_s$, $(t_a \geq t_0)$, and t_a and t_s can be equal to infinity. Then, an admissible uncertainty for the MAS (7.2) can be defined by the IQC inputs $\xi_{ik}(t)$.

Assume that the Lebesgue space $L_2[T_1, T_2]$ represents a class of functions $w_{i,k}(t)$: $[T_1, T_2] \to \mathbb{R}^{m_i}$ of finite energy (i.e., square-integrable), and for any $T_1 < T_2 \leq \infty$, $\|w_{i,k}(t)\|_{L_2}^2 = \int_{T_1}^{T_2} \|w_{i,k}(t)\|^2 \, dt < \infty$ is satisfied [24]. From (7.8), one has:

$$\|w_{ik}(t)\|_{L_2}^2 \leq \|\xi_{ik}(t)\|_{L_2}^2 + \delta_{ik} \tag{7.9}$$

namely,

$$\|w_{ik}(t)\|_{L_2} \leq \sqrt{\|\xi_{ik}(t)\|_{L_2}^2 + \delta_{ik}} \leq \|\xi_{ik}(t)\|_{L_2} + \delta_{wik} \tag{7.10}$$

where $0 < \delta_{w,ik} \leq \sqrt{\delta_{ik}}$ is an unknown constant. In addition, according to the equivalence of all norm in a finite dimensional vector space, the inequality (7.10) can be equivalent to

$$\|w_{ik}(t)\| \leq \|\xi_{ik}(t)\| + \delta_{w,ik}. \tag{7.11}$$

By combining (7.2) with (7.11), one can obtain

$$\begin{aligned}\sum_{k=1}^{p_i} \|B_{ik} w_{ik}(t)\| &\leq \sum_{k=1}^{p_i} \|B_{ik}\xi_{ik}(t)\| + \sum_{k=1}^{p_i} \|B_{ik}\|\delta_{w,ik} \\ &\leq \zeta_{1i}|u_i^F(t)| + \zeta_{2i}|v_i(t)| + \delta_{w,i}\end{aligned} \tag{7.12}$$

where

$$\zeta_{1i} = \sum_{k=1}^{p_i} \|B_{ik}\|\|G_{ik}\|, \quad \zeta_{2i} = \sum_{k=1}^{p_i} \|B_{ik}\|\|H_{ik}\|.$$

Furthermore, $\delta_{w,ik}$ is an unknown positive constant, and $\delta_{w,i} = \sum_{k=1}^{p_i} \|B_{ik}\|\delta_{w,ik}$.

7.2.3 Actuator Fault Model

This chapter tackles the concurrent issues of actuator efficiency loss and bias faults in the followers of an MAS. For agent i, when faults occur, the practical control input $u_i^F(t)$ can be represented as

$$u_i^F(t) = \rho_i(t, t_{\rho,i})u_i(t) + r_i(t, t_{r,i}), \quad i \in \mathscr{V}_N \tag{7.13}$$

where $\rho_i(t, t_{\rho,i})$ and $r_i(t, t_{r,i})$ indicate the actuator efficiency loss and bias faults, respectively, in the ith agent, where $t_{\rho,i}$ and $t_{r,i}$ are unknown fault time instants. The considered faulty modes are listed as follows:

$$\begin{cases} \rho_i(t, t_{\rho,i}) = 1, r_i(t, t_{r,i}) = 0, & \text{healthy actuator} \\ \rho_i(t, t_{\rho,i}) = 1, r_i(t, t_{r,i}) \neq 0, & \text{actuator bias fault} \\ 0 < \rho_i(t, t_{\rho,i}) \leq 1, r_i(t, t_{r,i}) = 0, & \text{loss of effectiveness} \\ 0 < \rho_i(t, t_{\rho,i}) \leq 1, r_i(t, t_{r,i}) \neq 0, & \text{loss of effectiveness with bias fault.} \end{cases}$$

Some other necessary assumptions will be provided here for the main results.

Assumption 7.2 The unknown parameters $\rho_i(t, t_{\rho,i})$ and $r_i(t, t_{r,i})$ referring to (7.13) satisfy the following constraints:

$$
\begin{aligned}
&0 < \underline{\rho}_i \leq \rho_i(t, t_{\rho,i}) \leq \bar{\rho}_i < \infty \\
&|r_i(t, t_{r,i})| \leq \bar{r}_i, \quad i \in \mathscr{V}_N.
\end{aligned} \tag{7.14}
$$

Assumption 7.3 ([6]) It is assumed that the constraint $\zeta_{1i} < \underline{\rho}_i / \bar{\rho}_i$ holds, where ζ_{1i} is defined in (7.12), while $\underline{\rho}_i$ and $\bar{\rho}_i$ are the upper and lower bounds of actuator efficiency factor $\rho_i(t, t_{\rho,i})$ in (7.13), respectively.

Remark 7.2 Two key points should be highlighted regarding the aforementioned assumptions. Firstly, conditions akin to Assumption 7.2 are prevalent in the literature on actuator *fault-tolerant* control, as seen in [3, 5–7, 9–12] and related works. However, the availability of the upper and lower bounds of the actuator efficiency factor $\rho_i(t, t_{\rho,i})$ is required in [3, 5, 6, 9–12], while an additional analytical effort to estimate both the upper and lower bounds of the actuator efficiency factor $\rho_i(t, t_{\rho,i})$ is indispensable for establishing the control strategy proposed by [7]. This consequently complicates the strategy design process and results in inconvenient usage in practice. The Secondly, Assumption 7.3 mirrors a condition previously stated in [6]. However, In contrast to [6], this chapter does not require the upper and lower bounds of the actuator efficiency factor $\rho_i(t, t_{\rho,i})$, enhancing the practicality of our proposed control strategies.

7.2.4 ANAS Model

Actuator *saturation* is a typical constraint encountered in real-world applications. This chapter focuses on studying the leader-following *complete consensus* issue under the influence of ANAS. The ANAS $Sat(u_i^F(t))$ is expressed as

$$
\operatorname{sat}(u_i^F(t)) = \begin{cases} u_{\max,i}, & \text{if } u_i^F(t) \geq b_{r,i} \\ h_{r,i}(u_i^F(t)), & \text{if } 0 \leq u_i^F(t) < b_{r,i} \\ h_{l,i}(u_i^F(t)), & \text{if } b_{l,i} < u_i^F(t) < 0 \\ u_{\min,i}, & \text{if } u_i^F(t) \leq b_{l,i}, \, i \in \mathscr{V}_N \end{cases} \tag{7.15}
$$

where $b_{l,i} < 0$ and $b_{r,i} > 0$ are the maximum and minimum bounds of *saturation* and $u_{\min,i} < 0$ and $u_{\max,i} > 0$ are the *saturation* constraint, while $h_{r,i}(u_i^F(t))$ and $h_{l,i}(u_i^F(t))$ are any bounded nonlinear functions. In the case where the faults have been considered, the error input produced by *saturation* can be denoted as below:

$$
\begin{aligned}
\Delta u_i(t) &\triangleq \operatorname{Sat}(u_i^F(t)) - u_i^F(t) \\
&= \begin{cases} u_{\max,i} - u_i^F(t), & \text{if } u_i^F(t) \geq b_{r,i} \\ h_{r,i}(u_i^F(t)) - u_i^F(t), & \text{if } 0 \leq u_i^F(t) < b_{r,i} \\ h_{l,i}(u_i^F(t)) - u_i^F(t), & \text{if } b_{l,i} < u_i^F(t) < 0 \\ u_{\min,i} - u_i^F(t), & \text{if } u_i^F(t) \leq b_{l,i}. \end{cases}
\end{aligned} \tag{7.16}
$$

Therefore, the MAS (7.2) is rewritten as

$$\begin{cases} \dot{x}_i(t) = v_i(t) \\ \dot{v}_i(t) = u_i^F(t) + \Delta u_i(t) + f_i(x_i, v_i, t) + \sum_{k=1}^{p_i} B_{ik} w_{ik}(t), i \in \mathscr{V}_N. \end{cases} \tag{7.17}$$

Remark 7.3 The actuator *saturations* models that have been considered in the past, which necessitate a linear relationship between $h_{r,i}(u_i^F(t))/h_{l,i}(u_i^F(t))$ and $u_i^F(t)$ as considered in [13, 15–18], are actually specific instances of our ANAS model defined in Eq. (7.15). Furthermore, our ANAS model, as outlined in Eq. (7.15), is more comprehensive than the previous model proposed by [19], which required a strict monotonicity condition for $h_{r,i}(u_i^F(t))$ and $h_{l,i}(u_i^F(t))$. These aspects render our ANAS model more applicable in practical scenarios, given that actuators are often subject to realistic nonlinear constraints.

7.3 Fully Distributed Adaptive Fault-Tolerant Pseudo-PID SMC Design

In this section, the technique of the pseudo-PID SMC method is integrated with adaptive control strategies to introduce two innovative fully distributed adaptive *fault-tolerant* control schemes that are designed to handle faults while accommodating no-saturation and ANAS cases.

7.3.1 No-Saturation Case

In this subsection, controllers $u_i(t)$ need to be designed for each follower that uses only local information such that the practical leader-following *complete consensus* problem of the MAS (7.1)–(7.2) with actuator faults, IQCs, and no *saturation*, (i.e., $\Delta u_i(t) = 0$) is solved. For arbitrary *initial conditions* $x_i(t_0)$ and $v_i(t_0)$, the following equations are satisfied:

$$x_i(t) - x_0(t) \to 0, \quad v_i(t) - v_0(t) \to 0, \quad i \in \mathscr{V}_N. \tag{7.18}$$

To achieve the above objective, a novel sliding-mode surface (i.e., pseudo-PID sliding-mode surface) is first proposed as follows:

$$s_i(t) = v_i(t) + \alpha_i e_{x,i}(t) + \int_0^t \beta_i e_{x,i}(\tau) d\tau, \quad i \in \mathscr{V}_N \tag{7.19}$$

with

$$\dot{s}_i(t) = \dot{v}_i(t) + \alpha_i e_{v,i}(t) + \beta_i e_{x,i}(t), \quad i \in \mathscr{V}_N \tag{7.20}$$

where α_i and β_i are positive constants. $\alpha_i e_{x,i}(t)$, $\int_0^t \beta_i e_{x,i}(\tau)d\tau$, and $v_i(t)$ can be regarded as the proportional, integral, and pseudo-derivative terms, respectively.

To obtain the result of Theorem 7.1, the nonlinear function in the system is constrained as follows:

Assumption 7.4 There exist nonnegative functions $\chi_f(x_i, v_i, t)$ and unknown nonnegative scalars $\kappa_i > 0$, satisfying

$$|f_i(x_i, v_i, t)| \le \kappa_i \chi_f(x_i, v_i, t), \quad i \in \mathscr{V}_N.$$

Remark 7.4 Assumption 7.4, which is commonly adopted in the literature on single-agent systems as exemplified by [25, 26], refers to the process of extracting crude information from a nonlinear function based on rudimentary model data [25, 26]. Assumption 7.4 is broader in scope compared to the one presented in [6], where the nonlinear function $f_i(x_i, v_i, t)$ in [6] is expected to fulfill a differential mean value property, and the nonlinear dynamics between the leader and the followers are required to be identical.

Then, based on (7.19), the fully distributed adaptive *fault-tolerant* control law is proposed below:

$$u_i(t) = -(k_i + \Delta k_i(t))s_i(t) \tag{7.21}$$

where

$$\Delta k_i(t) = \frac{\hat{\eta}_i(t)\Psi_i^2(t)}{\Psi_i(t)|s_i(t)| + \iota_i}$$
$$\Psi_i(t) = 1 + \chi_f(x_i, v_i, t) + |v_i(t)| + |e_{v,i}(t)| + |e_{x,i}(t)|$$

and the adaptive law is constructed as

$$\dot{\hat{\eta}}_i(t) = -\varphi_i\sigma_i(t)\hat{\eta}_i(t) + \frac{\varphi_i\Psi_i^2(t)s_i^2(t)}{\Psi_i(t)|s_i(t)| + \iota_i}, \quad \hat{\eta}_i(0) \ge 0. \tag{7.22}$$

Here, k_i, φ_i, and ι_i are all positive parameters. Moreover, $\sigma_i(t)$ are any uniform boundedness continuous functions expressed as

$$\sigma_i(t) > 0, \quad \lim_{t\to\infty}\int_{t_0}^{t} \sigma_i(\tau)d\tau \le \bar{\sigma}_i < +\infty, \quad i \in \mathscr{V}_N \tag{7.23}$$

with $\bar{\sigma}_i$ being unknown finite bounds.

Then, the theorem proposed here can ensure the practical leader-following *complete consensus* and boundedness of the system (7.1)-(7.2) in *finite time.*

Theorem 7.1 *Under Assumptions 7.1–7.4, for MAS (7.1)–(7.2) with no-saturation actuator, the practical leader-following complete consensus problem is solved under the adaptive controller (7.21)–(7.22). Moreover, the position and velocity tracking errors are guaranteed to be bounded in finite time.*

Proof To analyze the practical leader-following tracking *complete consensus* problem of the MAS (7.1)–(7.2), we construct the following Lyapunov function:

$$V(t) = \sum_{i=1}^{N} \left[\frac{1}{2} s_i^2(t) + \frac{1}{2\varphi_i(\underline{\rho}_i - \zeta_{1i}\bar{\rho}_i)} \tilde{\eta}_i^2(t) \right] \tag{7.24}$$

where $\varphi_i > 0,\ i \in \mathscr{V}_N$ are positive constants, and the estimation errors $\tilde{\eta}_i(t), i \in \mathscr{V}_N$ are expressed as

$$\tilde{\eta}_i(t) = \eta_i^* - (\underline{\rho}_i - \zeta_{1i}\bar{\rho}_i)\hat{\eta}_i(t), \quad i \in \mathscr{V}_N \tag{7.25}$$

where $\underline{\rho}_i$ and $\bar{\rho}_i$ are the upper and lower bounds of the actuator efficiency factor $\rho_i(t, t_{\rho,i})$ defined in (7.13), while ζ_{1i} are defined in (7.12) and η_i^* will be defined in (7.29).

Using Assumption 7.4 and substituting (7.12), (7.13), and (7.17) with $\Delta u_i(t) = 0$ into (7.20), it follows that

$$\begin{aligned} \dot{s}_i(t) &= \dot{v}_i(t) + \alpha_i e_{v,i}(t) + \beta_i e_{x,i}(t) \\ &= u_i^F(t) + f_i(x_i, v_i, t) + \sum_{k=1}^{p_i} B_{ik} w_{ik}(t) + \alpha_i e_{v,i}(t) + \beta_i e_{x,i}(t) \\ &\le \rho_i(t, t_{\rho,i})u_i(t) + r_i(t, t_{r,i}) + \kappa_i \chi_f(x_i, v_i, t) + \zeta_{1i}|\rho_i(t, t_{\rho,i})u_i(t)| \\ &\quad + \zeta_{2i}|v_i(t)| + \zeta_{1i}|r_i(t, t_{r,i})| + \delta_{w,i} + \alpha_i e_{v,i}(t) + \beta_i e_{x,i}(t). \end{aligned} \tag{7.26}$$

From Assumption 7.3, one has $\underline{\rho}_i - \zeta_{1i}\bar{\rho}_i > 0$. Then, substituting (7.21) into (7.26) yields

$$\begin{aligned} s_i(t)\dot{s}_i(t) &\le s_i(t)\{\rho_i(t, t_{\rho,i})u_i(t) + \zeta_{1i}\rho_i(t, t_{\rho,i})|u_i(t)| + r_i(t, t_{r,i}) + \kappa_i\chi_f(x_i, v_i, t) \\ &\quad + \zeta_{1i}|r_i(t, t_{r,i})| + \delta_{w,i} + \zeta_{2i}|v_i(t)| + \alpha_i e_{v,i}(t) + \beta_i e_{x,i}(t)\} \\ &\le -(\underline{\rho}_i - \zeta_{1i}\bar{\rho}_i)[(k_i + \Delta k_i(t))]s_i^2(t) + [r_i(t, t_{r,i}) + \kappa_i\chi_f(x_i, v_i, t) \\ &\quad + \zeta_{1i}|r_i(t, t_{r,i})| + \delta_{w,i} + \zeta_{2i}|v_i(t)| + \alpha_i e_{v,i}(t) + \beta_i e_{x,i}(t)]s_i(t) \\ &\le -(\underline{\rho}_i - \zeta_{1i}\bar{\rho}_i)[(k_i + \Delta k_i(t))]s_i^2(t) + \eta_i^*\Psi_i(t)|s_i(t)| \end{aligned} \tag{7.27}$$

where the following fact has been applied:

$$\begin{aligned}
&\big[r_i(t, t_{r,i}) + \kappa_i \chi_f(x_i, v_i, t) + \zeta_{1i}|r_i(t, t_{r,i})| + \delta_{w,i} + \zeta_{2i}|v_i(t)| + \alpha_i e_{v,i}(t) + \beta_i e_{x,i}(t)\big] s_i(t) \\
&\le \big[\bar{r}_i + \kappa_i \chi_f(x_i, v_i, t) + \zeta_{1i}\bar{r}_i + \delta_{w,i} + \zeta_{2i}|v_i(t)| + \alpha_i |e_{v,i}(t)| + \beta_i |e_{x,i}(t)|\big] |s_i(t)| \\
&\le \eta_i^* \Psi_i(t)|s_i(t)| \qquad (7.28)
\end{aligned}$$

with

$$\Psi_i(t) = 1 + \chi_f(x_i, v_i, t) + |v_i(t)| + |e_{v,i}(t)| + |e_{x,i}(t)| \qquad (7.29)$$

$$\eta_i^* = \max\{\bar{r}_i + \zeta_{1i}\bar{r}_i + \delta_{w,i}, \kappa_i, \zeta_{2i}, \alpha_i, \beta_i\}. \qquad (7.30)$$

From (7.29), one has $\Psi_i(t) \ge 0$, so in (7.22) that

$$\dot{\hat{\eta}}_i(t) + \varphi_i \sigma_i(t) \hat{\eta}_i(t) \ge 0.$$

This means that

$$\hat{\eta}_i(t) \ge e^{-\int_0^t \varphi_i \sigma_i(\tau) d\tau} \hat{\eta}_i(0).$$

For any $\hat{\eta}_i(0) \ge 0$, we thus conclude that the adaptive law $\hat{\eta}_i(t)$ will always be nonnegative, which implies that

$$\frac{\hat{\eta}_i(t)\Psi_i^2(t)}{\Psi_i(t)|s_i(t)| + \iota_i} \ge 0.$$

Consequently, by recalling the fact $\underline{\rho}_i - \zeta_{1i}\bar{\rho}_i > 0$ and the definition of $\Delta k_i(t)$ in (7.21), one can derive

$$\begin{aligned}
&-(\underline{\rho}_i - \zeta_{1i}\bar{\rho}_i)\Delta k_i(t) s_i^2(t) + \eta_i^* \Psi_i(t)|s_i(t)| \\
&= -(\underline{\rho}_i - \zeta_{1i}\bar{\rho}_i) \frac{\hat{\eta}_i(t)\Psi_i^2(t)s_i^2(t)}{\Psi_i(t)|s_i(t)| + \iota_i} + \eta_i^* \Psi_i(t)|s_i(t)| \\
&= \frac{-(\underline{\rho}_i - \zeta_{1i}\bar{\rho}_i)\hat{\eta}_i(t)\Psi_i^2(t)s_i^2(t)}{\Psi_i(t)|s_i(t)| + \iota_i} + \frac{\eta_i^*\Psi_i^2(t)s_i^2(t) + \eta_i^*\Psi_i(t)|s_i(t)|\iota_i}{\Psi_i(t)|s_i(t)| + \iota_i} \\
&\le [\eta_i^* - (\underline{\rho}_i - \zeta_{1i}\bar{\rho}_i)\hat{\eta}_i(t)] \frac{\Psi_i^2(t)s_i^2(t)}{\Psi_i(t)|s_i(t)| + \iota_i} + \eta_i^* \iota_i \qquad (7.31)
\end{aligned}$$

where the fact

$$\frac{\eta_i^*\Psi_i(t)|s_i(t)|\iota_i}{\Psi_i(t)|s_i(t)| + \iota_i} \le \eta_i^* \iota_i$$

has been used.

After the derivation to (7.25), one can get

$$\dot{\tilde{\eta}}_i(t) = -(\underline{\rho}_i - \zeta_{1i}\bar{\rho}_i)\dot{\hat{\eta}}_i(t).$$

Correspondingly, by using (7.22), it is derived that

$$
\begin{aligned}
-\frac{1}{\varphi_i}\tilde{\eta}_i(t)\dot{\hat{\eta}}_i(t) &= -\frac{1}{\varphi_i}\tilde{\eta}_i(t)\left(-\varphi_i\sigma_i(t)\hat{\eta}_i(t)+\frac{\varphi_i\Psi_i^2(t)s_i^2(t)}{\Psi_i(t)|s_i(t)|+\iota_i}\right)\\
&= \sigma_i(t)\tilde{\eta}_i(t)\hat{\eta}_i(t)-\tilde{\eta}_i(t)\frac{\Psi_i^2(t)s_i^2(t)}{\Psi_i(t)|s_i(t)|+\iota_i}\\
&= \frac{\sigma_i(t)}{(\underline{\rho}_i-\zeta_{1i}\bar{\rho}_i)}\tilde{\eta}_i(t)\left(\eta_i^*-\tilde{\eta}_i(t)-\frac{\Psi_i^2(t)s_i^2(t)}{\Psi_i^2(t)|s_i(t)|+\iota_i}\right)\\
&\leq \frac{\sigma_i(t)}{2(\underline{\rho}_i-\zeta_{1i}\bar{\rho}_i)}(\eta_i^{*2}-\tilde{\eta}_i^2(t))-\tilde{\eta}_i(t)\frac{\Psi_i^2(t)s_i^2(t)}{\Psi_i^2(t)|s_i(t)|+\iota_i}.
\end{aligned}
\tag{7.32}
$$

Then, using the deducing parts from (7.31) and (7.32), the time derivation of $V(t)$ in (7.24) is

$$
\begin{aligned}
\dot{V}(t) &= \sum_{i=1}^{N}[s_i(t)\dot{s}_i(t)-\frac{1}{\varphi_i}\tilde{\eta}_i(t)\dot{\hat{\eta}}_i(t)]\\
&= \sum_{i=1}^{N}\left[-(\underline{\rho}_i-\zeta_{1i}\bar{\rho}_i)k_is_i^2(t)-\frac{\sigma_i(t)}{2(\underline{\rho}_i-\zeta_{1i}\bar{\rho}_i)}\tilde{\eta}_i^2(t)+\eta_i^*\iota_i+\frac{\sigma_i(t)}{2(\underline{\rho}_i-\zeta_{1i}\bar{\rho}_i)}\eta_i^{*2}(t)\right]\\
&\leq -\varsigma_1V(t)+\varsigma_2
\end{aligned}
\tag{7.33}
$$

where the parameters ς_1 and ς_2 are positive, and given as follows:

$$
\begin{aligned}
\varsigma_1 &= \min_{i\in\mathscr{V}_N}\{2(\underline{\rho}_i-\zeta_{1i}\bar{\rho}_i)k_i,\varphi_i\min_{0\leq t\leq\infty}\sigma_i(t)\}\\
\varsigma_2 &= \sum_{i=1}^{N}[\eta_i^*\iota_i+\frac{\sup_{t\geq0}\sigma_i(t)}{2(\underline{\rho}_i-\zeta_{1i}\bar{\rho}_i)}\eta_i^{*2}(t)].
\end{aligned}
\tag{7.34}
$$

Multiplying both sides of (7.33) by $e^{\varsigma_1 t}$ yields

$$
\frac{d}{dt}(V(t)e^{\varsigma_1 t})\leq\varsigma_2e^{\varsigma_1 t}. \tag{7.35}
$$

By integrating (7.35) over $[t_0,t]$ and simultaneously multiplying it by $e^{-\varsigma_1 t}$, one has

$$
\begin{aligned}
0\leq V(t) &\leq \frac{\varsigma_2}{\varsigma_1}+\left(V(t_0)-\frac{\varsigma_2}{\varsigma_1}\right)e^{-\varsigma_1(t-t_0)}\\
&\leq V(t_0)+\frac{\varsigma_2}{\varsigma_1}
\end{aligned}
\tag{7.36}
$$

where

$$V(t_0) = \sum_{i=1}^{N} \left[\frac{1}{2} s_i^2(t_0) + \frac{1}{2\varphi_i(\underline{\rho}_i - \zeta_{1i}\bar{\rho}_i)} \tilde{\eta}_i^2(t_0) \right] \geq 0.$$

Let $V(T_m) = V_{\max}$ with $V_{\max}$ being a sufficiently large positive constant. Then, from (7.36), one can obtain

$$T_m = \frac{\ln \frac{V(T_m) - \frac{\varsigma_2}{\varsigma_1}}{V(t_0) - \frac{\varsigma_2}{\varsigma_1}}}{-\varsigma_1} + t_0 = \frac{\ln \frac{V_{\max} - \frac{\varsigma_2}{\varsigma_1}}{V(t_0) - \frac{\varsigma_2}{\varsigma_1}}}{-\varsigma_1} + t_0 \tag{7.37}$$

which implies that $V(t) \leq V_{\max}$ when $t \geq T_m$.

Recalling (7.24), one has

$$\begin{aligned} \frac{1}{2} \sum_{i=1}^{N} s_i^2(t) &\leq V(t) \\ \sum_{i=1}^{N} \frac{1}{2\varphi_i(\underline{\rho}_i - \zeta_{1i}\bar{\rho}_i)} \tilde{\eta}_i^2(t) &\leq V(t) \end{aligned} \tag{7.38}$$

that is

$$||s(t)|| \leq \sqrt{2V(t)}, \quad ||\tilde{\eta}(t)|| \leq \sqrt{2\mu V(t)}, \quad \mu = \min_{1 \leq i \leq N} \{\varphi_i(\underline{\rho}_i - \zeta_{1i}\bar{\rho}_i)\} \tag{7.39}$$

where

$$s(t) = [s_1(t), s_2(t), \cdots, s_N(t)]^T \tag{7.40}$$

$$\tilde{\eta}(t) = [\tilde{\eta}_1(t), \tilde{\eta}_2(t), \cdots, \tilde{\eta}_N(t)]^T. \tag{7.41}$$

Since $V(t) \leq V_{\max}$ for $t \geq T_m$, it further follows from (7.39) that

$$||s(t)|| \leq \sqrt{2V_{\max}} \tag{7.42}$$

$$||\tilde{\eta}(t)|| \leq \sqrt{2\mu V_{\max}}, \quad t \geq T_m, \tag{7.43}$$

which implies that, for all $t \geq T_m$, all signals in the system converge to an adjustable region Ω in *finite time*, where Ω is expressed as

$$\Omega = \left\{ s(t), \tilde{\eta}(t) \middle| ||s(t)|| \leq \sqrt{2V_{\max}}, ||\tilde{\eta}(t)|| \leq \sqrt{2\mu V_{\max}} \right\}$$

which can be adjusted by choosing the appropriate parameters ς_1 and ς_2 defined in (7.34).

From (7.19), the boundedness of $s_i(t)$ in *finite time* ensures that $v_i(t)$, $e_{x,i}(t)$, and $\int_0^t e_{x,i}(\tau)d\tau$ are bounded in *finite time*. Since $e_{x,i}(t) = e_{x,i}(0) + \int_0^t e_{v,i}(\tau)d\tau$ and $v_i(t)$ are bounded, $e_{v,i}(t)$ is also bounded. Therefore, it is clear from Lemma 7.1 that the practical *complete consensus* problem of the leader-following system can be solved in *finite time*. This proof is completed. □

Remark 7.5 Although the control strategy outlined in Theorem 7.1 ensures only practical leader-following complete consensus in cases with actuator faults and IQCs, it does not necessitate comprehensive network topology details as seen in prior studies [6, 9]. Instead, the scheme (7.21)–(7.22) relies solely on accessible local neighbor data. Furthermore, the proposed control scheme (7.21)–(7.22) achieves independence from the leader's acceleration by incorporating the pseudo-PID SMC technique.

Remark 7.6 Although the research on *fault-tolerant* control for MAS under IQCs as presented in [6] is noteworthy, three significant points deserve emphasis. Firstly, [6] involves numerous analytical approximations for unknown parameters, whereas the control scheme (7.21)–(7.22) requires only a single parameter estimation per agent, simplifying the structure and reducing computational costs in our approach outlined in Theorem 7.1. Secondly, in contrast to [6], our method does not necessitate the knowledge of the upper and lower bounds of the faults. Lastly, the interplay between actuator faults, IQCs, and ANASs complicates the controller design in this chapter, making it more intricate than the design presented in [6].

7.3.2 ANAS Case

Actuator *saturations*, a common phenomenon in real-world systems, can lead to performance deterioration or even system instability, as highlighted in [13]. In light of this, this section will introduce an innovative anti-windup compensator designed to mitigate the negative effects of ANASs.

On the other hand, although Assumption 7.4 has streamlined the controller design, it comes with a significant limitation: the requirement for rudimentary model information, which is often challenging to acquire. This difficulty can restrict the practical application of the method presented in Theorem 7.1. To address this challenge, FLSs will be employed to approximate the unknown nonlinear functions $f_i(x_i, v_i, t)$ from (7.2).

Furthermore, a novel anti-windup compensator is designed to eliminate the adverse impact caused by ANASs. The anti-windup compensator is constructed as follows:

$$\begin{cases} \dot{\lambda}_{1i}(t) = -c_{1i}\lambda_{1i}(t) + \lambda_{2i}(t) \\ \dot{\lambda}_{2i}(t) = -c_{2i}\lambda_{2i}(t) + \Delta u_i(t) \end{cases} \tag{7.44}$$

where c_{1i} and c_{2i} are positive parameters. The related lemma is stated below:

Lemma 7.2 *The anti-windup compensator (7.44) can attenuate the adverse impact of saturation by selecting the appropriate parameters c_{1i} and c_{2i}.*

Proof The expression of the compensator (7.44) can be represented as

$$\dot{\lambda}_i(t) = A_i\lambda_i(t) + B_i\Delta u_i(t) \tag{7.45}$$

where

$$A_i = \begin{bmatrix} -c_{1i} & 1 \\ 0 & -c_{2i} \end{bmatrix}, \quad B_i = \begin{bmatrix} 0 \\ 1 \end{bmatrix}.$$

The solution for $\lambda_i(t)$ is

$$\lambda_i(t) = e^{A_i t}\lambda_i(0) + \int_0^t e^{A_i(t-\tau)} B_i \Delta u_i(\tau)d\tau. \tag{7.46}$$

To ensure that $\lambda_{1i} \to 0$, $\lambda_{2i} \to 0$ at last. A_i should be Hurwitz, which needs $c_{1i} > 0$ and $c_{2i} > 0$. Therefore, according [27], there are two constants, $k_{0i} > 0$ and $\lambda_{0i} > 0$, such that

$$\|e^{A_i t}\| \le k_{0i}e^{-\lambda_{0i}t}.$$

Then, it follows from (7.46) that

$$\|\lambda_i(t)\| \le k_{0i}\|\lambda_{0i}(0)\|e^{-\lambda_{0i}t} + k_{0i}\int_0^t e^{-\lambda_{0i}(t-\tau)}|\Delta u_i(\tau)|d\tau. \tag{7.47}$$

Then, using the following Schwartz inequality [27]

$$\int_0^t |a(\tau)b(\tau)|d\tau \le \left(\int_0^t a^2(\tau)d\tau\right)^{\frac{1}{2}} \left(\int_0^t b^2(\tau)d\tau\right)^{\frac{1}{2}}$$

we can deduce from (7.47) that

$$\begin{aligned}
\|\lambda_i(t)\| &\le k_{0i}\|\lambda_i(0)\|e^{-\lambda_{0i}t} + k_{0i}\left(\int_0^t e^{-2\lambda_{0i}(t-\tau)d\tau}\right)^{\frac{1}{2}} \left(\int_0^t \Delta u_i^2(\tau)d\tau\right)^{\frac{1}{2}} \\
&= k_{0i}\|\lambda_i(0)\|e^{-\lambda_{0i}t} + k_{0i}\left(\frac{1}{2\lambda_{0i}}(1-e^{-2\lambda_{0i}t})\right)^{\frac{1}{2}} \left(\int_0^t \Delta u_i^2(\tau)d\tau\right)^{\frac{1}{2}} \\
&\le k_{0i}\|\lambda_i(0)\|e^{-\lambda_{0i}t} + \frac{k_{0i}}{\sqrt{2\lambda_{0i}}}\left(\int_0^t \Delta u_i^2(\tau)d\tau\right)^{\frac{1}{2}}
\end{aligned}$$

which means that $\lambda_i(t)$ converges exponentially to a region connected to the *saturation*. When there is no *saturation*, $\lambda_i(t)$ converges to zero (i.e., $\Delta u_i(t) = 0$). Furthermore, for the deep *saturation* case, $\Delta u_i(t)$ will be large, which means a large λ_{0i}. Then, c_{1i} and c_{2i} should be large enough.

Then, redefine the tracking errors as follows:

$$e_{x,i}(t) = \sum_{j \in N_i} a_{ij}(x_i(t) - x_j(t)) - \lambda_{1i}(t) \tag{7.48}$$

$$e_{v,i}(t) = \sum_{j \in N_i} a_{ij}(v_i(t) - v_j(t)) - \dot{\lambda}_{1i}(t). \tag{7.49}$$

When the effects of ANASs are fully eliminated, then $\Delta u_i(t) \to 0$. The limits

$$\lim_{t\to\infty} \lambda_{1i}(t) \to 0, \quad \lim_{t\to\infty} \lambda_{2i}(t) \to 0$$

hold due to the Hurwitz property of A_i. Thus, when

$$\lim_{t\to\infty} e_{x,i}(t) \to 0, \quad \lim_{t\to\infty} e_{v,i}(t) \to 0$$

a practical leader-following *complete consensus* can be guaranteed.

Then, by adopting the same pseudo-PID sliding-mode surface (7.19), the control method composed of the anti-windup compensator is redesigned as follows:

$$u_i(t) = -(k_i + \Delta k_i(t))s_i(t) - \frac{1}{2}s_i(t)\theta_i(t)\xi_{1,i}^T(v_i)\xi_{1,i}(v_i) \tag{7.50}$$

where $\Delta k_i(t)$ and $\hat{\eta}_i(t)$ are the same as the definition in Theorem 7.1, while

$$\Psi_i(t) = 1 + |v_i(t)| + |e_{v,i}(t)| + |e_{x,i}(t)| + |\dot{\lambda}_{2i}(t)| + c_{2i}|\lambda_{2i}(t)|. \tag{7.51}$$

The update law for $\hat{\eta}_i(t)$ is defined as in (7.22), while the law for $\theta_i(t)$ is defined as follows:

$$\dot{\theta}_i(t) = \frac{\phi_i}{2}\xi_i^T(v_i)\xi_i(v_i)s_i^2(t) - \phi_i\theta_i(t), \quad \theta_i(0) \geq 0 \tag{7.52}$$

where ϕ_i is a positive parameter.

Then, the following theorem establishes the bounded stability of the leader-following MAS (7.1)–(7.2) under the interplay of actuator faults, IQCs, and ANASs.

Theorem 7.2 *Under Assumptions 7.1–7.3, for MAS (7.1)–(7.2) subject to mutual coupling among actuator faults, IQCs, and ANASs, the practical leader-following complete consensus problem is solved under adaptive controller (7.50) and update laws (7.22) and (7.52). In addition, the position and velocity tracking errors are guaranteed to be bounded in finite time.*

Proof The proof of this theorem will be simplified since most of the proof is the same as that of Theorem 7.1. To solve the practical leader-following tracking *complete consensus* problem of system (7.1)–(7.2) subject to mutual coupling between actuator faults, IQCs, and ANASs, first, the derivative of $s_i(t)$ is expressed as

$$\begin{aligned}\dot{s}_i(t) &= \dot{v}_i(t) + \alpha_i \dot{e}_{x,i}(t) + \beta_i e_{x,i}(t)\\ &= u_i^F(t) + \Delta u_i(t) + f_i(x_i, v_i, t) + \sum_{k=1}^{p_i} B_{ik} w_{ik}(t) + \alpha_i e_{v,i}(t) + \beta_i e_{x,i}(t)\\ &\leq \rho_i(t, t_{\rho,i}) u_i(t) + \zeta_{1i}\rho_i(t, t_{\rho,i})|u_i(t)| + W_i^{*T}\xi_i(v_i) + \varepsilon_i(t) + r_i(t, t_{r,i}) + \delta_{w,i}\\ &\quad + \zeta_{1i}|r_i(t, t_{r,i})| + \zeta_{2i}|v_i(t)| + \alpha_i e_{v,i}(t) + \beta_i e_{x,i}(t) + \dot{\lambda}_{2i}(t) + c_{2i}\lambda_{2i}(t)\end{aligned} \tag{7.53}$$

where the unknown nonlinear function $f_i(x_i, v_i, t)$ is approximated by FLSs and the estimation error $\varepsilon_i(t)$ satisfies

$$|\varepsilon_i(t)| \leq \bar{\varepsilon}_i(t) < \infty \tag{7.54}$$

where $\bar{\varepsilon}_i(t)$ are unknown positive parameters.

Then, the following facts can be derived:

$$W_i^{*T}\xi_i(v_i)s_i(t) \leq \frac{\underline{\rho}_i - \zeta_{1i}\bar{\rho}_i}{2}\theta_i^* \xi_i^T(v_i)\xi_i(v_i)s_i^2(t) + \frac{1}{2(\underline{\rho}_i - \zeta_{1i}\bar{\rho}_i)} \tag{7.55}$$

$$\begin{aligned}&[\varepsilon_i(t) + r_i(t, t_{r,i}) + \zeta_{1i}|r_i(t, t_{r,i})| + \delta_{w,i} + \zeta_{2i}|v_i(t)| + \alpha_i e_{v,i}(t)\\ &\quad + \beta_i e_{x,i}(t) + \dot{\lambda}_{2i}(t) + c_{2i}\lambda_{2i}(t)]s_i(t)\\ &\leq [\bar{\varepsilon}_i + \bar{r}_i + \zeta_{1i}\bar{r}_i + \delta_{w,i} + \zeta_{2i}|v_i(t)| + \alpha_i|e_{v,i}(t)| + \beta_i|e_{x,i}(t)|\\ &\quad + |\dot{\lambda}_{2i}(t)| + c_{2i}|\lambda_{2i}(t)|]|s_i(t)|\\ &\leq \eta_i^*\Psi_i(t)|s_i(t)|\end{aligned} \tag{7.56}$$

where

$$\begin{aligned}\Psi_i(t) &= 1 + |v_i(t)| + |e_{v,i}(t)| + |e_{x,i}(t)| + |\dot{\lambda}_{2i}(t)| + |\lambda_{2i}(t)|]\\ \theta_i^* &= \|W_i^*\|^2.\end{aligned} \tag{7.57}$$

By substituting the adaptive laws (7.52), we can obtain

$$
\begin{aligned}
&-\frac{(\underline{\rho}_i - \zeta_{1i}\bar{\rho}_i)}{\phi_i}\tilde{\theta}_i(t)\dot{\theta}_i(t)\\
&= \tilde{\theta}_i(t)((\underline{\rho}_i - \zeta_{1i}\bar{\rho}_i)\theta_i(t) - \frac{(\underline{\rho}_i - \zeta_{1i}\bar{\rho}_i)}{2}\xi_i^T(v_i)\xi_i(v_i)s_i^2(t))\\
&\leq -\frac{(\underline{\rho}_i - \zeta_{1i}\bar{\rho}_i)}{2}\tilde{\theta}_i(t)\xi_i^T(v_i)\xi_i(v_i)s_i^2(t) - \frac{(\underline{\rho}_i - \zeta_{1i}\bar{\rho}_i)}{2}\tilde{\theta}_i^2(t) + \frac{(\underline{\rho}_i - \zeta_{1i}\bar{\rho}_i)\theta_i^{*2}}{2}.
\end{aligned}
\tag{7.58}
$$

Moreover, a new Lyapunov function is expressed as

$$
V(t) = \sum_{i=1}^{N}\left[\frac{1}{2}s_i^2(t) + \frac{\underline{\rho}_i - \zeta_{1i}\bar{\rho}_i}{2\phi_i}\tilde{\theta}_i^2(t) + \frac{1}{2\varphi_i(\underline{\rho}_i - \zeta_{1i}\bar{\rho}_i)}\tilde{\eta}_i^2(t)\right] \tag{7.59}
$$

where $\tilde{\theta}_i(t) = \theta_i^* - \theta_i(t)$ is the estimation error. Considering the new Lyapunov function defined above, by referring to the proof of Theorem 7.1, the rest of the proof is omitted here for simplicity. The proof is completed. □

Remark 7.7 By summarizing Theorem 7.2, two key points should be highlighted. Firstly, Theorem 7.2 incorporates FLSs, which is advantageous as it is not contingent on the quantity of rules, thereby potentially reducing computational complexity. However, the control structure of Theorem 7.2 remains more intricate compared to that of Theorem 7.1. Consequently, in practical applications, the selection of the control strategy should be tailored to the specific circumstances. In systems where crude information can be obtained, the control strategy from Theorem 7.1 may be more suitable; otherwise, Theorem 7.2 should be considered. Secondly, while [19] focuses on a single system with ANAS, its application is limited by the assumption that the nonlinear function must be strictly monotonically increasing. In contrast, this chapter employs a novel anti-windup compensator to address the ANAS conditions without relying on this assumption, enhancing the practical applicability of our proposed compensator.

7.4 Simulation Examples

In this part, an MAS with one leader and five followers (i.e., $N = 5$) is considered to illustrate the practicality of the approaches in Theorems 7.1 and 7.2 and simultaneously verify the advantage of the designed anti-windup compensator.

Figure 7.1 shows the topology of the MAS. The nonlinear continuous functions $f_i(x_i, v_i, t)$ are expressed as

$$
f_i(x_i, v_i, t) = -\sin(x_i(t)) - 0.25v_i(t) + 1.5\cos(2.5t) + \sin(v_i(t))v_i^2(t)
$$

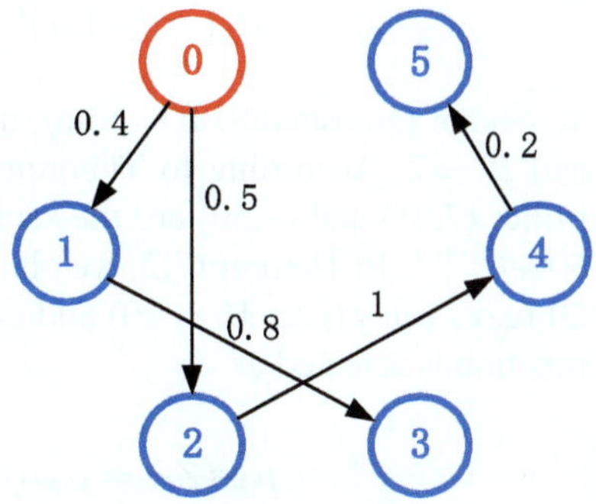

Fig. 7.1 Communication topology

where $1 \leq i \leq 5$. In addition, the acceleration trajectory of the leader is selected as

$$a_0(t) = \begin{cases} 0.05t\ km/h^2, & 1h \leq t < 5h \\ 0.25\ km/h^2, & 5h \leq t < 10h \\ -0.05t + 0.75\ km/h^2, & 10h \leq t < 15h \\ 0\ km/h^2, & \text{otherwise.} \end{cases}$$

The input ANAS constrained within an interval is given as follows:

$$\text{sat}(u_i^F(t)) = \begin{cases} 10, & u_i^F(t) \geq 10 \\ 0.9u_i^F(t), & 2 \leq u_i^F(t) < 10 \\ 1 - 0.2\sin(0.1u_i^F(t)), & 0 \leq u_i^F(t) < 2 \\ 1 + 0.1\cos(0.1u_i^F(t)), & -1 \leq u_i^F(t) < 0 \\ 0.8u_i^F(t), & -4 \leq u_i^F(t) < -1 \\ -4, & u_i^F(t) < -4. \end{cases}$$

Let the loss of effectiveness in the fault model be

$$\rho_i(t, t_{\rho,i}) = \begin{cases} 0.5 + 0.1\cos(0.1t), & 5 \leq t < 10 \\ 0.5 + 0.3\cos(0.1t), & 10 \leq t < 25 \\ 1, & \text{otherwise} \end{cases}$$

and the bias fault be

$$r_i(t, t_{r,i}) = 0.2\cos(0.01t).$$

Furthermore, the parameters of the controller and initial value of the agent are given below. In (7.2), the value of IQCs is set by $p_i = 1$,

$$\psi_{i,1} = \sin(\xi_{i,1}(t)), \quad B_{i,1} = [-0.5, 0.5 \times i]$$

$$H_{i,1} = [-1, i]^T, \quad G_{i,1} = [-0.2, 0.2 \times i]^T.$$

To design the pseudo-PID sliding-mode surface, we select the parameters $\alpha_i = 10$ and $\beta_i = 2$. According to Theorems 7.1 and 7.2, some of the parameters in controller (7.21) and (7.50) are the same: $\varphi_i = 0.01$, $\iota_i = 0.001$, $k_i = 50$, and $\sigma_i(t) = 500e^{-0.001t}$. In Theorem 7.2, we obtain $\phi_i = 2$ from (7.52), and for FLSs, we define 20 fuzzy rules (i.e., $M_i = 20$) and one input, $Z_i = v_i$. Then, the fuzzy membership function is selected as

$$\mu_{F_1^l}(Z_i) = \mu_{F_1^l}(v_i) = \exp\left[-\left(\frac{v_i + \frac{(3-l)\pi}{6}}{\frac{\pi}{12}} \right)^2 \right]$$

where $i = 1, 2, \cdots, 5$ and $l_i = 1, 2, \cdots, 20$. To construct the anti-windup compensator, we select $c_{1i} = 5, c_{2i} = 30, \lambda_{1i}(0) = 1$, and $\lambda_{2i}(0) = 0$. Moreover, some *initial conditions* are chosen: $\hat{\eta}_i(0) = 1$, $s_i(0) = 1$, and $\theta_i(0) = 1$.

7.4.1 No-Saturation Case

In this section, the simulation will verify the finite-time convergence of various signals of the MAS (7.1)–(7.2) by employing the distributed *fault-tolerant* strategy in Theorem 7.1 with and without ANASs. Figure 7.2 shows the simulation result under the controller in Theorem 7.1 in the absence of *saturation*. The position and velocity tracking curves with controller (7.21) are represented by Fig. 7.2a and b. The tracking error trajectory curves are depicted in Fig. 7.2c. It is evident that, under the presence of IQCs and actuator faults, the position and velocity errors achieve bounded stability within a *finite time*. The sliding-mode surface $s_i(t)$, control input $u_i(t)$, and parameter estimates $\eta_i(t)$ are revealed in Fig. 7.2d–f, respectively. Observations indicate that all system signals remain bounded. To demonstrate the effectiveness of the controller from Theorem 7.2 in mitigating the effects of *saturation*, ANAS *saturation* is introduced to the controller proposed in Theorem 7.1 for comparative purposes, and the simulation results are represented in Fig. 7.3. Obviously, Fig. 7.3c indicates a marked deterioration in tracking control performance, with a significant reduction in the convergence rate of the tracking error. The entire system is susceptible to *saturation* and even instability.

7.4.2 ANAS Case

The simulation of the system with ANASs applied in the control scheme (7.50) and adaptive laws (7.22) and (7.52) in Theorem 7.2 is shown in Fig. 7.4. The position and velocity tracking curves with controller (7.50) are shown in Fig. 7.4a and b, respectively. Figure 7.4c displays the position and velocity tracking errors under input *saturations*, evidencing that practical leader-following *complete consensus* can

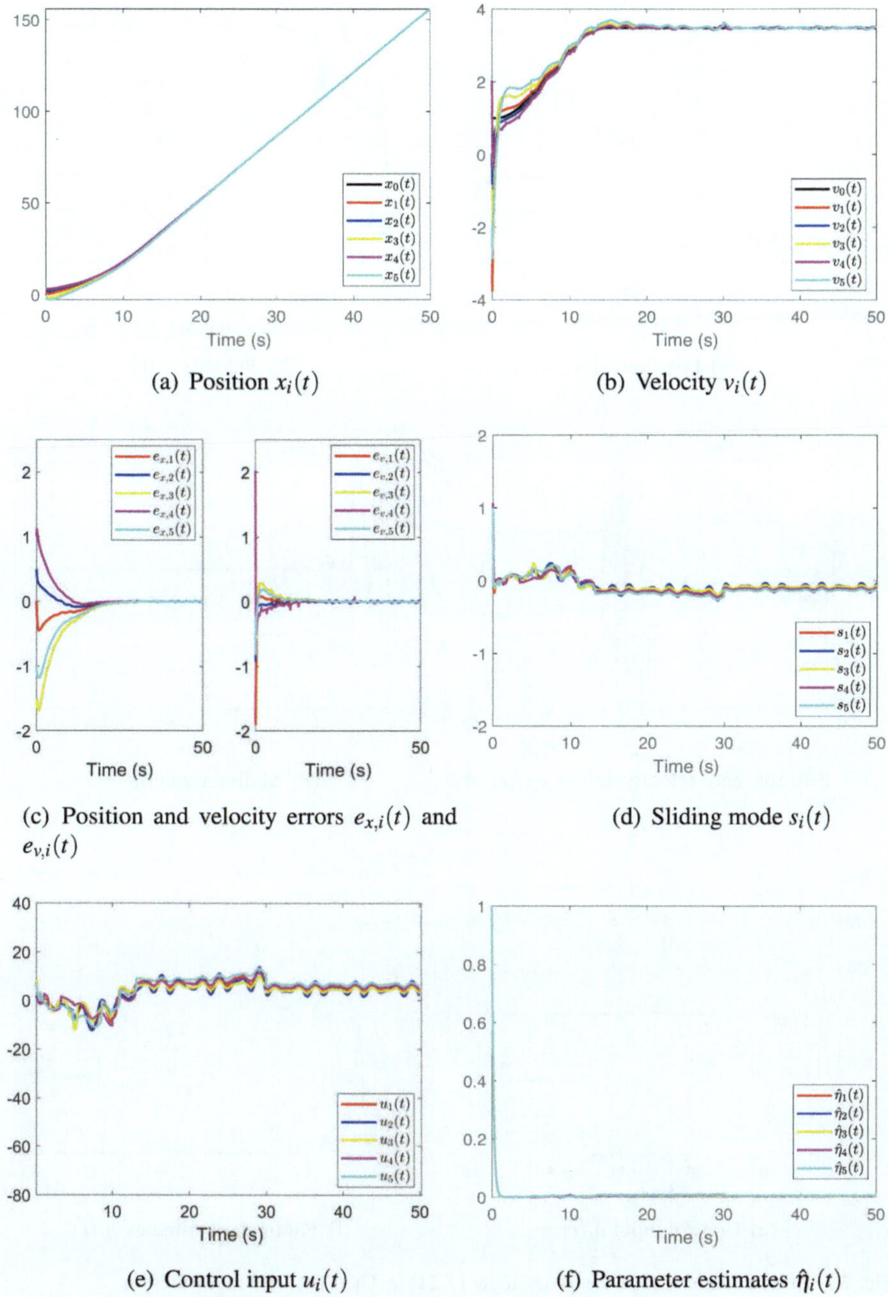

(a) Position $x_i(t)$

(b) Velocity $v_i(t)$

(c) Position and velocity errors $e_{x,i}(t)$ and $e_{v,i}(t)$

(d) Sliding mode $s_i(t)$

(e) Control input $u_i(t)$

(f) Parameter estimates $\hat{\eta}_i(t)$

Fig. 7.2 Simulation results of the MAS under (7.21) in Theorem 7.1 with no saturation

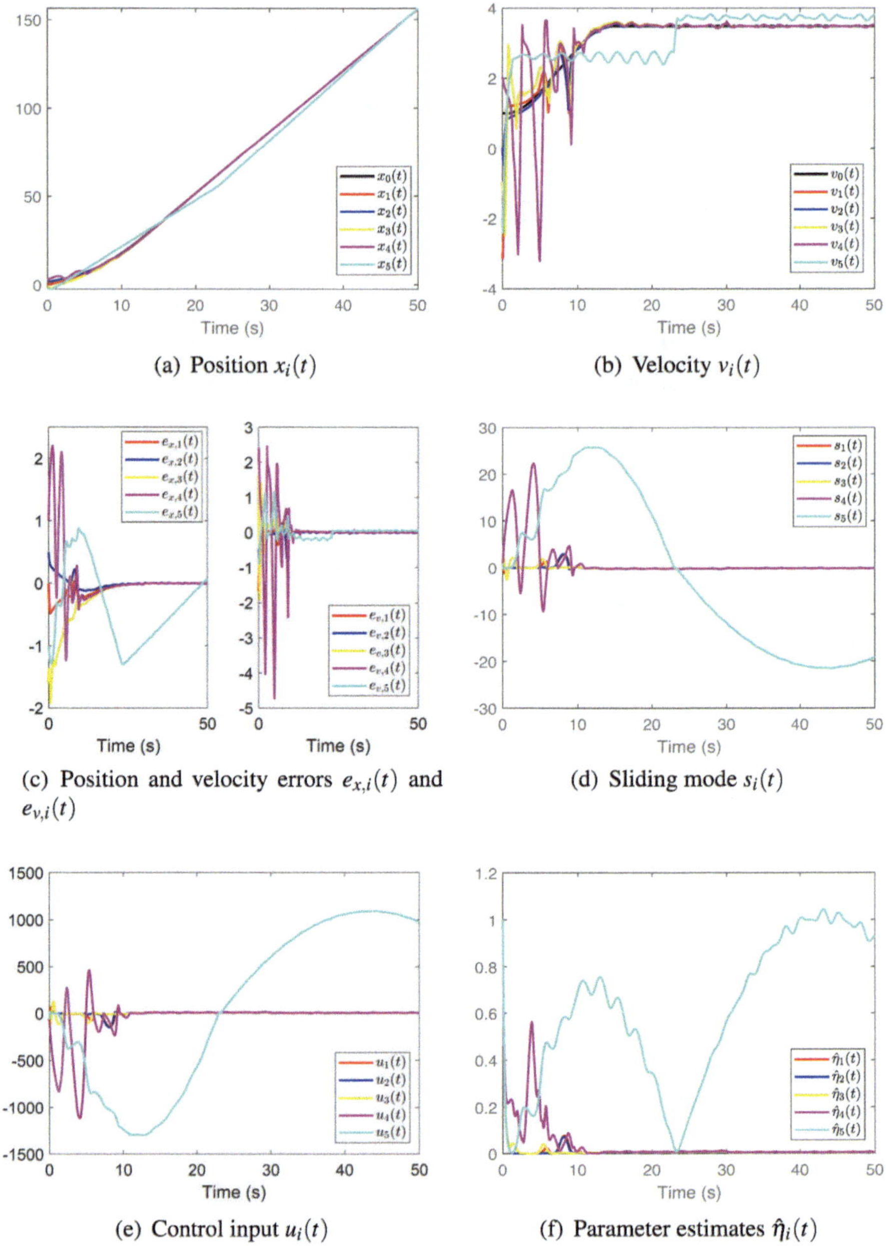

(a) Position $x_i(t)$

(b) Velocity $v_i(t)$

(c) Position and velocity errors $e_{x,i}(t)$ and $e_{v,i}(t)$

(d) Sliding mode $s_i(t)$

(e) Control input $u_i(t)$

(f) Parameter estimates $\hat{\eta}_i(t)$

Fig. 7.3 Simulation results of the MAS under (7.21) in Theorem 7.1 with saturation

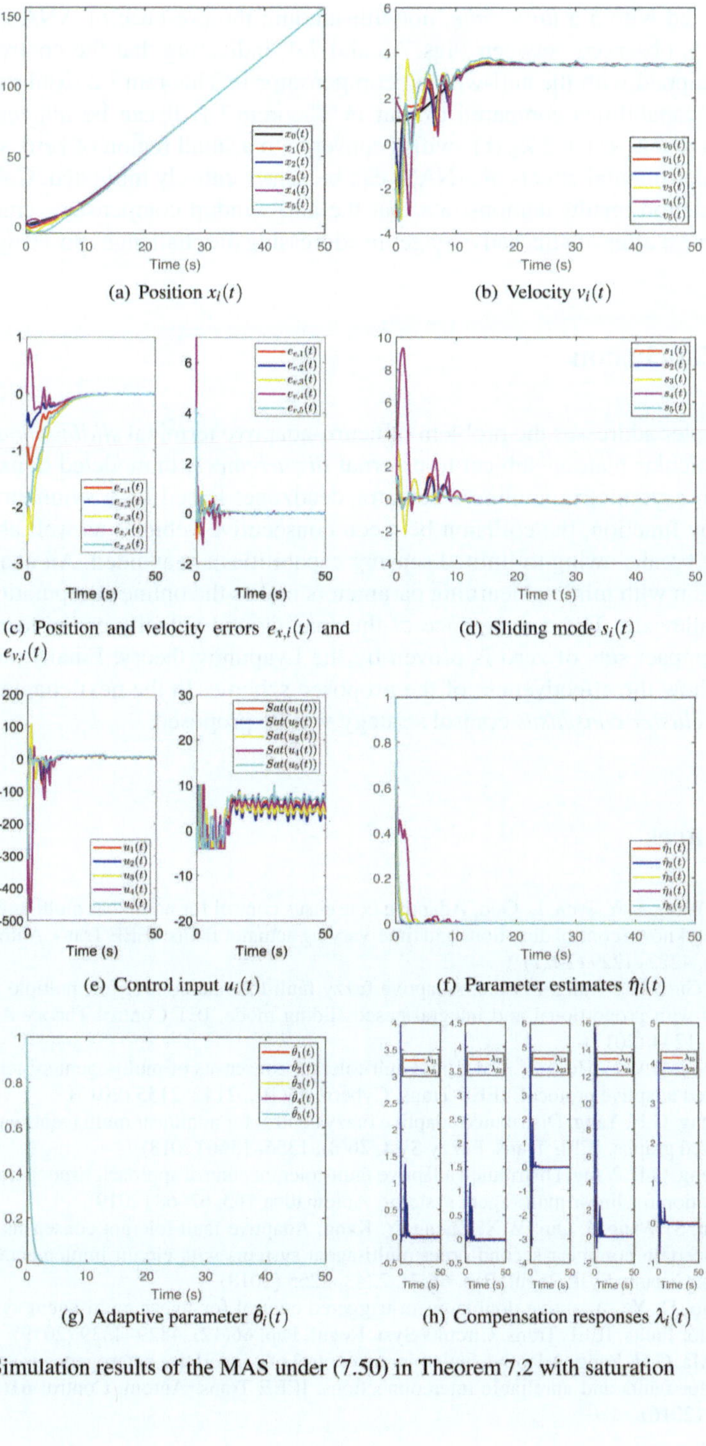

(a) Position $x_i(t)$

(b) Velocity $v_i(t)$

(c) Position and velocity errors $e_{x,i}(t)$ and $e_{v,i}(t)$

(d) Sliding mode $s_i(t)$

(e) Control input $u_i(t)$

(f) Parameter estimates $\hat{\eta}_i(t)$

(g) Adaptive parameter $\hat{\theta}_i(t)$

(h) Compensation responses $\lambda_i(t)$

Fig. 7.4 Simulation results of the MAS under (7.50) in Theorem 7.2 with saturation be achieved within a *finite time*, notwithstanding the presence of ANASs. A stark contrast is observed between Figs. 7.3 and 7.4, indicating that the control mechanism equipped with the anti-windup compensator in Theorem 7.2 exhibits superior tracking capabilities compared to that in Theorem 7.1. It can be noticeable from Fig. 7.4h that $\lambda_{1i}(t)$ and $\lambda_{2i}(t)$ swiftly converge to a small region of zero, signifying that the detrimental effects of ANASs can be nearly entirely mitigated. Collectively, the simulation results demonstrate that the anti-windup compensator employed in Theorem 7.2 offers distinct advantages in addressing the challenges posed by ANASs.

7.5 Conclusion

This chapter addresses the problem of neuro-adaptive terminal *sliding mode control* for a vehicular platoon subject to external *disturbances*, unmodeled dynamics and unknown asymmetric nonlinear actuator deadzone. Based on a symmetric barrier Lyapunov function, the collision between consecutive vehicles as well as the connectivity breaks owing to limited sensing capabilities are avoided. An adaptive NN mechanism with minimal learning parameters makes the online computation burden greatly alleviate. The convergence of the spacing and velocity errors to arbitrarily small compact sets of zero is proven by the Lyapunov theory. Finally, simulation results show the effectiveness of the proposed scheme. In the next chapter, a *fault-tolerant cluster consensus* control strategy will be proposed.

References

1. C.L. Wang, C.Y. Wen, L. Guo, Adaptive consensus control for nonlinear multi-agent systems with unknown control directions and time-varying actuator faults. IEEE Trans. Autom. Control **66**(9), 4222–4229 (2021)
2. X.G. Guo, J.L. Wang, F. Liao, Adaptive fuzzy fault-tolerant control for multiple high-speed trains with proportional and integral-based sliding mode. IET Control Theory Appl. **11**(8), 1234–1244 (2017)
3. S. Chen, D.W.C. Ho, L. Li, M. Liu, Fault-tolerant consensus of multi-agent system with distributed adaptive protocol. IEEE Trans. Cybern. **45**(10), 2142–2155 (2015)
4. C. Deng, G.H. Yang, Distributed adaptive fuzzy control for nonlinear multiagent systems under directed graphs. IEEE Trans. Fuzzy Syst. **26**(3), 1356–1366 (2018)
5. C. Deng, G.H. Yang, Distributed adaptive fault-tolerant control approach to cooperative output regulation for linear multi-agent systems. Automatica **103**, 62–68 (2019)
6. X. Jin, S. Wang, J. Qin, W.X. Zheng, Y. Kang, Adaptive fault-tolerant consensus for a class of uncertain nonlinear second-order multi-agent systems with circuit implementation. IEEE Trans. Circuits Syst. Regul. Pap. **65**(7), 2243–2255 (2018)
7. S. Luo, D. Ye, Adaptive double event-triggered control for linear multi-agent systems with actuator faults. IEEE Trans. Circuits Syst. Regul. Pap. **66**(12), 4829–4839 (2019)
8. H.J. Ma, G.H. Yang, Adaptive fault tolerant control of cooperative heterogeneous systems with actuator faults and unreliable interconnections. IEEE Trans. Autom. Control **61**(11), 3240–3255 (2016)

9. J. Qin, G. Zhang, W.X. Zheng, Y. Kang, Adaptive sliding mode consensus tracking for second-order nonlinear multiagent systems with actuator faults. IEEE Trans. Cybern. **49**(5), 1605–1615 (2019)
10. X. Wang, G.H. Yang, Adaptive reliable coordination control for linear agent networks with intermittent communication constraints. IEEE Trans. Control Netw. Syst. **5**(3), 1120–1131 (2018)
11. D. Ye, M.M. Chen, H.J. Yang, Distributed adaptive event-triggered fault-tolerant consensus of multiagent systems with general linear dynamics. IEEE Trans. Cybern. **49**(3), 757–767 (2019)
12. Z. Zuo, J. Zhang, Y. Wang, Adaptive fault-tolerant tracking control for linear and Lipschitz nonlinear multi-agent systems. IEEE Trans. Ind. Electron. **62**(6), 3923–3931 (2015)
13. X.G. Guo, J.L. Wang, F. Liao, R.S.H. Teo, CNN-based distributed adaptive control for vehicle-following platoon with input saturation. IEEE Trans. Intell. Transp. Syst. **19**(10), 3121–3132 (2018)
14. X.G. Guo, J.J. Zhao, H.J. Li, J.L. Wang, F. Liao, Y. Chen, Novel auxiliary saturation compensation design for neuroadaptive NTSM tracking control of high speed trains with actuator saturation. J. Franklin Inst. **357**(3), 1582–1602 (2020)
15. J. Fu, G. Wen, T. Huang, Z. Duan, Consensus of multi-agent systems with heterogeneous input saturation levels. IEEE Trans. Circuits Syst. Express Briefs **66**(6), 1053–1057 (2019)
16. M. Rehan, C.K. Ahn, M. Chadli, Consensus of one-sided Lipschitz multi-agents under input saturation. IEEE Trans. Circuits Syst. Express Briefs **67**(4), 745–749 (2020)
17. L. Shi, Y. Li, Z. Lin, Semi-global leader-following output consensus of heterogeneous multi-agent systems with input saturation. Int. J. Robust Nonlinear Control **28**(16), 4916–4930 (2018)
18. X. Yi, T. Yang, J. Wu, K.H. Johansson, Distributed event-triggered control for global consensus of multi-agent systems with input saturation. Automatica **100**, 1–9 (2019)
19. M. Chen, S.S. Ge, B. Ren, Adaptive tracking control of uncertain MIMO nonlinear systems with input constraints. Automatica **47**(3), 452–465 (2011)
20. X.G. Guo, J.L. Wang, F. Liao, R.S.H. Teo, Neuroadaptive quantized pid sliding-mode control for heterogeneous vehicular platoon with unknown actuator deadzone. Int. J. Robust Nonlinear Control **29**(1), 188–208 (2019)
21. Z. Ding, Z. Li, Distributed adaptive consensus control of nonlinear output-feedback systems on directed graphs. Automatica **72**, 46–52 (2016)
22. Y. Lv, Z. Li, Z. Duan, Distributed PI control for consensus of heterogeneous multiagent systems over directed graphs. IEEE Trans. Syst. Man Cybern. Syst. **50**(4), 1602–1609 (2020)
23. A. Megretski, A. Rantzer, System analysis via integral quadratic constraints. IEEE Trans. Autom. Control **42**(6), 819–830 (1997)
24. I.R. Petersen, V.A. Ugrinovskii, A.V. Savkin, *Robust Control Design Using H_∞ Methods*. Springer Science & Business Media, London (2000)
25. Y. Song, X. Huang, C. Wen, Tracking control for a class of unknown nonsquare MIMO non-affine systems: a deep-rooted information based robust adaptive approach. IEEE Trans. Autom. Control **61**(10), 3227–3233 (2016)
26. Y. Song, X. Huang, C. Wen, Robust adaptive fault-tolerant PID control of MIMO nonlinear systems with unknown control direction. IEEE Trans. Ind. Electron. **64**(6), 4876–4884 (2017)
27. P.A. Ioannou, J. Sun, *Robust Adaptive Control*, vol. 1, Prentice-Hall Upper Saddle River, NJ (1996)

Chapter 8
Fault-Tolerant Cluster Consensus of MASs with Actuator Fault and IQCs

Abstract In this chapter, the issue of achieving *cluster consensus* in a heterogeneous second-order leader-following multi-agent system (MAS) with nonlinear dynamics, actuator faults, and integral quadratic constraints (IQCs) is explored within a directed network topology. Two adaptive *fault-tolerant* pinning control strategies are proposed, one with fixed pinning gains and the other with adaptive pinning gains, to ensure *cluster consensus* within a *finite time* frame using only local topological information. An adaptive input compensation mechanism is introduced to counteract the negative effects of actuator faults, with the estimation of only a single parameter required for each agent, thereby reducing computational costs. Moreover, the use of pinning control, rather than a full control scheme, is shown to enhance the cost-effectiveness of the strategies for large-scale MASs.

8.1 Introduction

Chapters 3–7 introduce the *complete consensus* control strategy of multi-agent system (MAS) in complex environment. However, the practical application scenario of multi-objective and multi-task limits the application of the designed *complete consensus* strategy. *Cluster consensus* in MASs has become more prevalent than traditional *complete consensus* for executing a single task [1–7]. *cluster consensus* has garnered significant attention over recent years due to its extensive applicability in task distribution and social learning networks across various environments and objectives within wireless sensor networks [8–12]. Broadly speaking, the existing literatures offer a multitude of effective control strategies, which can be broadly categorized into pinning control methods [9, 10, 13, 14] and fully equipped control methods [15, 16]. Compared with fully equipped control strategies, pinning control methods stand out for their efficiency, as they target only a subset of nodes within the system, leading to a substantial reduction in communication expenses and computational intricacy, and conserving substantial physical resources for MASs, as highlighted in [13, 14]. However, selecting the suitable nodes to pin in the pinning control strategy is a complex issue. On the basis of achieving satisfactory control performance, a small

X.-G. Guo et al., *Anti-disturbance Intrusion- and Fault-Tolerant Control of Multi-agent Systems*, https://doi.org/10.1007/978-981-96-9614-7_8

number of pinned nodes can economize resources for the MAS. In addition, the majority of existing research on pinning control has not addressed the implications of actuator faults. However, in practical terms, even minor actuator defects can severely degrade system performance or even trigger instability [17–24]. Therefore, the endeavor to maintain the integrity and security of MASs with unexpected actuator faults through a reduced number of pinning nodes is a research domain of considerable practical and theoretical value.

In general, *disturbances* are considered inevitable in many practical applications, and it is difficult to establish a general mathematical model for a physical system subjected to external *disturbances* and sensor measurement noise [25–28]. It is generally believed that *disturbances* and uncertainties in the system are the two main underlying causes of the instability problem [26]. Especially in the field of MASs, which are interconnected and coupled, the impact of uncertainties and external *disturbances* on one agent can ripple through to others. Therefore, characterizing the properties of uncertainties and external *disturbances* and mitigating their adverse effects is a crucial issue. The integral quadratic constraint (IQC) is recognized for its effectiveness in capturing the characteristics of uncertainties and external *disturbances* [29, 30]. Consequently, it is logical to develop adaptive *fault-tolerant* pinning control strategies aimed at achieving *cluster consensus* in MASs that are simultaneously subject to IQCs and actuator faults, even though the interplay between IQCs and actuator faults significantly complicates the design of such controllers.

This chapter is dedicated to addressing the complex issue of adaptive *fault-tolerant* pinning control for heterogeneous second-order MASs that are affected by IQCs, nonlinear dynamics, and actuator faults. The focus is on systems with a directed communication structure that includes a directed spanning tree. The chapter presents both fixed and adaptive pinning control strategies to achieve *cluster consensus* in a *finite time* frame. The primary challenges include: (1) Developing a pinning control mechanism for a *directed graph* that requires a small number of fixed control nodes. (2) Creating an adaptive *fault-tolerant* control approach that does not necessitate the estimation of an excessive number of parameters for each agent. (3) Implementing *fault-tolerant* control for an MAS where there is significant interplay between the IQCs and actuator faults. The key contributions of this chapter are outlined below:

1. Fewer fixed pinning nodes: In contrast to the *complete consensus* control strategy in the previous chapters, this chapter designs a *cluster consensus* control scheme based on the pinning control technology for MASs with multiple leaders. Moreover, different from the approach in [31] requires pinning of all nodes and the method in [32] involves random pinning nodes, this chapter advocates for pinning only a select group of fixed nodes. This approach minimizes the unpredictability associated with pinning control and indicates that the proposed strategies can decrease computational complexity and resource consumption.
2. Novel adaptive *fault-tolerant* control schemes: In comparison to the methodologies presented in [29], the benefits of the schemes crafted in this chapter can be highlighted as follows: (1) The case of *complete consensus* addressed in [29] is treated as a particular instance of *cluster consensus* in this chapter. (2) This chapter

opts for a pinning control approach, diverging from the fully equipped control approach utilized in [29]. (3) Distinguished from [29], which entails the estimation of a multitude of unknown parameters for each agent, the adaptive input compensation mechanism proposed in this chapter necessitates the estimation of only a single parameter per agent. These advantages render the strategies designed in this chapter both structurally streamlined and computationally efficient.

3. Finite-time *cluster consensus*: The strategies developed in this chapter are designed to ensure that *cluster consensus* achieves bounded stability within a *finite time* frame, diverging from the method in [33], which only ensures bounded stability for *cluster consensus*. When compared with the pinning control approach to *cluster consensus* detailed in [10], the inclusion of nonlinear dynamics, IQCs, and actuator faults in this chapter's model broadens its applicability. Furthermore, the intricate coupling between IQCs and actuator faults compounds the design challenges, making it more demanding than what is faced in [10].

8.2 Problem Statement and Preliminaries

8.2.1 System Model

Consider a second-order nonlinear MAS consisting of N followers, labelled as $1, 2, \cdots, N$ and divided into s subgroups. The motion of the ith follower in the σ_ith group is expressed as follows:

$$\begin{aligned}
\dot{x}_i(t) &= v_i(t), \quad i = 1, 2, \cdots, N \\
\dot{v}_i(t) &= f_{\sigma_i}(t, x_i, v_i) + u_i^F(t) + \sum_{k=1}^{p_i} B_{i,k} w_{i,k}(t), \quad \sigma_i = 1, 2, \cdots, s \\
\xi_{i,k}(t) &= H_{i,k} v_i(t) + G_{i,k} u_i^F(t), \quad k = 1, 2, \cdots, p_i
\end{aligned} \tag{8.1}$$

where $x_i(t) \in \mathbb{R}$ and $v_i(t) \in \mathbb{R}$ are the position and velocity of the ith agent, respectively; $f_{\sigma_i}(x_i, v_i, t)$ is a nonlinear differentiable function, representing the nonlinear dynamics. $u_i^F(t)$ is the actuator signal of the ith agent that has faulty; $\xi_{i,k}(t) \in \mathbb{R}^{q_i}$ and $w_{i,k}(t) \in \mathbb{R}^{m_i}$ denote the input and output of the IQC, respectively; $B_{i,k}$, $H_{i,k}$, and $G_{i,k}$ are real constant matrices with corresponding dimensions.

In the σ_ith subgroup, the motion of the virtual leader is described by

$$\dot{x}_{\sigma_i}^r(t) = v_{\sigma_i}^r(t) \tag{8.2}$$

$$\dot{v}_{\sigma_i}^r(t) = f_{\sigma_i}(t, x_{\sigma_i}^r, v_{\sigma_i}^r), \quad \sigma_i = 1, 2, \cdots, s \tag{8.3}$$

in which $x_{\sigma_i}^r(t)$ and $v_{\sigma_i}^r(t)$ stand for the position and velocity of the leader, respectively; $f_{\sigma_i}(t, x_{\sigma_i}^r, v_{\sigma_i}^r)$ represents the nonlinear dynamics, which may be heterogeneous for different subgroups.

In practice, actuators may become faulty. In this chapter, bias faults and efficiency faults are considered. For the ith agent, the actuator fault model is described as follows:

$$u_i^F(t) = \begin{cases} u_i(t), & \text{no fault} \\ \rho_i(t)u_i(t) + r_i(t), & \text{occuring fault} \end{cases} \tag{8.4}$$

where $u_i(t)$ is the control signal of the ith agent; $\rho_i(t)$ and $r_i(t)$ are time-varying unknown functions, which denote the efficiency factor and bias fault of the ith agent actuator, respectively.

Assumption 8.1 The *directed graph* of the leader-following MASs (8.1) and (8.2) has a directed spanning tree.

Assumption 8.2 The efficiency factor and bias fault are bounded and satisfy the following relations:

$$0 < \underline{\rho}_i \leq \rho_i(t) \leq \overline{\rho}_i < \infty \tag{8.5}$$

$$|r_i(t)| \leq \overline{r}_i, \quad i = 1, 2, \cdots, N \tag{8.6}$$

where $\underline{\rho}_i$ and $\overline{\rho}_i$ are the upper and lower bounds of the efficiency factor, respectively; $\overline{r}_i$ is the upper bound of the bias fault.

Remark 8.1 In the actuator fault model (8.4), the variables $\rho_i(t)$ and $r_i(t)$, which are unknown and change over time, are crucial in identifying the type of actuator fault. Here's a detail of the fault classification: fault-free ($\rho_i(t) = 1, r_i(t) = 0$); bias fault ($\rho_i(t) = 1, r_i(t) \neq 0$); efficiency loss fault ($\rho_i(t) \neq 1, r_i(t) = 0$); efficiency loss and bias faults ($\rho_i(t) \neq 1, r_i(t) \neq 0$). Furthermore, the efficiency factor in this chapter can exceed 1, which is not typically considered in many studies, such as [29, 34, 35]. An efficiency factor greater than 1 signifies an excessive fault, which is more common in circuits and sensors, such as excessive sensor offset faults [36].

8.2.2 *IQC Constraints*

Assume that the relationship between the IQC input $\xi_{i,k}(t)$ and output $w_{i,k}(t)$ in the follower model (8.1) is described as:

$$w_{i,k}(t) = \psi_{i,k}\left(t, \xi_{i,k}(\cdot)|_{t_0}^{t}\right), \quad i \in \mathscr{V}, k = 1, 2, \cdots, p_i \tag{8.7}$$

in which $\psi_{i,k}(\cdot, \cdot)$ stands for nonlinear time-varying dynamic constraints.

Based on Chap. 7, one can obtain

$$\sum_{k=1}^{p_i} \|B_{i,k}w_{i,k}(t)\| \le \sum_{k=1}^{p_i} \|B_{i,k}\xi_{i,k}(t)\| + \sum_{k=1}^{p_i} \|B_{i,k}\|\delta_{wi,k}$$
$$\le \zeta_{1i}|u_i^F(t)| + \zeta_{2i}|v_i(t)| + \delta_{di}, \quad i \in \mathscr{V} \tag{8.8}$$

where

$$\zeta_{1i} = \sum_{k=1}^{p_i} \|B_{i,k}\|\|G_{i,k}\| < \frac{\underline{\rho}_i}{\overline{\rho}_i}, \quad \zeta_{2i} = \sum_{k=1}^{p_i} \|B_{i,k}\|\|H_{i,k}\|, \quad \delta_{di} = \sum_{k=1}^{p_i} \|B_{i,k}\|\delta_{wi,k}. \tag{8.9}$$

Remark 8.2 The IQC output $w_{i,k}(t)$, as mentioned in the inequality (8.8), is influenced by the control signal $u_i^F(t)$, the constant δ_{di}, and the follower velocity $v_i(t)$. This indicates that the IQC output can be derived from various sources. Compared to the uncertainty modeling methods in [37, 38], the IQCs model is more practical and applicable to real-world scenarios. Furthermore, the assumption $\zeta_{1i} < \underline{\rho}_i/\overline{\rho}_i$ has been proposed in [25, 29].

In this chapter, the nonlinear differentiable function $f_{\sigma_i}(t, x_i, v_i)$ is characterized based on the differential mean value theorem, i.e., Lemma 2.10.

Assumption 8.3 If $f(t, x(t), v(t))$ is a continuously nonlinear differentiable function, there exist unknown bounded real constants $\beta \ge 0$ and $\eta \ge 0$ such that

$$\left\|\frac{\partial f(t, x(t), v(t))}{\partial x}\right\| \le \beta, \quad \left\|\frac{\partial f(t, x(t), v(t))}{\partial v}\right\| \le \eta. \tag{8.10}$$

Therefore, according to Lemma 2.10, it gives

$$|f(t, x(t), v(t)) - f(t, x(t_0), v(t_0))| \le \beta|x(t) - x(t_0)| + \eta|v(t) - v(t_0)|. \tag{8.11}$$

Furthermore, the values of β and η can be determined by (8.10).

Remark 8.3 In cases where there is no interaction term between $x(t)$ and $v(t)$, Lemma 2.10 is applicable for representing continuously differentiable nonlinear functions. This lemma is widely used in various physical control systems, including anthropomorphic robotic arms [39] and multiple helicopters with 3-degree-of-freedom [40]. Furthermore, the differential mean value theorem can be utilized to locally linearize nonlinear systems. As per several studies, the Lipschitz condition (8.11) significantly mitigates the conservativeness associated with the nonlinearity $f(t, x(t), v(t))$ [29].

Drawing from Definition 2.2, the aim of this chapter is to develop adaptive *fault-tolerant* pinning control strategies $u_i(t)$ for each individual agent, leveraging state data from neighboring agents. The goal is to ensure that within the same subgroup, the position and velocity states of all agents remain uniformly bounded, and the state errors will reach a small set in a *finite time* period.

8.3 Adaptive Fault-Tolerant Control with Fixed Pinning Gain

First, a new adaptive *fault-tolerant* pinning control scheme with a fixed pinning gain is designed. This scheme introduces adaptive input compensation to combat the effects of actuator faults. The proposed adaptive *fault-tolerant* pinning control scheme is depicted in Fig. 8.1. Following this, an analysis for *cluster consensus* of MASs is performed.

8.3.1 Design of Adaptive Fault-Tolerant Pinning Controller

By combining the pinning control strategy with the adaptive *fault-tolerant* compensation method, a *cluster consensus* controller can be proposed as

$$u_i(t) = u_{1i}(t) + u_{2i}(t) + u_{3i}(t) \tag{8.12}$$

$$u_{1i}(t) = -\gamma \sum_{j=1}^{N} l_{ij} x_j - \gamma \sum_{j=1}^{N} l_{ij} v_j \tag{8.13}$$

$$u_{2i}(t) = \begin{cases} -b_i \tilde{x}_i - b_i \tilde{v}_i + \mathscr{P}_i, & i \in \tilde{\mathscr{V}}_{\sigma_i} \\ -b_i \tilde{x}_i - b_i \tilde{v}_i, & i \in \mathscr{V}_{\sigma_i} \backslash \tilde{\mathscr{V}}_{\sigma_i} \text{ and } \mathscr{D}_i = 0 \\ 0, & \text{otherwise} \end{cases} \tag{8.14}$$

$$u_{3i}(t) = -k_i(t)(\tilde{x}_i + \tilde{v}_i) \tag{8.15}$$

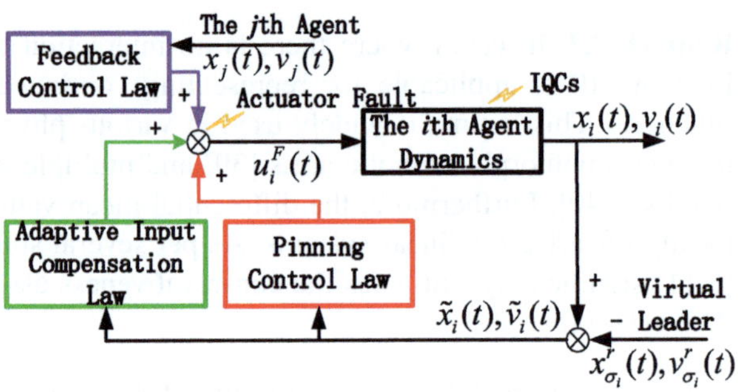

Fig. 8.1 Schematic of the designed control scheme

where $\tilde{\mathscr{V}}_{\sigma_i}$ is the subset of the σ_ith subgroup $\mathscr{V}_{\sigma_i}$, which denotes that the agents in subset $\tilde{\mathscr{V}}_{\sigma_i}$ can receive information from other subgroups' agents; $\mathscr{D}_i = 0$ implies that the ith agent can not get information from other agents. In addition,

$$\tilde{x}_i = x_i - x^r_{\sigma_i}, \quad \sigma_i = 1, 2, \cdots, s \tag{8.16}$$

$$\tilde{v}_i = v_i - v^r_{\sigma_i}, \quad i = 1, 2, \cdots, N \tag{8.17}$$

$$\mathscr{P}_i = \gamma \sum_{j=1}^{N} l_{ij} x^r_{\sigma_j} + \gamma \sum_{j=1}^{N} l_{ij} v^r_{\sigma_j} \tag{8.18}$$

where $\gamma > 0$ denotes the coupling strength and b_i $(1 \leq i \leq N)$ the pinning gain of the ith agent; $b_i > 0$ if the ith agent is chosen as pinning node, otherwise $b_i = 0$. The adaptive input compensation $k_i(t)$ and adaptive update law are as follows:

$$k_i(t) = \frac{\hat{\chi}_i(t)\Psi_i^2(t)}{\Psi_i(t)|\tilde{x}_i + \tilde{v}_i| + \iota_i} \tag{8.19}$$

$$\dot{\hat{\chi}}_i(t) = -\varphi_i \varsigma_i(t)\hat{\chi}_i(t) + \frac{\varphi_i \Psi_i^2(t)(\tilde{x}_i + \tilde{v}_i)^2}{\Psi_i(t)|\tilde{x}_i + \tilde{v}_i| + \iota_i} \tag{8.20}$$

$$\Psi_i(t) = 1 + |v_i(t)| \tag{8.21}$$

where $\hat{\chi}_i(0) \geq 0$; ι_i and φ_i are positive constants; $\varsigma_i(t)$ is a positive continuously bounded function with $\lim_{t\to\infty} \int_{t_0}^{t} \varsigma_i(\tau)\, d\tau \leq \overline{\varsigma}_i < +\infty$, in which $\overline{\varsigma}_i$ is an unknown positive constant.

Remark 8.4 It should be noted that the control input $u_i(t)$ in (8.12) is composed of three distinct components: the feedback component $u_{1i}(t)$, the pinning control component $u_{2i}(t)$, and the input compensation component $u_{3i}(t)$. Both the feedback component $u_{1i}(t)$ and the input compensation component $u_{3i}(t)$ are designed to ensure uniform performance in the presence of IQCs and actuator faults.

Remark 8.5 The fully equipped control approach demands the development of a tracking controller for every agent [15, 16], whereas the pinning control approach necessitates the design of tracking controllers for a select few agents. Consequently, the fully equipped control approach is a particular instance of the pinning control approach. It is evident that, considering the computational burden, equipment, and communication resources, the pinning control approach is more economical for large-scale MASs compared to the fully equipped control approach.

Remark 8.6 In the pinning control part $u_{2i}(t)$, if the ith agent $(i \in \mathscr{V}_{\sigma_i})$ can not receive information from other subgroups $(\mathscr{V} \backslash \mathscr{V}_{\sigma_i})$, then $\sum_{j=1}^{N} l_{ij} x_{\sigma_j} = 0$ and $\sum_{j=1}^{N} l_{ij} v_{\sigma_j} = 0$. If the ith agent can receive information from an agent in another subgroup $i \in \tilde{\mathscr{V}}_{\sigma_i}$, the pinning control strategy $u_{2i}(t)$ needs $\sum_{j=1}^{N} l_{ij} x^r_{\sigma_j}$ and

$\sum_{j=1}^{N} l_{ij} v_{\sigma_j}^r$ to eliminate the effects of other agents from other subgroups. Additionally, nodes in graph $\mathscr{G}$ that have an in-degree of zero are special as they are unable to gather information from other nodes (i.e., $\mathscr{D}i = 0$). Hence, these nodes must be the pinning nodes. Unlike the control strategy in [31] where all nodes are pinning or [32] where some nodes are arbitrarily considered as pinning nodes, only the fixed nodes require pinning in the controller $u_{2i}(t)$. The controllers designed for all nodes in [31] can be viewed as a particular case of the pinning controller (8.12).

Remark 8.7 With respect to the control law (8.12), a distinct advantage is worth mentioning. In the analysis of *fault-tolerant* consensus for MASs amidst actuator faults and IQCs, a multitude of unknown parameters are required to be estimated in [29, 41]. The greater the number of parameters to be estimated, the more intricate the control scheme becomes and the more arduous the computation. The approach proposed in this chapter requires the estimation of only one parameter per agent via the adaptive update law (8.19).

Remark 8.8 The adaptive update law (8.19) presented in this chapter relies solely on the state information of the agent and its virtual leader. The predetermined parameters φ_i, ι_i, and the time-varying functions $\varsigma_i(t)$ consume minimal computational resources. Moreover, the control performance can be tuned by modifying these parameters. As a result, the adaptive update law (8.19) in this chapter is straightforward to implement in practical systems.

8.3.2 Cluster Consensus Analysis

The following theorem is a formal presentation of the main results of this section.

Theorem 8.1 *Consider the MAS (8.1) and (8.2) under Assumptions 8.1–8.3. By applying the adaptive fault-tolerant pinning controller (8.12), the cluster consensus can be guaranteed in finite time if the following conditions are established:*

$$\lambda_{\max}(\tilde{L}_l) < -\frac{\alpha}{\gamma} \tag{8.22}$$

$$b_i > \frac{\lambda_{\max}(B - C M_l^{-1} C^T)}{\tilde{\rho}_l} \tag{8.23}$$

where

$$\tilde{L} = \frac{-\tilde{\rho}\mathscr{L} - \mathscr{L}^T \tilde{\rho}}{2}$$

with $\mathscr{L}$ being the Laplacian matrix of the directed graph, and $\tilde{\rho}$ is defined in (8.25); $\tilde{L}_l$ is a matrix that removes the first l column-rows of $\tilde{L}$; moreover

$$M = \alpha I_N + \gamma \tilde{L}, \quad \alpha = \max\{p, q+1\}$$
$$p = \max_{1\le i\le N} \left\{\frac{3\beta_i + \eta_i}{2}\right\}, \quad q = \max_{1\le i\le N} \left\{\frac{3\eta_i + \beta_i}{2}\right\}$$

where β_i and η_i are unknown bounded positive constants; B, C, and M_l are defined in Lemma 2.11.

Proof First, based on the follower system (8.1), virtual leader system (8.2), and fault model (8.4), the state error system of the MAS is given by

$$\begin{aligned} \dot{\tilde{x}}_i(t) &= \tilde{v}_i(t) \\ \dot{\tilde{v}}_i(t) &= \rho_i(t)u_i(t) + r_i(t) + \sum_{k=1}^{p_i} B_{i,k} w_{i,k}(t) + f_{\sigma_i}(t, x_i, v_i) - f_{\sigma_i}(t, x^r_{\sigma_i}, v^r_{\sigma_i}). \end{aligned} \tag{8.24}$$

Let

$$\begin{aligned} \tilde{x} &= [\tilde{x}_1^T, \tilde{x}_2^T, \cdots, \tilde{x}_N^T]^T \\ \tilde{v} &= [\tilde{v}_1^T, \tilde{v}_2^T, \cdots, \tilde{v}_N^T]^T \\ \mathscr{B} &= \text{diag}\{b_1, b_2, \cdots, b_l, 0, \cdots, 0\} \end{aligned}$$

with $b_i > 0$. Then, the following Lyapunov function is constructed.

$$V(t) = \frac{1}{2}(\tilde{x}^T, \tilde{v}^T)\Omega(\tilde{x}^T, \tilde{v}^T)^T + \sum_{i=1}^{N} \frac{1}{2\varphi_i \tilde{\rho}_i}\tilde{\chi}_i^2(t) \tag{8.25}$$

where

$$\tilde{\rho}_i = (\underline{\rho}_i - \zeta_{1i}\overline{\rho}_i), \quad \tilde{\rho} = \text{diag}\{\tilde{\rho}_1, \tilde{\rho}_2, \cdots, \tilde{\rho}_N\}$$

$$\Omega = \begin{bmatrix} \gamma(\tilde{\rho}\mathscr{L} + \mathscr{L}^T\tilde{\rho}) + 2\tilde{\rho}\mathscr{B} I_N \\ I_N & I_N \end{bmatrix}$$

and $\tilde{\chi}_i(t)$ is described in (8.31). Note that Ω is a positive definite matrix when appropriate values are chosen for γ and b_i to guarantee that

$$\gamma(\tilde{\rho}\mathscr{L} + \mathscr{L}^T\tilde{\rho}) + 2\tilde{\rho}\mathscr{B} - I_N > 0$$

holds based on Lemmas 2.8 and 2.1.

From (8.24), the derivative of $V(t)$ is as follows:

$$
\begin{aligned}
\dot{V}(t) &= \tilde{x}^T[\gamma(\tilde{\rho}\mathscr{L} + \mathscr{L}^T\tilde{\rho}) + 2\tilde{\rho}\mathscr{B}]\tilde{v} + \tilde{v}^T\dot{\tilde{v}} + \tilde{v}^T\tilde{v} + \tilde{x}^T\dot{\tilde{v}} - \sum_{i=1}^{N}\frac{1}{\varphi_i}\tilde{\chi}_i(t)\dot{\hat{\chi}}_i(t) \\
&= \tilde{x}^T[\gamma(\tilde{\rho}\mathscr{L} + \mathscr{L}^T\tilde{\rho}) + 2\tilde{\rho}\mathscr{B}]\tilde{v} + \tilde{v}^T\tilde{v} + \sum_{i=1}^{N}(\tilde{x}_i + \tilde{v}_i)\rho_i(t)u_i + \sum_{i=1}^{N}(\tilde{x}_i + \tilde{v}_i)r_i(t) \\
&\quad + \sum_{i=1}^{N}(\tilde{x}_i + \tilde{v}_i)(f_{\sigma_i}(t, x_i, v_i) - f_{\sigma_i}(t, x_{\sigma_i}^r, v_{\sigma_i}^r)) - \sum_{i=1}^{N}\frac{1}{\varphi_i}\tilde{\chi}_i(t)\dot{\hat{\chi}}_i(t) \\
&\quad + \sum_{i=1}^{N}(\tilde{x}_i + \tilde{v}_i)\sum_{k=1}^{p_i}B_{i,k}w_{i,k}(t).
\end{aligned} \tag{8.26}
$$

To simplify the notation, let $s_i = \tilde{x}_i + \tilde{v}_i$. Combining (8.8) with (8.26) leads to the following expression:

$$
\begin{aligned}
\dot{V}(t) \le\; & \tilde{x}^T[\gamma(\tilde{\rho}\mathscr{L} + \mathscr{L}^T\tilde{\rho}) + 2\tilde{\rho}\mathscr{B}]\tilde{v} + \tilde{v}^T\tilde{v} + \sum_{i=1}^{N}s_i(f_{\sigma_i}(t, x_i, v_i) - f_{\sigma_i}(t, x_{\sigma_i}^r, v_{\sigma_i}^r)) \\
& + \sum_{i=1}^{N}s_i\rho_i(t, t_{\rho i})u_i + \sum_{i=1}^{N}[\zeta_{2i}|v_i(t)| + (1 + \zeta_{1i})\overline{r}_i + \delta_{di}]|s_i| \\
& + \sum_{i=1}^{N}\zeta_{1i}\overline{\rho}_i|u_i||s_i| - \sum_{i=1}^{N}\frac{1}{\varphi_i}\tilde{\chi}_i(t)\dot{\hat{\chi}}_i(t).
\end{aligned} \tag{8.27}
$$

According to (8.9), one has $\underline{\rho}_i - \zeta_{1i}\overline{\rho}_i > 0$. In addition, from (8.19), one can get

$$
\dot{\hat{\chi}}_i(t) + \varphi_i\varsigma_i(t)\hat{\chi}_i(t) \ge 0 \tag{8.28}
$$

namely

$$
\hat{\chi}_i(t) \ge e^{-\varphi_i\int_{t_0}^{t}\varsigma_i(\tau)d\tau}\hat{\chi}_i(0) \tag{8.29}
$$

which means that the adaptive update law $\hat{\chi}_i(t)$ will be nonnegative when $\hat{\chi}_i(0) \ge 0$ is chosen. Mathematically, it follows from (8.8) that

$$
\begin{aligned}
\tilde{u}_1 \triangleq & \sum_{i=1}^{N}[s_i\rho_i(t)u_i + \zeta_{1i}\overline{\rho}_i|u_i||s_i| - \frac{1}{\varphi_i}\tilde{\chi}_i(t)\dot{\hat{\chi}}_i(t)] + \sum_{i=1}^{N}[(\zeta_{2i}|v_i(t)| + (1+\zeta_{1i})\overline{r}_i + \delta_{di}]|s_i| \\
= & -\sum_{i=1}^{N} s_i\rho_i(t)[\gamma\sum_{j=1}^{N} l_{ij}s_j + (b_i + k_i(t))s_i] + \sum_{i=1}^{N}[(\zeta_{2i}|v_i(t)| + (1+\zeta_{1i})\overline{r}_i + \delta_{di}]|s_i| \\
& + \sum_{i=1}^{N}\zeta_{1i}\overline{\rho}_i|\gamma s_i\sum_{j=1}^{N} l_{ij}s_j + (b_i + k_i(t))s_i^2| - \sum_{i=1}^{N}\frac{1}{\varphi_i}\tilde{\chi}_i(t)\dot{\hat{\chi}}_i(t) \\
\leq & -\sum_{i=1}^{N}(\underline{\rho}_i - \zeta_{1i}\overline{\rho}_i)|\gamma s_i\sum_{j=1}^{N} l_{ij}s_j + (b_i + k_i(t))s_i^2| + \sum_{i=1}^{N}[\chi_i^*\Psi_i(t)|s_i| - \frac{1}{\varphi_i}\tilde{\chi}_i(t)\dot{\hat{\chi}}_i(t)]
\end{aligned}
\tag{8.30}
$$

where

$$
\chi_i^* = \max_{i=1}^{N}\{\zeta_{2i}, (1+\zeta_{1i})\overline{r}_i) + \delta_{di}\} \tag{8.31}
$$

$$
\tilde{\chi}_i(t) = \chi_i^* - (\underline{\rho}_i - \zeta_{1i}\overline{\rho}_i)\hat{\chi}_i(t). \tag{8.32}
$$

By combining the adaptive input compensation $k_i(t)$ with adaptive update law $\hat{\chi}_i(t)$ in (8.19), one can obtain

$$
\begin{aligned}
& -\frac{1}{\varphi_i}\tilde{\chi}_i(t)\dot{\hat{\chi}}_i(t) \\
= & -\frac{1}{\varphi_i}\tilde{\chi}_i(t)\Big(-\varphi_i\varsigma_i(t)\hat{\chi}_i(t) + \frac{\varphi_i\Psi_i^2(t)s_i^2(t)}{\Psi_i(t)|s_i(t)| + \iota_i}\Big) \\
= & \frac{\varsigma_i(t)}{\underline{\rho}_i - \zeta_{1i}\overline{\rho}_i}\tilde{\chi}_i(t)(\chi_i^* - \tilde{\chi}_i(t)) - \tilde{\chi}_i(t)\frac{\Psi_i^2(t)s_i^2(t)}{\Psi_i(t)|s_i(t)| + \iota_i} \\
\leq & -\frac{\bar{\varsigma}_i}{2(\underline{\rho}_i - \zeta_{1i}\overline{\rho}_i)}\tilde{\chi}_i^2(t) - \tilde{\chi}_i(t)\frac{\Psi_i^2(t)s_i^2(t)}{\Psi_i(t)|s_i(t)| + \iota_i} + \frac{\bar{\varsigma}_i}{2(\underline{\rho}_i - \zeta_{1i}\overline{\rho}_i)}\chi_i^{*2}
\end{aligned}
\tag{8.33}
$$

and

$$
\begin{aligned}
& -(\underline{\rho}_i - \zeta_{1i}\overline{\rho}_i)k_i(t)s_i^2(t) + \chi_i^*\Psi_i(t)|s_i| \\
= & -(\underline{\rho}_i - \zeta_{1i}\overline{\rho}_i)\frac{\hat{\chi}_i(t)\Psi_i(t)s_i(t)^2}{\Psi_i(t)|s_i(t)| + \iota_i} + \chi_i^*\Psi_i(t)|s_i(t)| \\
\leq & \tilde{\chi}_i(t)\frac{\Psi_i^2(t)s_i^2(t)}{\Psi_i(t)|s_i(t)| + \iota_i} + \chi_i^*\iota_i.
\end{aligned}
\tag{8.34}
$$

Combining (8.33) with (8.34), it follows from (8.30) that

$$\tilde{u}_1 \leq -\sum_{i=1}^{N}(\underline{\rho}_i - \zeta_{1i}\overline{\rho}_i)|\gamma s_i \sum_{j=1}^{N} l_{ij}s_j + b_i s_i^2| - \sum_{i=1}^{N} \frac{\bar{\varsigma}_i}{2(\underline{\rho}_i - \zeta_{1i}\overline{\rho}_i)}\tilde{\chi}_i^2(t) + \sum_{i=1}^{N} \frac{\bar{\varsigma}_i}{2(\underline{\rho}_i - \zeta_{1i}\overline{\rho}_i)}\chi_i^{*2} + \sum_{i=1}^{N}\chi_i^* \iota_i. \tag{8.35}$$

Based on Assumption 8.3 and Lemma 2.7, it is derived that

$$\begin{aligned}\sum_{i=1}^{N}\tilde{x}_i^T(f_{\sigma_i}(t, x_i, v_i) - f_{\sigma_i}(t, x_{\sigma_i}^r, v_{\sigma_i}^r)) &\leq \sum_{i=1}^{N}\tilde{x}_i^T(\beta_i|\tilde{x}_i| + \eta_i|\tilde{v}_i|) \\ &\leq \sum_{i=1}^{N}\left(\frac{2\beta_i + \eta_i}{2}\tilde{x}_i^T\tilde{x}_i + \frac{\eta_i}{2}\tilde{v}_i^T\tilde{v}_i\right). \end{aligned}\tag{8.36}$$

Similarly, one can obtain the following expression:

$$\sum_{i=1}^{N}\tilde{v}_i^T(f_{\sigma_i}(t, x_i, v_i) - f_{\sigma_i}(t, x_{\sigma_i}^r, v_{\sigma_i}^r)) \leq \sum_{i=1}^{N}\left(\frac{2\eta_i + \beta_i}{2}\tilde{v}_i^T\tilde{v}_i + \frac{\beta_i}{2}\tilde{x}_i^T\tilde{x}_i\right). \tag{8.37}$$

It follows from (8.36) and (8.37) that

$$\begin{aligned}&\sum_{i=1}^{N}(\tilde{x}_i + \tilde{v}_i)^T(f_{\sigma_i}(t, x_i, v_i) - f_{\sigma_i}(t, x_{\sigma_i}^r, v_{\sigma_i}^r)) \\ &\leq \sum_{i=1}^{N}\left(\frac{3\beta_i + \eta_i}{2}\tilde{x}_i^T\tilde{x}_i + \frac{3\eta_i + \beta_i}{2}\tilde{v}_i^T\tilde{v}_i\right) \\ &\leq \max_{1\leq i\leq N}\left\{\frac{3\beta_i + \eta_i}{2}\right\}\tilde{x}^T\tilde{x} + \max_{1\leq i\leq N}\left\{\frac{3\eta_i + \beta_i}{2}\right\}\tilde{v}^T\tilde{v}. \end{aligned}\tag{8.38}$$

Substituting (8.35) and (8.38) into (8.27) results in the following term:

$$\begin{aligned}\dot{V}(t) \leq\, &\tilde{x}^T(\alpha I_N - \gamma\tilde{\rho}\mathscr{L} - \tilde{\rho}\mathscr{B})\tilde{x} + \tilde{v}^T(\alpha I_N - \gamma\tilde{\rho}\mathscr{L} - \tilde{\rho}\mathscr{B})\tilde{v} \\ &- \sum_{i=1}^{N}\frac{\bar{\varsigma}_i}{2(\underline{\rho}_i - \zeta_{1i}\overline{\rho}_i)}\tilde{\chi}_i^2(t) + \sum_{i=1}^{N}\chi_i^*\iota_i + \sum_{i=1}^{N}\frac{\bar{\varsigma}_i}{2(\underline{\rho}_i - \zeta_{1i}\overline{\rho}_i)}\chi_i^{*2}. \end{aligned}\tag{8.39}$$

By defining $\tilde{L} = \frac{-\tilde{\rho}\mathscr{L} - \mathscr{L}^T\tilde{\rho}}{2}$, (8.39) can be simplified as follows:

$$\begin{aligned}\dot{V}(t) &\le \tilde{x}^T(\alpha I_N + \gamma\tilde{L} - \tilde{\rho}\mathscr{B})\tilde{x} + \tilde{v}^T(\alpha I_N + \gamma\tilde{L} - \tilde{\rho}\mathscr{B})\tilde{v} \\ &\quad - \sum_{i=1}^{N}\frac{\bar{\varsigma}_i}{2(\underline{\rho}_i - \zeta_{1i}\bar{\rho}_i)}\tilde{\chi}_i^2(t) + \sum_{i=1}^{N}\chi_i^*\iota_i + \sum_{i=1}^{N}\frac{\bar{\varsigma}_i}{2(\underline{\rho}_i - \zeta_{1i}\bar{\rho}_i)}\chi_i^{*2}\end{aligned} \tag{8.40}$$

where $\alpha I_N + \gamma\tilde{L} - \tilde{\rho}\mathscr{B}$ is a symmetric matrix.

Considering Lemma 2.11, when the conditions (8.22) and (8.23) in Theorem 8.1 are satisfied, $\alpha I_N + \gamma\tilde{L} - \tilde{\rho}\mathscr{B}$ is negative definite. Consequently, the following expression holds:

$$\begin{aligned}\dot{V}(t) &\le \lambda_{\min}(M)[\tilde{x}^T, \tilde{v}^T][\tilde{x}^T, \tilde{v}^T]^T - \sum_{i=1}^{N}\frac{\bar{\varsigma}_i}{2(\underline{\rho}_i - \zeta_{1i}\bar{\rho}_i)}\tilde{\chi}_i^2(t) \\ &\quad + \sum_{i=1}^{N}\chi_i^*\iota_i + \sum_{i=1}^{N}\frac{\bar{\varsigma}_i}{2(\underline{\rho}_i - \zeta_{1i}\bar{\rho}_i)}\chi_i^{*2} \\ &\le -z_1V(t) + z_2\end{aligned} \tag{8.41}$$

where

$$\begin{aligned}M &= \text{diag}\{\alpha I_N + \gamma\tilde{L} - \tilde{\rho}\mathscr{B}, \alpha I_N + \gamma\tilde{L} - \tilde{\rho}\mathscr{B}\} \\ z_1 &= \min\left\{\frac{-\lambda_{\min}(M)}{\lambda_{\max}(\Omega)}, \vartheta\right\} > 0, \quad \vartheta = \min_{1\le i\le N}\bar{\varsigma}_i\varphi_i \\ z_2 &= \sum_{i=1}^{N}\chi_i^*\iota_i + \sum_{i=1}^{N}\frac{\bar{\varsigma}_i}{2(\underline{\rho}_i - \zeta_{1i}\bar{\rho}_i)}\chi_i^{*2}.\end{aligned}$$

Multiplying the two sides of the inequality (8.41) by e^{z_1t} and integrating it over the interval $[t_0, t]$ and multiplying the result by e^{-z_1t} results in the following relation:

$$0 \le V(t) \le \frac{z_2}{z_1} + \left(V(t_0) - \frac{z_2}{z_1}\right)e^{-z_1(t-t_0)} \le V(t_0) + \frac{z_2}{z_1} \tag{8.42}$$

in which

$$V(t_0) = \frac{1}{2}\left(\tilde{x}^T(t_0), \tilde{v}^T(t_0)\right)\Omega\left(\tilde{x}^T(t_0), \tilde{v}^T(t_0)\right)^T + \sum_{i=1}^{N}\frac{1}{2\varphi_i\bar{\rho}_i}\tilde{\chi}_i^2(t_0) \ge 0. \tag{8.43}$$

Suppose that there exists a moment T_m such that $V(T_m) = V_m$ holds, where $V_m > 0$ is a sufficiently large constant. Based on (8.42), the following expression is obtained:

$$T_m = \frac{\ln \frac{V(T_m)-\frac{z_2}{z_1}}{V(t_0)-\frac{z_2}{z_1}}}{-z_1} + t_0 = \frac{\ln \frac{V_m-\frac{z_2}{z_1}}{V(t_0)-\frac{z_2}{z_1}}}{-z_1} + t_0 \tag{8.44}$$

which implies that $V(t) \leq V_m$ if $t \geq T_m$.

From the Lyapunov function (8.25), one can get

$$\frac{1}{2}(\tilde{x}^T, \tilde{v}^T)\Omega(\tilde{x}^T, \tilde{v}^T)^T \leq V(t) \tag{8.45}$$

$$\sum_{i=1}^{N} \frac{1}{2\varphi_i \tilde{\rho}_i} \tilde{\chi}_i^2(t) \leq V(t) \tag{8.46}$$

namely

$$\|\varpi(t)\| \leq \sqrt{2\varepsilon V(t)} \tag{8.47}$$

$$\|\tilde{\chi}(t)\| \leq \sqrt{2\varkappa V(t)} \tag{8.48}$$

where

$$\begin{aligned} \varpi(t) &= (\tilde{x}^T, \tilde{v}^T)^T, \quad \tilde{\chi}(t) = [\tilde{\chi}_1(t), \tilde{\chi}_2(t), \cdots, \tilde{\chi}_N(t)]^T \\ \varepsilon &= \frac{1}{\lambda_{\min}(\Omega)}, \quad \varkappa = \min_{1 \leq i \leq N} \{\varphi_i \tilde{\rho}_i\}. \end{aligned} \tag{8.49}$$

From (8.42) and (8.44), it follows that

$$\begin{aligned} \|\varpi(t)\| &\leq \sqrt{2\varepsilon V_m} \\ \|\tilde{\chi}(t)\| &\leq \sqrt{2\varkappa V_m}, \quad t \geq T_m. \end{aligned} \tag{8.50}$$

This implies that all signals of this system converge in Euclidean norm to a small set

$$\Xi = \{\varpi(t), \tilde{\chi}(t) | \|\varpi(t)\| \leq \sqrt{2\varepsilon V_m}, \|\tilde{\chi}(t)\| \leq \sqrt{2\varkappa V_m}\}$$

in *finite time*. It should be noted that Ξ can vary when the values of z_1 and z_2 are adjusted. In addition, it is worth noting that the parameter z_2 can be kept as small as possible by reducing the parameters $\overline{\varsigma}_i$ and ι_i in the adaptive compensation (8.19) when other parameter values are fixed; the smaller z_2 is, the smaller the set Ξ is. However, too small a value of z_2 may increase the amplitude of the adaptive compensation and cause the control input to change dramatically. Thus, selecting the system parameters is a tradeoff problem.

The boundedness of ϖ in *finite time* ensures that the position error $\tilde{x}_i$ and velocity error $\tilde{v}_i$ are also bounded in *finite time*. Consequently, by Definition 2.2, the MAS will achieve *cluster consensus* in *finite time*. The proof is completed. □

Remark 8.9 The adaptive *fault-tolerant* issue involving IQCs for individual agents was studied in [25], utilizing *sliding mode control* techniques. The analysis of *cluster consensus* for MASs in this chapter is more demanding than the single-agent case in [25]. In contrast to the *sliding mode control* strategies used in [25], Theorem 8.1 employs graph theory and pinning control techniques. It is evident that the problem tackled in this chapter is not a simple extension of the problem in [25].

Remark 8.10 In Theorem 8.1, an adaptive *fault-tolerant* pinning control strategy with fixed gains is designed to ensure *cluster consensus* of the MAS (8.1) and (8.2). However, the manual tuning of fixed pinning gains has a notable drawback: it typically fails to accommodate varying conditions such as external *disturbances*, uncertainties, and actuator faults. To address this limitation, the subsequent section will incorporate an adaptive control approach that is both rapidly responsive and capable of automatic updates [13]. This approach is designed to overcome the deficiencies associated with fixed pinning gains by dynamically adjusting to the system's changing requirements.

8.4 Adaptive Fault-Tolerant Control with Adaptive Pinning Gain

In this section, to address the limitations imposed by fixed pinning gains, an adaptive *fault-tolerant* pinning control strategy with adaptive pinning gains is introduced; this approach allows for an automatic and rapidly response. Following this, the *cluster consensus* analysis of MASs is detailed.

8.4.1 Design of Adaptive Fault-Tolerant Pinning Controller

For the controller with adaptive pinning gains, the controller structure and adaptive input compensation are as those presented in (8.12) and (8.19), respectively, except for the part $u_{2i}(t)$, which is as follows:

$$u_{2i}(t) = \begin{cases} -c_i\tilde{x}_i - h_i\tilde{v}_i + \mathscr{P}_i, & i \in \tilde{\mathscr{V}}_{\sigma_i} \\ -c_i\tilde{x}_i - h_i\tilde{v}_i, & i \in \mathscr{V}_{\sigma_i}\backslash\tilde{\mathscr{V}}_{\sigma_i} \text{ and } \mathscr{D}_i = 0 \\ 0, & \text{otherwise} \end{cases} \tag{8.51}$$

where $\tilde{x}_i$, $\tilde{v}_i$, and $\mathscr{P}_i$ are defined in (8.16). The adaptive pinning gains c_i $(c_i \geq 0)$ and h_i $(h_i \geq 0)$ are automatically updated as follows:

$$\begin{aligned} \dot{c}_i &= -\mu_{1i}\pi_{1i}(t)c_i + \mu_{1i}\tilde{x}_i^T\tilde{x}_i \\ \dot{h}_i &= -\mu_{2i}\pi_{2i}(t)h_i + \mu_{2i}\tilde{v}_i^T\tilde{v}_i \end{aligned} \tag{8.52}$$

where $c_i(0) \geq 0$, $h_i(0) \geq 0$, μ_{mi} is a positive constant, and $\pi_{mi}(t)$ denotes positive continuously bounded functions with

$$\lim_{t\to\infty}\int_{t_0}^{t}\pi_{mi}(\tau)\,d\tau \leq \overline{\pi}_{mi} < +\infty.$$

Furthermore, $\overline{\pi}_{mi}$ is the upper bound of $\pi_{mi}(t)$, $m = 1, 2$.

8.4.2 Cluster Consensus Analysis

Based on the graph theory and adaptive compensation techniques, we establish the following theorem.

Theorem 8.2 *Consider the MAS (8.1) and (8.2) under Assumptions 8.1–8.3. By applying the adaptive fault-tolerant pinning controller (8.12) and (8.19) with (8.51) and the adaptive pinning update law (8.52), cluster consensus can be guaranteed in finite time if the following condition is established.*

$$\lambda_{\max}(\tilde{L}_l) < -\frac{\alpha}{\gamma} \tag{8.53}$$

where $\tilde{L}_l$ and α correspond to those in Theorem 8.1.

Proof First, the following diagonal matrices are defined based on the state error function of the MAS (8.24):

$$\begin{aligned} \Delta &= \mathrm{diag}\{c_1, c_2, \cdots, c_l, 0, \cdots, 0\} \\ \Theta &= \mathrm{diag}\{h_1, h_2, \cdots, h_l, 0, \cdots, 0\} \\ \tilde{\Delta} &= \mathrm{diag}\{\tilde{c}_1, \tilde{c}_2, \cdots, \tilde{c}_l, 0, \cdots, 0\} \\ \tilde{\Theta} &= \mathrm{diag}\{\tilde{h}_1, \tilde{h}_2, \cdots, \tilde{h}_l, 0, \cdots, 0\} \end{aligned} \tag{8.54}$$

where $\tilde{c}_l$ and $\tilde{h}_l$ are positive constant estimates of the adaptive pinning gains c_i and h_i, respectively. In addition, as in the analysis of (8.30), Δ and Θ are nonnegative.

Subsequently, the following Lyapunov Function is constructed:

$$V(t) = \frac{1}{2}(\tilde{x}^T, \tilde{v}^T)\Omega(\tilde{x}^T, \tilde{v}^T)^T + \sum_{i=1}^{N} \frac{1}{2\varphi_i \tilde{\rho}_i} \tilde{\chi}_i^2(t)$$

$$+ \frac{1}{2}\sum_{i=1}^{l} \frac{\tilde{\rho}_i}{\mu_{1i}}(c_i - \tilde{c}_i)^2 + \frac{1}{2}\sum_{i=1}^{l} \frac{\tilde{\rho}_i}{\mu_{2i}}(h_i - \tilde{h}_i)^2 \tag{8.55}$$

where

$$\Omega = \begin{bmatrix} \gamma(\tilde{\rho}\mathscr{L} + \mathscr{L}^T \tilde{\rho}) + \tilde{\rho}\Delta + \tilde{\rho}\Theta & I_N \\ I_N & I_N \end{bmatrix}$$

$\tilde{\chi}_i(t)$ and $\tilde{\rho}_i$ are defined in (8.25). Because γ and $\tilde{\rho}$ are positive, and Δ and Θ are nonnegative, according to Lemmas 2.8 and 2.1, if

$$\gamma(\tilde{\rho}\mathscr{L} + \mathscr{L}^T \tilde{\rho}) + \tilde{\rho}\Delta + \tilde{\rho}\Theta - I_N > 0$$

holds by choosing the appropriate value of γ, Ω can be a positive definite matrix.

From (8.24), the derivative of V (t) is as follows:

$$\dot{V}(t) = \tilde{x}^T[\gamma(\tilde{\rho}\mathscr{L} + \mathscr{L}^T \tilde{\rho}) + \tilde{\rho}\Delta + \tilde{\rho}\Theta]\tilde{v} + \tilde{v}^T\dot{\tilde{v}} + \tilde{v}^T\tilde{v} + \sum_{i=1}^{l} \frac{\tilde{\rho}_i}{\mu_{1i}}(c_i - \tilde{c}_i)\dot{c}_i$$

$$+ \sum_{i=1}^{l} \frac{\tilde{\rho}_i}{\mu_{2i}}(h_i - \tilde{h}_i)\dot{h}_i + \tilde{x}^T\dot{\tilde{v}} - \sum_{i=1}^{N} \frac{1}{\varphi_i}\tilde{\chi}_i(t)\dot{\tilde{\chi}}_i(t)$$

$$= \tilde{x}^T[\gamma(\tilde{\rho}\mathscr{L} + \mathscr{L}^T \tilde{\rho}) + \tilde{\rho}\Delta + \tilde{\rho}\Theta]\tilde{v} + \tilde{v}^T\tilde{v}$$

$$+ \sum_{i=1}^{N} s_i(f_{\sigma_i}(t, x_i, v_i) - f_{\sigma_i}(t, x_{\sigma_i}^r, v_{\sigma_i}^r)) + \sum_{i=1}^{N} s_i \sum_{k=1}^{p_i} B_{i,k} w_{i,k}(t)$$

$$- \sum_{i=1}^{N} \frac{1}{\varphi_i}\tilde{\chi}_i(t)\dot{\tilde{\chi}}_i(t) + \sum_{i=1}^{l} \tilde{\rho}_i(c_i - \tilde{c}_i)(-\pi_{1i}(t)c_i + \tilde{x}_i^T\tilde{x}_i)$$

$$+ \sum_{i=1}^{l} \tilde{\rho}_i(h_i - \tilde{h}_i)(-\pi_{2i}(t)h_i + \tilde{v}_i^T\tilde{v}_i) + \sum_{i=1}^{N} s_i\rho_i(t)u_i + \sum_{i=1}^{N} s_i r_i(t). \tag{8.56}$$

Substituting (8.8) into (8.56) leads to the following expression:

$$\begin{aligned}\dot{V}(t) \leq & \tilde{x}^T[\gamma(\tilde{\rho}\mathscr{L}+\mathscr{L}^T\tilde{\rho})+\tilde{\rho}\Delta+\tilde{\rho}\Theta]\tilde{v}+\tilde{v}^T\tilde{v} \\ & +\sum_{i=1}^{N} s_i(f_{\sigma_i}(t,x_i,v_i)-f_{\sigma_i}(t,x_{\sigma_i}^r,v_{\sigma_i}^r))-\sum_{i=1}^{N}\frac{1}{\varphi_i}\tilde{\chi}_i(t)\dot{\hat{\chi}}_i(t) \\ & +\sum_{i=1}^{N} s_i\rho_i(t)u_i+\sum_{i=1}^{N}\zeta_{1i}\overline{\rho}_i|u_i||s_i|+\sum_{i=1}^{N}\big[\zeta_{2i}|v_i(t)|+(1+\zeta_{1i})\overline{r}_i+\delta_{di}\big]|s_i| \\ & +\sum_{i=1}^{l}\tilde{\rho}_i(c_i-\tilde{c}_i)\big(-\pi_{1i}(t)c_i+\tilde{x}_i^T\tilde{x}_i\big)+\sum_{i=1}^{l}\tilde{\rho}_i(h_i-\tilde{h}_i)\big(-\pi_{2i}(t)h_i+\tilde{v}_i^T\tilde{v}_i\big).\end{aligned} \tag{8.57}$$

Similar to (8.30), one can obtain the following expression:

$$\begin{aligned}\tilde{u}_2 \triangleq & \sum_{i=1}^{N} s_i\rho_i(t)u_i+\sum_{i=1}^{N}\zeta_{1i}\overline{\rho}_i|u_i||s_i|+\sum_{i=1}^{N}[\zeta_{2i}|v_i(t)|+(1+\zeta_{1i})\overline{r}_i+\delta_{di}]|s_i| \\ \leq & -\sum_{i=1}^{N}(\underline{\rho}_i-\zeta_{1i}\overline{\rho}_i)\Big|s_i[\gamma\sum_{j=1}^{N}l_{ij}s_j+c_i\tilde{x}_i+h_i\tilde{v}_i+k_i(t)s_i]\Big| \\ & +\sum_{i=1}^{N}[\chi_i^*\Psi_i(t)|s_i|-\frac{1}{\varphi_i}\tilde{\chi}_i(t)\dot{\hat{\chi}}_i(t)].\end{aligned} \tag{8.58}$$

Combining (8.33) with (8.34), it follows from (8.58) that

$$\begin{aligned}\tilde{u}_2 \leq & -\sum_{i=1}^{N}(\underline{\rho}_i-\zeta_{1i}\overline{\rho}_i)\Big|s_i(\gamma\sum_{j=1}^{N}l_{ij}s_j+c_i\tilde{x}_i+h_i\tilde{v}_i)\Big|-\sum_{i=1}^{N}\frac{\bar{\varsigma}_i}{2(\underline{\rho}_i-\zeta_{1i}\overline{\rho}_i)}\tilde{\chi}_i^2(t) \\ & +\sum_{i=1}^{N}\frac{\bar{\varsigma}_i}{2(\underline{\rho}_i-\zeta_{1i}\overline{\rho}_i)}\chi_i^{*2}+\sum_{i=1}^{N}\chi_i^*\iota_i.\end{aligned} \tag{8.59}$$

Substituting (8.38) and (8.59) into (8.57) yields that

$$\begin{aligned}\dot{V}(t) \leq & \alpha\tilde{x}^T\tilde{x}+\alpha\tilde{v}^T\tilde{v}-\tilde{x}^T\gamma\tilde{\rho}\mathscr{L}\tilde{x}-\tilde{v}^T\gamma\tilde{\rho}\mathscr{L}\tilde{v}-\tilde{x}^T\tilde{\rho}\Delta\tilde{x}-\tilde{v}^T\tilde{\rho}\Theta\tilde{v} \\ & +\sum_{i=1}^{l}\tilde{\rho}_i(c_i-\tilde{c}_i)(-\pi_{1i}(t)c_i+\tilde{x}_i^T\tilde{x}_i)+\sum_{i=1}^{l}\tilde{\rho}_i(h_i-\tilde{h}_i)(-\pi_{2i}(t)h_i+\tilde{v}_i^T\tilde{v}_i) \\ & -\sum_{i=1}^{N}\frac{\bar{\varsigma}_i}{2(\underline{\rho}_i-\zeta_{1i}\overline{\rho}_i)}\tilde{\chi}_i^2(t)+\sum_{i=1}^{N}\chi_i^*\iota_i+\sum_{i=1}^{N}\frac{\bar{\varsigma}_i}{2(\underline{\rho}_i-\zeta_{1i}\overline{\rho}_i)}\chi_i^{*2}.\end{aligned} \tag{8.60}$$

Since $\pi_{1i}(t) > 0$ and $\pi_{2i}(t) > 0$, one can obtain

$$\begin{aligned}
&\tilde{\rho}_i(c_i - \tilde{c}_i)(-\pi_{1i}(t)c_i) \\
&= \tilde{\rho}_i\pi_{1i}(t)[-\frac{1}{2}c_i^2 - \frac{1}{2}(c_i - \tilde{c}_i)^2 + \frac{1}{2}\tilde{c}_i^2] \\
&\leq -\frac{1}{2}\tilde{\rho}_i\pi_{1i}(t)(c_i - \tilde{c}_i)^2 + \frac{1}{2}\tilde{\rho}_i\overline{\pi}_{1i}\tilde{c}_i^2
\end{aligned} \tag{8.61}$$

and

$$\tilde{\rho}_i(h_i - \tilde{h}_i)(-\pi_{2i}(t)h_i) \leq -\frac{1}{2}\tilde{\rho}_i\pi_{2i}(t)(h_i - \tilde{h}_i)^2 + \frac{1}{2}\tilde{\rho}_i\overline{\pi}_{2i}\tilde{h}_i^2. \tag{8.62}$$

According to (8.61) and (8.62), (8.60) can be simplified as follows:

$$\begin{aligned}
\dot{V}(t) \leq\, & \tilde{x}^T(\alpha I_N + \gamma\tilde{L} - \tilde{\rho}\tilde{\Delta})\tilde{x} + \tilde{v}^T(\alpha I_N + \gamma\tilde{L} - \tilde{\rho}\tilde{\Theta})\tilde{v} - \frac{1}{2}\sum_{i=1}^{l}\tilde{\rho}_i\pi_{1i}(t)(c_i - \tilde{c}_i)^2 \\
& + \frac{1}{2}\sum_{i=1}^{l}\tilde{\rho}_i\overline{\pi}_{1i}\tilde{c}_i^2 - \frac{1}{2}\sum_{i=1}^{l}\tilde{\rho}_i\pi_{2i}(t)(h_i - \tilde{h}_i)^2 + \frac{1}{2}\sum_{i=1}^{l}\tilde{\rho}_i\overline{\pi}_{2i}\tilde{h}_i^2 \\
& - \sum_{i=1}^{N}\frac{\varsigma_i}{2(\underline{\rho}_i - \zeta_{1i}\overline{\rho}_i)}\tilde{\chi}_i^2(t) + \sum_{i=1}^{N}\chi_i^*\iota_i + \sum_{i=1}^{N}\frac{\varsigma_i}{2(\underline{\rho}_i - \zeta_{1i}\overline{\rho}_i)}\chi_i^{*2}.
\end{aligned} \tag{8.63}$$

The condition (8.22) of Theorem 8.1 holds when $\tilde{c}_i$ and $\tilde{h}_i$ are sufficiently large. Hence, when the condition (8.53) of Theorem 8.2 is satisfied, $\alpha I_N + \gamma\tilde{L} - \tilde{\rho}\tilde{\Delta}$ and $\alpha I_N + \gamma\tilde{L} - \tilde{\rho}\tilde{\Theta}$ are negative definite matrices. Therefore, the following expression can be obtained:

$$\dot{V}(t) \leq -z_1V(t) + z_2 \tag{8.64}$$

where

$$\begin{aligned}
z_1 &= \min\left\{\frac{-\lambda_{\min}(M)}{\lambda_{\max}(\Omega)}, \min_{1\leq i\leq N}\mu_{1i}\pi_{1i}(t), \min_{1\leq i\leq N}\mu_{2i}\pi_{2i}(t), \min_{1\leq i\leq N}\varsigma_i\varphi_i\right\} > 0 \\
z_2 &= \frac{1}{2}\sum_{i=1}^{l}\tilde{\rho}_i\overline{\pi}_{1i}\tilde{c}_i^2 + \frac{1}{2}\sum_{i=1}^{l}\tilde{\rho}_i\overline{\pi}_{2i}\tilde{h}_i^2 + \sum_{i=1}^{N}\frac{\varsigma_i}{2(\underline{\rho}_i - \zeta_{1i}\overline{\rho}_i)}\chi_i^{*2} + \sum_{i=1}^{N}\chi_i^*\iota_i \\
M &= \mathrm{diag}\{\alpha I_N + \gamma\tilde{L} - \tilde{\rho}\tilde{\Delta}, \alpha I_N + \gamma\tilde{L} - \tilde{\rho}\Theta\}.
\end{aligned}$$

The rest of the analysis can be easily done in the form of the proof of Theorem 8.1, and is hence omitted here. The proof is completed. □

Remark 8.11 As opposed to the manual setting of fixed pinning gains in Theorems 8.1 and 8.2 features adaptive pinning gains that are automatically updated to address different scenarios, including external *disturbances*, uncertainties, and actuator faults. Moreover, this chapter attains a more comprehensive form of *cluster consensus*, i.e., dynamic *cluster consensus*, compared to the static version discussed in [13].

8.5 Simulation Examples

In this section, two simulation examples are offered to validate the potency of the two pinning control methods advanced in this chapter. Each of these simulations features a system consisting of five agents ($N = 5$) divided into two distinct subgroups, G_1 and G_2. The topological structure of the MAS is illustrated in Fig. 8.2. The existence of a directed spanning tree in the *directed graph* makes that Assumption 8.1 is hold.

8.5.1 *Fixed Pinning Gain*

The dynamics of the two subgroups G_1 and G_2 are heterogeneous. For the first subgroup G_1, the dynamics of the agents are as follows:

$$\begin{aligned} \dot{x}_i(t) &= v_i(t), \quad i = 1, 3, 4 \\ \dot{v}_i(t) &= -\sin(x_i) - 0.25v_i + 1.5\cos(2.5t) + u_i^F(t) + \sum_{k=1}^{p_i} B_{i,k} w_{i,k}(t) \end{aligned} \tag{8.65}$$

while for the second subgroup G_2, the dynamics of the agents are as follows:

$$\begin{aligned} \dot{x}_i(t) &= v_i(t), \quad i = 2, 5 \\ \dot{v}_i(t) &= 0.5\sin(2t) - 0.5v_i + u_i^F(t) + \sum_{k=1}^{p_i} B_{i,k} w_{i,k}(t) \end{aligned} \tag{8.66}$$

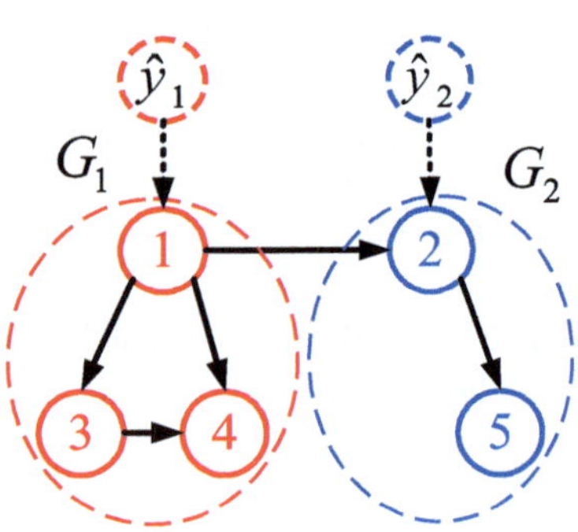

Fig. 8.2 Two subgroup with heterogeneous dynamics

where the nonlinear differentiable functions of the virtual leader $\hat{y}_1$ and $\hat{y}_2$ in the first and second subgroup are described by

$$\begin{aligned} f_1(t, x_1^r, v_1^r) &= -\sin(x_1^r) - 0.25v_1^r + 1.5\cos(2.5t) \\ f_2(t, x_2^r, v_2^r) &= 0.5\sin(2t) - 0.5v_2^r. \end{aligned} \tag{8.67}$$

When Assumption 8.3 holds, one can obtain

$$\begin{aligned} &|f_1(t, x_i, v_i) - f_1(t, x_1^r, v_1^r)| \le |x_i - x_1^r| + 0.25|v_i - v_1^r|, i \in \{1, 3, 4\} \\ &|f_2(t, x_i, v_i) - f_2(t, x_2^r, v_2^r)| \le 0.5|v_i - v_2^r|, i \in \{2, 5\} \end{aligned} \tag{8.68}$$

which implies that $p = 1.625, q = 0.875$, and $\alpha = 1.875$ according to Assumption 8.3.

It can be seen from Fig. 8.2 that the 1st agent in the G_1 subgroup can not receive information from the G_2 subgroup; in addition, its in-degree is zero, namely $i = 1 \in \mathscr{V}_1 \backslash \tilde{\mathscr{V}}_1$ and $\mathscr{D}_1 = 0$. The 2nd agent in the G_2 subgroup can receive information from the G_1 subgroup, namely $i = 2 \in \tilde{\mathscr{V}}_2$. Thus, according to the pinning control strategy and discussions in Remark 8.6, the 1st and 2nd agents can be chosen as the pinning nodes.

The outputs of the IQC are chosen as $w_{i,1} = \sin(\xi_{i,1})$, while the matrix gains of the IQCs are assumed to be

$$B_{i,1} = [-0.5, 0.5i], \quad H_{i,1} = [-1, i]^T, \quad G_{i,1} = [-0.12, 0.12i]^T$$

the time-varying actuator bias faults $r_i(t)$ and actuator efficiency factors $\rho_i(t)$ are as follows:

$$\begin{cases} r_i(t) = 3\sin(0.1i \times t) \\ \rho_1(t) = 1.5 + 0.1\cos(0.1t) \\ \rho_2(t) = 0.8 - 0.1\sin(t), \quad i = 1, 2, \cdots, 5 \\ \rho_3(t) = 0.6 + 0.2\sin(1.2t), \quad 5s \le t < 25s \\ \rho_4(t) = 0.3 + 0.1\sin(1.2t) \\ \rho_5(t) = 0.4 - 0.1\cos(0.5t). \end{cases}$$

According to the definition of $\tilde{\rho}$, the efficiency parameter matrix is $\tilde{\rho} = \text{diag}\{0.57, 0.50, 0.15, 0.35, 0.45\}$. By using the condition (8.22) in Theorem 8.1, one can get $\gamma \ge 41.6213$. Thus, in accordance with the condition (8.23), the coupling strength and fixed pinning gains are chosen as follows: $\gamma = 50$, $d_1 = d_2 = 22$, and $d_3 = d_4 = d_5 = 0$. In addition, the following parameters are chosen: $\iota_i = 0.05$, $\varphi_i = 0.001$, $\varsigma_i(t) = 20e^{-t}$, and $\zeta_{1i} = 0.1\sqrt{1 + i^2}$, where $i = 1, 2, \cdots, 5$.

Figure 8.3a and b demonstrate that the adaptive *fault-tolerant* pinning controller with fixed gains (8.12) as outlined in Theorem 8.1 is capable of achieving *fault-tolerant* cluster consensus for the MAS (8.65)–(8.66) with a *directed graph* that includes a directed spanning tree. This achievement is despite the presence of non-

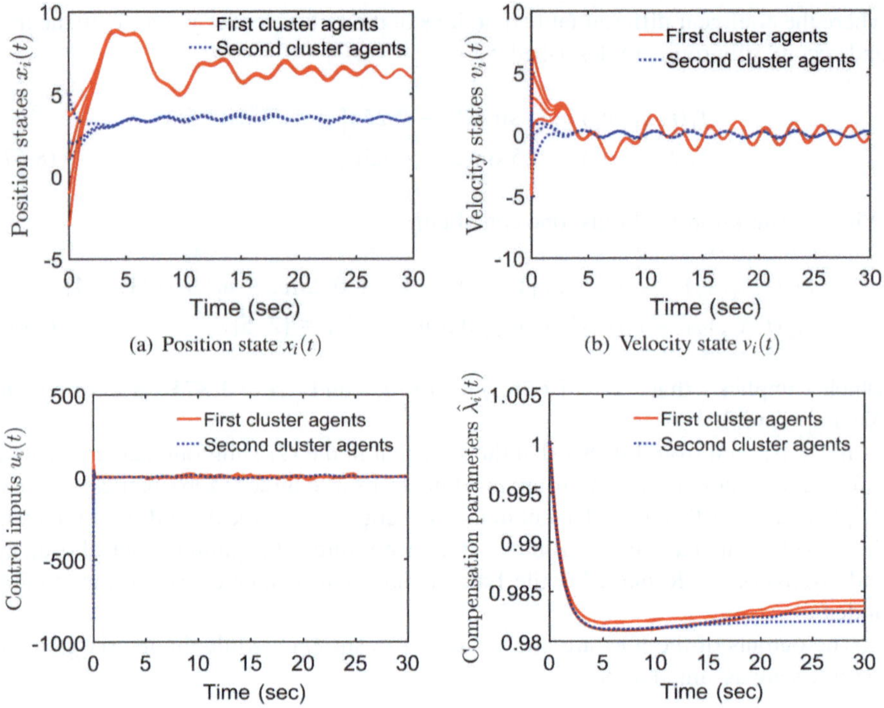

(a) Position state $x_i(t)$ (b) Velocity state $v_i(t)$

(c) Control input $u_i(t)$ (d) Compensation parameter

Fig. 8.3 Simulation results of controller with fixed pinning gains of Theorem 8.1

linear dynamics, IQCs, and actuator faults. Fig. 8.3c displays the trajectories of the control inputs for all agents. The change trajectories of the compensation parameters of the system agents are shown in Fig. 8.3d. It is evident that the compensation parameters converge and stabilize in *finite time*. Moreover, in accordance with the adaptive law (8.19), the convergence of compensation parameter also confirms the convergence of the state error system, and thus confirms the MAS achieves *cluster consensus*.

To gain a clearer understanding of how actuator faults affect the system's stability, Fig. 8.4a and b present, respectively, the position and velocity state trajectories when the pinning leader-following approach of [10] is applied to the same MAS (8.65)–(8.66) under the same influence of nonlinear dynamics, IQCs, and actuator faults. It is observed that the presence of actuator faults and IQCs can lead to system instability. Moreover, by comparing Fig. 8.4 with Fig. 8.3a and b, it becomes evident that the strategy proposed in this chapter offers a more practical solution compared to the pinning leader-following approach of [10].

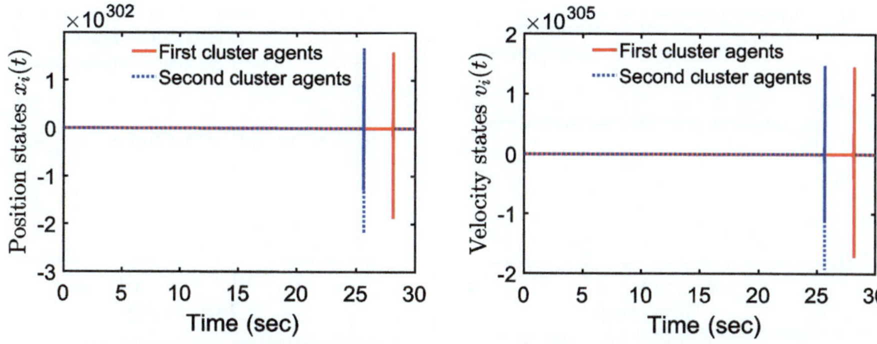

(a) Position state $x_i(t)$ (b) Velocity state $v_i(t)$

Fig. 8.4 Simulation results based on controller of [10]

8.5.2 Adaptive Pinning Gain

In this subsection, the efficacy of the strategy utilizing adaptive pinning control is confirmed. In this example, the MAS and system parameters are consistent with those in the previous example; the first and second agents continue to be designated as the pinning nodes.

The simulation results depicted in Fig. 8.5a and b demonstrate that the adaptive *fault-tolerant* pinning controller (8.12) and (8.19) with (8.51) and the adaptive pinning update law (8.52) of Theorem 8.2 can also achieve *fault-tolerant cluster consensus* for the MAS (8.65)–(8.66) under the effects of nonlinear dynamics, IQCs, and actuator faults. Figure 8.5c and d illustrate the control inputs and compensation parameters for each system agents, respectively. Additionally, Fig. 8.5e and f display the adaptive pinning gain trajectories of the first and second agents, respectively.

By comparing the results in Fig. 8.5 with those in Fig. 8.3, one can discern that the consensus performance of the control strategy of the control approach detailed in Theorem 8.1 is similar to that of Theorem 8.2. This indicates that the adaptive pinning control with variable gains can self-update and maintain a similar level of tracking performance as the fixed pinning control with static gains.

For comparative analysis, the state consensus error index is applied to evaluate the *fault-tolerant* control scheme discussed in this chapter against the scheme reported in [19], and the comparison result illustrated in Fig. 8.6.

$$J(t) = \frac{1}{N}\sqrt{\sum_{i=1}^{N}\left(x_i(t) - x^r_{\sigma_i}(t)\right)^2 + \sum_{i=1}^{N}\left(v_i(t) - v^r_{\sigma_i}(t)\right)^2}.$$

It should be observed that while both control schemes are capable of achieving *cluster consensus*, the *fault-tolerant* control strategy designed in this chapter outperforms the method from [19] in terms of state consensus performance.

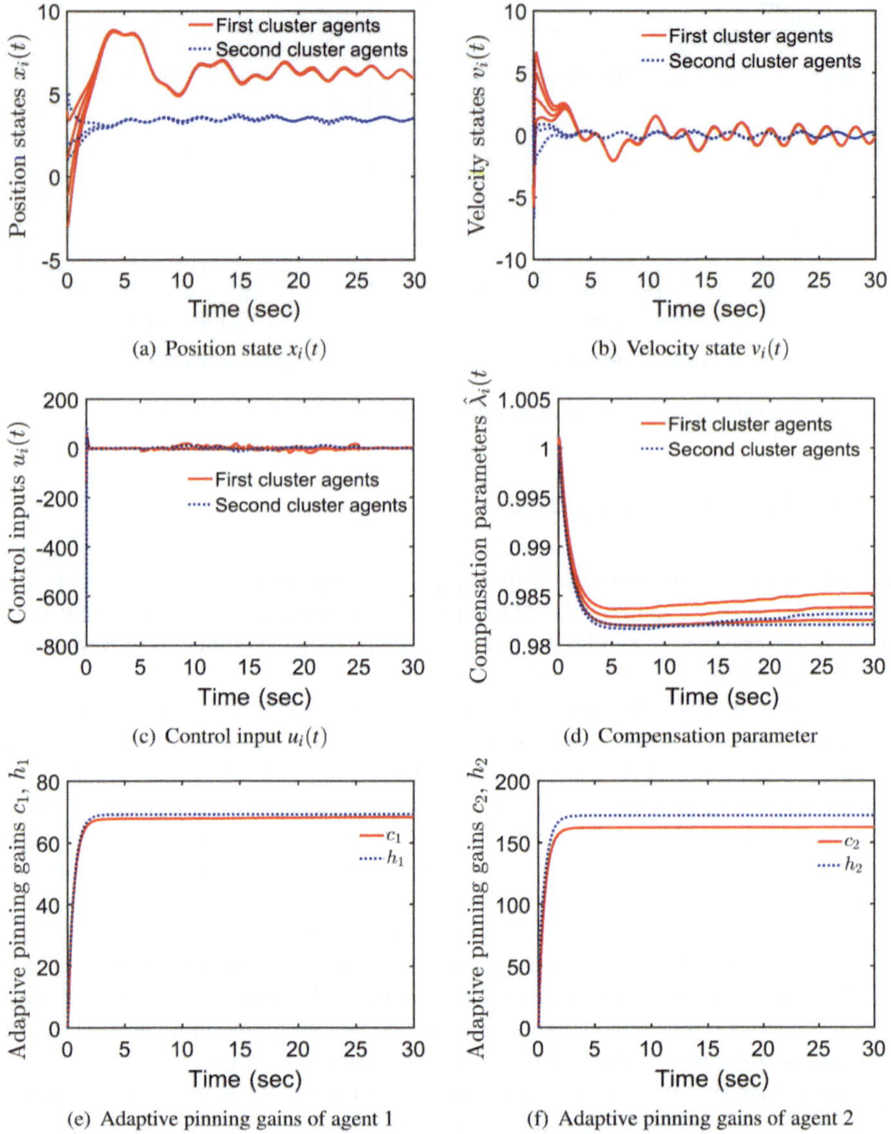

(a) Position state $x_i(t)$

(b) Velocity state $v_i(t)$

(c) Control input $u_i(t)$

(d) Compensation parameter

(e) Adaptive pinning gains of agent 1

(f) Adaptive pinning gains of agent 2

Fig. 8.5 Simulation results of controller with adaptive pinning gains of Theorem 8.2

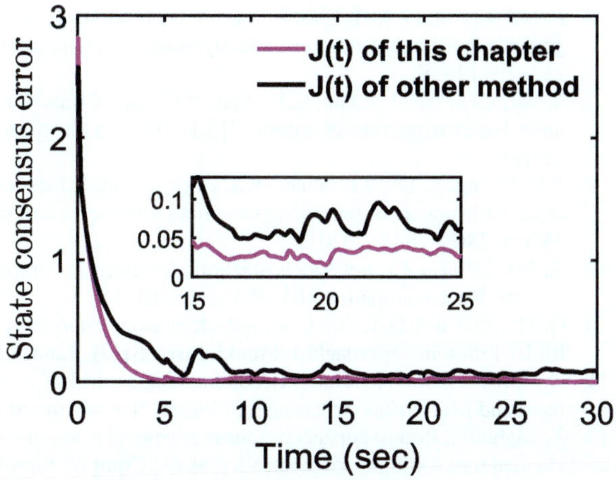

Fig. 8.6 Comparison between the fault-tolerant controller in Theorem 8.2 and the method in [19]

8.6 Conclusion

This chapter proposes two adaptive *fault-tolerant* pinning control strategies, one with fixed gains and the other with adaptive gains, for heterogeneous MASs organized under a *directed graph* that possesses a directed spanning tree. Both strategies are designed to achieve *cluster consensus* for the MASs within a *finite time* frame, notwithstanding the presence of actuator faults and IQCs. These strategies involve a single parameter that necessitates estimation. Furthermore, the chapter establishes two sufficient conditions to guarantee that both position and velocity errors remain bounded. The chapter concludes with simulation examples that underscore the efficacy and benefits of the proposed control strategies. The subsequent chapter will focus on the pursuit of cluster consensus under adversarial conditions.

References

1. R. Agha, M. Rehan, C.K. Ahn, G. Mustafa, S. Ahmad, Adaptive distributed consensus control of one-sided Lipschitz nonlinear multiagents. IEEE Trans. Syst. Man Cybern. Syst. **49**(3), 568–579 (2019)
2. X.G. Guo, P.M. Liu, H.J. Li, J.L. Wang, C.K. Ahn, Cluster synchronization of heterogeneous nonlinear multi-agent systems with actuator faults and IQCs through adaptive fault-tolerant pinning control. Inf. Sci. **575**, 289–305 (2021)
3. A.H. Hu, J.H. Park, J.D. Cao, Group consensus of multi-agent networks with hybrid interactions. Neurocomputing **404**, 267–275 (2020)
4. X.B. Li, Z.H. Yu, Z.W. Li, N.Q. Wu, Group consensus via pinning control for a class of heterogeneous multi-agent systems with input constraints. Inf. Sci. **542**, 247–262 (2021)
5. E. Shahamatkhah, M. Tabatabaei, Leader-following group consensus of discrete-time fractional-order double-integrator multi-agent systems. ISA Trans. **106**, 262–270 (2020)

6. L. Shi, J.L. Shao, M.T. Cao, H. Xia, Asynchronous group consensus for discrete-time heterogeneous multi-agent systems under dynamically changing interaction topologies. Inf. Sci. **463**, 282–293 (2018)
7. Y. Xu, Z.G. Wu, Y.J. Pan, C.K. Ahn, H.C. Yan, Consensus of linear multiagent systems with input-based triggering condition. IEEE Trans. Syst. Man Cybern. Syst. **49**(11), 2308–2317 (2019)
8. Y.H. Cheng, L. Shi, J.L. Shao, W.X. Zheng, Sampled-data scaled group consensus for second-order multi-agent systems with switching topologies and random link failures. J. Frankl. Inst. **357**(5), 2868–2881 (2020)
9. Q. Ma, J.W. Lu, Cluster synchronization for directed complex dynamical networks via pinning control. Neurocomputing **101**, 354–360 (2013)
10. Q. Ma, Z. Wang, G.Y. Miao, Second-order group consensus for multi-agent systems via pinning leader-following approach. J. Frankl. Inst. **351**(3), 1288–1300 (2014)
11. D.M. Xie, L. Shi, F.C. Jiang, Group tracking control of second-order multi-agent systems with fixed and Markovian switching topologies. Neurocomputing **281**, 37–46 (2018)
12. Z. Yaghoubi, Robust cluster consensus of general fractional-order nonlinear multi agent systems via adaptive sliding mode controller. Math. Comput. Simul. **172**, 15–32 (2020)
13. X.G. Guo, H.J. Li, J.J. Zhao, W.W. Che, Static group synchronization of second-order multi-agent systems via pinning control. J. Frankl. Inst. **356**(9), 4842–4858 (2019)
14. X.Z. Jin, T. He, J.W. Xia, D. Wang, W. Guan, Adaptive general pinned synchronization of a class of disturbed complex networks. Commun. Nonlinear Sci. Numer. Simul. **67**, 658–669 (2019)
15. J.Y. Yu, Y. Shi, Scaled group consensus in multiagent systems with first/second-order continuous dynamics. IEEE Trans. Cybern. **48**(8), 2259–2271 (2018)
16. J.Y. Zhan, X. Li, Cluster consensus in networks of agents with weighted cooperative-competitive interactions. IEEE Trans. Circuits Syst. II, Exp. Briefs **65**(2), 241–245 (2017)
17. C. Deng, G.H. Yang, Distributed adaptive fault-tolerant control approach to cooperative output regulation for linear multi-agent systems. Automatica **103**, 62–68 (2019)
18. X. Wang, G.H. Yang, Fault-tolerant consensus tracking control for linear multiagent systems under switching directed network. IEEE Trans. Cybern. **50**(5), 1921–1930 (2020)
19. Y.J. Wang, Y.D. Song, F.L. Lewis, Robust adaptive fault-tolerant control of multiagent systems with uncertain nonidentical dynamics and undetectable actuation failures. IEEE Trans. Ind. Electron. **62**(6), 3978–3988 (2015)
20. C.L. Wang, C.Y. Wen, L. Guo, Adaptive consensus control for nonlinear multi-agent systems with unknown control directions and time-varying actuator faults. IEEE Trans. Autom. Control **66**(9), 4222–4229 (2021)
21. Y. Xu, Z.G. Wu, Distributed adaptive event-triggered fault-tolerant synchronization for multi-agent systems. IEEE Trans. Ind. Electron. **68**(2), 1537–1547 (2021)
22. L. Zhao, G.H. Yang, End to end communication rate-based adaptive fault tolerant control of multi-agent systems under unreliable interconnections. Inf. Sci. **460**, 331–345 (2018)
23. L. Zhao, G.H. Yang, Adaptive fault-tolerant control for nonlinear multi-agent systems with DoS attacks. Inf. Sci. **526**, 39–53 (2020)
24. W. Zou, C.K. Ahn, Z. Xiang, Fuzzy-approximation-based distributed fault-tolerant consensus for heterogeneous switched nonlinear multiagent systems. IEEE Trans. Fuzzy Syst. **29**(10), 2916–2925 (2021)
25. X.G. Guo, C.K. Ahn, Adaptive fault-tolerant pseudo-PID sliding-mode control for high-speed train with integral quadratic constraints and actuator saturation. IEEE Trans. Intell. Transp. Syst. **22**(12), 7421–7431 (2021)
26. X.G. Guo, J.L. Wang, F. Liao, R.S.H. Teo, Distributed adaptive sliding mode control strategy for vehicle-following systems with nonlinear acceleration uncertainties. IEEE Trans. Veh. Technol. **66**(2), 981–991 (2017)
27. M. Mohammadi, M.M. Arefi, P. Setoodeh, O. Kaynak, Optimal tracking control based on reinforcement learning value iteration algorithm for time-delayed nonlinear systems with external disturbances and input constraints. Inf. Sci. **554**, 84–98 (2021)

28. W.C. Zou, Z.R. Xiang, C.K. Ahn, Mean square leader–following consensus of second-order nonlinear multiagent systems with noises and unmodeled dynamics. IEEE Trans. Syst. Man Cybern. Syst. **49**(12), 2478–2486 (2019)
29. X. Jin, S. Wang, J. Qin, W.X. Zheng, Y. Kang, Adaptive fault-tolerant consensus for a class of uncertain nonlinear second-order multi-agent systems with circuit implementation. IEEE Trans. Circuits Syst. Regul. Pap. **65**(7), 2243–2255 (2018)
30. A. Megretski, A. Rantzer, System analysis via integral quadratic constraints. IEEE Trans. Autom. Control **42**(6), 819–830 (1997)
31. Y.Z. Feng, S.Y. Xu, B.Y. Zhang, Group consensus control for double-integrator dynamic multiagent systems with fixed communication topology. Int. J. Robust Nonlinear Control **24**(3), 532–547 (2014)
32. X. Wu, Y. Liu, J. Zhou, Pinning adaptive synchronization of general time-varying delayed and multi-linked networks with variable structures. Neurocomputing **147**, 492–499 (2015)
33. X.Y. Li, J.Z. Wang, Fault-tolerant tracking control for a class of nonlinear multi-agent systems. Syst. Control Lett. **135**, Art. No. 104576 (2020)
34. Q. Song, Y.D. Song, Data-based fault-tolerant control of high-speed trains with traction/braking notch nonlinearities and actuator failures. IEEE Trans. Neural Netw. **22**(12), 2250–2261 (2011)
35. Q. Song, Y.D. Song, PI-like fault-tolerant control of nonaffine systems with actuator failures. Acta Automatica Sinica **38**(6), 1033–1040 (2012)
36. A. Theissler, Detecting known and unknown faults in automotive systems using ensemble-based anomaly detection. Knowl. Based Syst. **123**, 163–173 (2017)
37. J. Back, J. Kim, A disturbance observer based practical coordinated tracking controller for uncertain heterogeneous multi-agent systems. Int. J. Robust Nonlinear Control **25**(14), 2254–2278 (2015)
38. X.J. Li, G.H. Yang, FLS-based adaptive synchronization control of complex dynamical networks with nonlinear couplings and state-dependent uncertainties. IEEE Trans. Cybern. **46**(1), 171–180 (2016)
39. A. Das, F.L. Lewis, Cooperative adaptive control for synchronization of second-order systems with unknown nonlinearities. Int. J. Robust Nonlinear Control **21**(13), 1509–1524 (2011)
40. Y.J. Zhou, X.H. Yu, C.Y. Sun, W.W. Yu, Robust synchronisation of second-order multi-agent system via pinning control. IET Contr. Theory Appl. **9**(5), 775–783 (2015)
41. J. Long, W. Wang, J.S. Huang, J. Zhou, K.X. Liu, Distributed adaptive control for asymptotically consensus tracking of uncertain nonlinear systems with intermittent actuator faults and directed communication topology. IEEE Trans. Cybern. **51**(8), 4050–4061 (2021)

Chapter 9
Intrusion-Tolerant Cluster Consensus of MASs Under Periodic DoS Attacks

Abstract In this chapter, an event-triggered pinning control scheme for *cluster consensus* is introduced, designed for nonlinear *multi-agent systems* (MASs) with a directed topology, under the influence of periodic denial-of-service (DoS) attacks. This scheme eliminates the requirement for the in-degree balance condition in MASs. Subsequently, the dynamics of the state error under the periodic DoS attacks are formulated. Additionally, an event-triggered mechanism (ETM) that is based on sampled data is employed to reduce unnecessary resource expenditure by the system.

9.1 Introduction

Cluster consensus, a unique form of consensus behavior, signifies that agents within the same cluster will ultimately achieve consensus, whereas agents in distinct clusters will follow distinct state trajectories [1, 2]. Given the broad applicability of *cluster consensus* in various practical scenarios, there has been a surge in research on clustering control methods for *multi-agent systems* (MASs) [1, 3–6]. However, many of these clustering control strategies have typically introduced the in-degree balance condition to simplify consensus analysis [4–6], which may limit the applicability of these strategies to real-world systems. Consequently, for MASs that do not satisfy the in-degree balance condition, there is a pressing need to devise a control approach that guarantees the achievement of *cluster consensus*.

As discussed in Chap. 4, interconnect a variety of infrastructures via wired or wireless networks, enhancing the convenience of physical processes and human-computer interaction [7]. However, the incorporation of wireless networks compromises the closed nature of traditional control systems. Consequently, the system becomes vulnerable to malicious *cyber attacks* from the networks [8]. Denial-of-service (DoS) attacks are a prevalent form of *cyber attacks* [9–12], with the objective of severing communication channels and obstructing signal transmission, ultimately leading to a degradation or loss of *cluster consensus* performance. Different from the observer-based consensus control scheme in Chap. 4, this chapter employs linear matrix inequality (LMI) technique to ensure the *cluster consensus* of MASs

X.-G. Guo et al., *Anti-disturbance Intrusion- and Fault-Tolerant Control of Multi-agent Systems*, https://doi.org/10.1007/978-981-96-9614-7_9

performance during *cyber attacks*. On the other hand, the existing results on *cluster consensus* are predicated on an ideal network environment [1, 3–6]. Moreover, the sampling control scheme can lead to excessive consumption of the system's computing and communication resources, particularly when the system achieve *cluster consensus* [13]. The event-triggered mechanism (ETM), which selects the time instants of control updates and sampling actions based on pre-specified trigger conditions, offers a solution to reduce such resource consumption [14–16]. Therefore, it is of significant value to investigate the issue of secure *cluster consensus* based on the ETM under the threat of DoS attacks.

Following the preceding analysis, this chapter focuses on the secure *cluster consensus* problem of nonlinear MASs under periodic DoS attacks, and introduces an event-triggered pinning control method. The key contributions of this chapter are summarized as:

(1) Periodic DoS Attacks: This chapter establishes the error system for MASs affected by periodic DoS attacks and provides a secure *cluster consensus* criterion under this *cybre attacks.* In addition, different from the standard cluster consensus analysis method in Chap. 8, it is a great challenge to use the delay-based analysis method to study *cluster consensus* in this chapter.
(2) Pinning controller: Different from the time-based communication in Chap. 8, this chapter proposes a centralized ETM under the cluster consensus framework to save communication resources. Furthermore, this chapter provides a method for designing the ETM parameter matrices using the LMI technique. These contributions help address the challenges posed by periodic DoS attacks in maintaining *cluster consensus* performance, ensuring both system security and resource efficiency.

9.2 Problem Formulation

9.2.1 Second-Order MAS Model

Consider a nonlinear MAS with second-order dynamics consisting of N follower agents and s distinct virtual leader agents. In addition, these follower agents are divided into s disjoint clusters. The dynamic of follower agent i is expressed as

$$\begin{aligned} \dot{x}_i(t) &= v_i(t) \\ \dot{v}_i(t) &= f_{\sigma_i}(t, x_i, v_i) + u_i(t). \end{aligned} \tag{9.1}$$

Furthermore, the leader agent of cluster σ_i is denoted by

$$\begin{aligned} \dot{x}^r_{\sigma_i}(t) &= v^r_{\sigma_i}(t) \\ \dot{v}^r_{\sigma_i}(t) &= f_{\sigma_i}(t, x_{\sigma_i}, v_{\sigma_i}) \end{aligned} \tag{9.2}$$

where $x_i(t) \in \mathbb{R}$ and $v_i(t) \in \mathbb{R}$ are the states of of follower agent i; $x^r_{\sigma_i}(t) \in \mathbb{R}$ and $v^r_{\sigma_i}(t) \in \mathbb{R}$ are the states of the leader agent of cluster σ_i; $u_i(t)$ is the control input; and $f_{\sigma_i}(t, \cdot, \cdot)$ denotes a nonlinear function.

Assumption 9.1 The Lipschitz condition is holds for the nonlinear function $f_{\sigma_i}(t, \cdot, \cdot)$ in (9.1) and (9.2), i.e., there are $\rho_i > 0$ and $\rho_i > 0$ such that the following condition holds:

$$|f_i(t, x^r_{\sigma_i}, v^r_{\sigma_i}) - f_i(t, x_i, v_i)| \le \rho_i |x^r_{\sigma_i} x_i| + \rho_i |v^r_{\sigma_i} - v_i| \tag{9.3}$$

where $\forall x_i, v_i, x^r_{\sigma_i}, v^r_{\sigma_i} \in \mathbb{R}$.

Assumption 9.2 ([17]) The directed communication structure of the MAS (9.1) encompasses a directed spanning tree.

9.2.2 *Periodic DoS Attack Model and ETM*

As a prevalent form of *cyber attacks*, the objective of DoS attacks is to sever communication and obstruct the conveyance of signals among adjacent agents. In this chapter, we will focus on the following model for periodic DoS attacks:

$$\mathscr{S}_{\text{DoS}} = \begin{cases} 1, & t \in [nT, nT + T_{\text{off}}) \\ 0, & t \in [nT + T_{\text{off}}, (n+1)T) \end{cases} \tag{9.4}$$

where the sequence $\{nT\}$ denotes the moments when the DoS attacks cease, and T_{off} represents the interval during which the DoS attacks are inactive. Within the interval $[nT, nT + T_{\text{off}})$, $\mathscr{S}_{\text{DoS}} = 1$ signifies that the network environment is functioning normally, allowing for the normal transmission of signals. Conversely, for $t \in [nT + T_{\text{off}}, (n+1)T)$, $\mathscr{S}_{\text{DoS}} = 0$ indicates that all signal transmissions are halted, resulting in the actual control input being effectively zero.

In this chapter, it is assumed that T and T_{off} are ascertainable, with the dormant period T_{off} being permitted to fluctuate over time. Moreover, a lower bound is presumed to exist for $T_{\text{off}} \ge T^{\min}_{\text{off}} > 0$.

In this chapter, a synthetical ETM based on the sampled data is presented to mitigate both communication burden and the effects of DoS attacks. The ETM is proposed as:

$$t^h_{k+1,n} = \inf_{\varphi > 0} \{t^h_{k,n} + \varphi h \in [nT, nT + T_{\text{off}}) | \mathscr{H}(t^h_{k,n} + \varphi h) \ge 0\} \cup \{nT\} \tag{9.5}$$

$$\begin{aligned} \mathscr{H}(t) = {} & [\tilde{x}(t) - \tilde{x}(t^h_{k,n})]^T W_1 [\tilde{x}(t) - \tilde{x}(t^h_{k,n})] - \omega_1 \tilde{x}^T(t^h_{k,n}) W_1 \tilde{x}^T(t^h_{k,n}) \\ & + [\tilde{v}(t) - \tilde{v}(t^h_{k,n})]^T W_2 [\tilde{v}(t) - \tilde{v}(t^h_{k,n})] \\ & - \omega_2 \tilde{v}^T(t^h_{k,n}) W_2 \tilde{v}^T(t^h_{k,n}) \end{aligned} \tag{9.6}$$

where

$$\tilde{x}(t) = [\tilde{x}_1(t), \tilde{x}_2(t), \cdots, \tilde{x}_N(t)], \quad \tilde{x}_i(t) = x_i(t) - x^r_{\sigma_i}(t)$$
$$\tilde{v}(t) = [\tilde{v}_1(t), \tilde{v}_2(t), \cdots, \tilde{v}_N(t)], \quad \tilde{v}_i(t) = v_i(t) - v^r_{\sigma_i}(t).$$

Furthermore, $\varphi \in \mathbb{N}$; $t^h_{k,n}$ presents the kth triggering instant in the interval $[nT, nT + T_{\text{off}})$; $\mathscr{H}(t)$ serves as a conditional function; $0 < h < T$ represents the sampling interval; W_1 and W_2 are positive definite matrices that need to be designed, and $\omega_1 \in (0, 1)$ and $\omega_2 \in (0, 1)$ are pre-specified parameters.

Employing the designed sampled-data-based ETM (9.6), the issue of Zeno behavior is inherently circumvented.

The purpose of this chapter is to establish a pinning-based event-triggered control strategy, aimed at the achievement of *cluster consensus* in an MAS with a directed topology under periodic DoS attacks.

9.3 Event-Triggered Pinning Control Strategy

In this section, the development of a pinning controller is undertaken to fulfill the control objectives. Building upon this, the state error system is established, and an analysis of the exponential stability of this system is conducted. Furthermore, a solution for determining the ETM parameter matrices is presented using the LMI approach.

9.3.1 *Design of Pinning Controller*

The pinning control law is proposed as

$$u_i(t) = \begin{cases} u_{1i}(t^h_{k,n}) + u_{2i}(t^h_{k,n}), & t \in [t^h_{k,n}, t^h_{k+1,n}) \cap [nT, nT + T_{\text{off}}) \\ 0, & t \in [nT + T_{\text{off}}, (n+1)T) \end{cases} \tag{9.7}$$

$$u_{1i}(t^h_{k,n}) = -\gamma \sum_{j=1}^{N} l_{ij} x_j(t^h_{k,n}) - \gamma \sum_{j=1}^{N} l_{ij} v_j(t^h_{k,n}) \tag{9.8}$$

$$u_{2i}(t^h_{k,n}) = \begin{cases} -b_i \tilde{x}_i(t^h_{k,n}) - b_i \tilde{v}_i(t^h_{k,n}) + \mathscr{U}_i, & i \in \tilde{\mathscr{V}}_{\sigma_i} \\ -b_i \tilde{x}_i(t^h_{k,n}) - b_i \tilde{v}_i(t^h_{k,n}), & \mathscr{D}_i = 0 \\ 0, & \text{otherwise} \end{cases} \tag{9.9}$$

where

$$\mathscr{U}_i = \gamma \sum_{j=1}^{N} l_{ij} x_{\sigma_i}^r (t_{k,n}^h) + \gamma \sum_{j=1}^{N} l_{ij} v_{\sigma_i}^r (t_{k,n}^h)$$

$$k(n) = \max\{k \in \mathbb{N} | t_{k,n}^h < nT + T_{\text{off}}\}.$$

In addition, $\gamma > 0$ presents the control gain, and $k \in \mathscr{K}(n) \triangleq \{0, 1, \cdots, k(n)\}$.

Remark 9.1 It should be noted that in the pinning control signal $u_{2i}(t_{k,n}^h)$, there are two categories of nodes that require pinning. The first category includes the root node of the topology graph, which is characterized by $\mathscr{D}_i = 0$. The second category comprises nodes capable of receiving information from nodes in other clusters, denoted by $i \in \tilde{\mathscr{V}}_{\sigma_i}$.

Remark 9.2 Unlike [4–6], the clustering control scheme delineated in this chapter effectively removes the constraint that the system (9.1) must satisfy the in-degree balance condition by introducing the term $\mathscr{U}_i$. This approach allows for greater flexibility and applicability of the control strategy in scenarios where the traditional in-degree balance condition may not be met or is difficult to satisfy, thereby enhancing the system's robustness and adaptability in the face of complex network interactions and potential disruptions.

9.3.2 State Error System

To facilitate the explanation, Let

$$\mathscr{R}_{k,n} \triangleq [t_{k,n}^h, t_{k+1,n}^h), \quad \mathscr{P}_{1,n} \triangleq [nT, nT + T_{\text{off}})$$

$$\mathscr{P}_{2,n} \triangleq [nT + T_{\text{off}}, (n+1)T), \quad k \in \mathscr{K}(n).$$

Then, $\mathscr{R}_{k,n}$ can be reformulated as

$$\mathscr{R}_{k,n} = \bigcup_{\nu=1}^{\delta_{k,n}} [t_{k,n}^h + (\nu - 1)h, t_{k,n}^h + \nu h) \bigcup [t_{k,n}^h + \delta_{k,n} h, t_{k+1,n}^h)$$

$$\delta_{k,n} = \sup\{\nu \in \mathbb{N} | t_{k,n}^h + \nu h < t_{k+1,n}^h\}.$$

Next, let

$$\begin{aligned}
\mathscr{L}_{k,n}^{\nu} &= [t_{k,n}^h + (\nu - 1)h, t_{k,n}^h + \nu h) \\
\mathscr{L}_{k,n}^{\delta_{k,n}} &= [t_{k,n}^h + \delta_{k,n} h, t_{k+1,n}^h) \\
\Pi_{k,n}^{\nu} &\triangleq \mathscr{L}_{k,n}^{\nu} \bigcap \mathscr{P}_{1,n}.
\end{aligned}$$

Subsequently, it follows that

$$\mathscr{P}_{1,n} = \bigcup_{k=0}^{k(n)} \bigcup_{\nu=1}^{\delta_{k,n}+1} \Pi_{k,n}^{\nu}.$$

Further, we define three separate piecewise functions

$$\eta_{k,n}(t) = \begin{cases} t - t_{k,n}^{h}, & t \in \Pi_{k,n}^{1} \\ t - t_{k,n}^{h} - h, & t \in \Pi_{k,n}^{2} \\ \vdots \\ t - t_{k,n}^{h} - \delta_{k,n} h, & t \in \Pi_{k,n}^{\delta_{k,n}+1} \end{cases} \tag{9.10}$$

$$\varkappa_{i,k,n}(t) = \begin{cases} 0, & t \in \Pi_{k,n}^{1} \\ \tilde{x}_i(t_{k,n}^{h}) - \tilde{x}_i(t_{k,n}^{h} + h), & t \in \Pi_{k,n}^{2} \\ \vdots \\ \tilde{x}_i(t_{k,n}^{h}) - \tilde{x}_i(t_{k,n}^{h} + \delta_{k,n} h), & t \in \Pi_{k,n}^{\delta_{k,n}+1} \end{cases} \tag{9.11}$$

$$\kappa_{i,k,n}(t) = \begin{cases} 0, & t \in \Pi_{k,n}^{1} \\ \tilde{v}_i(t_{k,n}^{h}) - \tilde{v}_i(t_{k,n}^{h} + h), & t \in \Pi_{k,n}^{2} \\ \vdots \\ \tilde{v}_i(t_{k,n}^{h}) - \tilde{v}_i(t_{k,n}^{h} + \delta_{k,n} h), & t \in \Pi_{k,n}^{\delta_{k,n}+1}. \end{cases} \tag{9.12}$$

Based on these piecewise functions, the sampled state error $\tilde{x}_i(t_{k,n})$ and $\tilde{v}_i(t_{k,n})$ at the triggering instant can be represented as

$$\begin{aligned} \tilde{x}_i(t_{k,n}) &= \tilde{x}_{i,k,n}(t) + \varkappa_{i,k,n}(t) \\ \tilde{v}_i(t_{k,n}) &= \tilde{v}_{i,k,n}(t) + \kappa_{i,k,n}(t) \end{aligned} \tag{9.13}$$

where

$$\begin{aligned} \tilde{x}_{i,k,n}(t) &= \tilde{x}_i(t - \eta_{k,n}(t)) \\ \tilde{v}_{i,k,n}(t) &= \tilde{v}_i(t - \eta_{k,n}(t)), \quad t \in \mathscr{R}_{k,n} \bigcap \mathscr{P}_{1,n}. \end{aligned}$$

Then, define the following notations:

$$\begin{aligned} \tilde{x}_{k,n}(t) &= [\tilde{x}_{1,k,n}(t), \tilde{x}_{2,k,n}(t), \cdots, \tilde{x}_{N,k,n}(t)] \\ \tilde{v}_{k,n}(t) &= [\tilde{v}_{1,k,n}(t), \tilde{v}_{2,k,n}(t), \cdots, \tilde{v}_{N,k,n}(t)] \\ \varkappa_{k,n}(t) &= [\varkappa_{1,k,n}(t), \varkappa_{2,k,n}(t), \cdots, \varkappa_{N,k,n}(t)] \\ \kappa_{k,n}(t) &= [\kappa_{1,k,n}(t), \kappa_{2,k,n}(t), \cdots, \kappa_{N,k,n}(t)]. \end{aligned}$$

Based on the above definitions, the following state error system is established

$$\begin{cases} \dot{\tilde{x}}_i(t) &= \begin{cases} \tilde{v}_i, & t \in \mathscr{R}_{k,n} \bigcap \mathscr{P}_{1,n} \\ \tilde{v}_i, & t \in \mathscr{P}_{2,n} \end{cases} \\ \dot{\tilde{v}}_i(t) &= \begin{cases} f_{\sigma_i}(t, x_i, v_i) - f_{\sigma_i}(t, x^r_{\sigma_i}, v^r_{\sigma_i}) + \tilde{\mathscr{J}}_i, & t \in \mathscr{R}_{k,n} \bigcap \mathscr{P}_{1,n} \\ f_{\sigma_i}(t, x_i, v_i) - f_{\sigma_i}(t, x^r_{\sigma_i}, v^r_{\sigma_i}), & t \in \mathscr{P}_{2,n} \end{cases} \\ \tilde{x}_i(t) &= \mathscr{I}_x(t), \quad t \in [-h, 0) \\ \tilde{v}_i(t) &= \mathscr{I}_v(t), \quad t \in [-h, 0) \end{cases} \tag{9.14}$$

where

$$\begin{aligned} \tilde{\mathscr{J}}_i = &-\gamma \sum_{j=1}^{N} l_{ij}(\tilde{x}_{j,k,n}(t) + \varkappa_{j,k,n}(t)) - \gamma \sum_{j=1}^{N} l_{ij}(\tilde{v}_{j,k,n}(t) + \kappa_{j,k,n}(t)) \\ &- b_i(\tilde{x}_{j,k,n}(t) + \varkappa_{j,k,n}(t)) - b_i(\tilde{v}_{j,k,n}(t) + \kappa_{j,k,n}(t)) \end{aligned} \tag{9.15}$$

$\mathscr{I}_x(t)$ and $\mathscr{I}_v(t)$ are *initial conditions* with $\mathscr{I}_x(0) \triangleq \mathscr{I}_{x0}$ satisfying $\mathscr{I}_v(0) \triangleq \mathscr{I}_{v0}$.

9.3.3 Cluster Consensus Analysis

According to [12], the Lyapunov function is defined as

$$V(t) = \begin{cases} V_1(t), & t \in [\mathscr{C}_{1,n}, \mathscr{C}_{2,n}) \\ V_2(t), & t \in [\mathscr{C}_{2,n}, \mathscr{C}_{1,n+1}) \end{cases} \tag{9.16}$$

where $\mathscr{C}_{1,n} = nT, \mathscr{C}_{2,n} = nT + T_{\text{off}}$,

$$\begin{aligned} V_\varsigma(t) = &[\tilde{x}^T(t), \tilde{v}^T(t)]\Omega_\varsigma[\tilde{x}^T(t), \tilde{v}^T(t)]^T + \int_{t-h}^{t} \tilde{x}^T(\tau) f_\varsigma(\tau) S_\varsigma \tilde{x}(\tau)\, d\tau \\ &+ \int_{t-h}^{t} \tilde{v}^T(\tau) f_\varsigma(\tau) H_\varsigma \tilde{v}(\tau)\, d\tau + \int_{-h}^{0}\int_{t+\tau}^{t} \dot{\tilde{x}}^T(\theta) f_\varsigma(\theta)(R_\varsigma + Z_\varsigma)\dot{\tilde{x}}(\theta)\, d\theta\, d\tau \\ &+ \int_{-h}^{0}\int_{t+\tau}^{t} \dot{\tilde{v}}^T(\theta) f_\varsigma(\theta)(P_\varsigma + Q_\varsigma)\dot{\tilde{v}}(\theta)\, d\varpi\, d\tau \end{aligned} \tag{9.17}$$

where $\varsigma \in \{1, 2\}$, $\tilde{x}(t) = [\tilde{x}_1(t), \tilde{x}_2(t), \cdots, \tilde{x}_N(t)]$, $\tilde{v} = [\tilde{v}_1(t), \tilde{v}_2(t), \cdots, \tilde{v}_N(t)]$, $S_\varsigma, H_\varsigma, R_\varsigma, Z_\varsigma, P_\varsigma, Q_\varsigma$ are positive define matrices, $\alpha_\varsigma > 0$,

$$f_1(\cdot) = e^{-2\alpha_1(t-\cdot)}, \quad f_2(\cdot) = e^{2\alpha_2(t-\cdot)}$$

$$\mathscr{B} = \text{diag}\{b_1, b_2, \ldots, b_l, 0, \ldots, 0\}$$

$$\Omega_1 = \begin{bmatrix} \gamma(\mathscr{L} + \mathscr{L}^T) + 2\mathscr{B} & I_N \\ I_N & I_N \end{bmatrix}, \quad \Omega_2 = \psi \begin{bmatrix} I_N & 0 \\ 0 & I_N \end{bmatrix}.$$

It should be observed that the Lyapunov functions $V(t)$ may exhibit discontinuities at the boundaries, and at these points, the derivative of the function is taken as the right-hand derivative.

To facilitate an intuitive analysis of the cluster consensus, we present the following lemma.

Lemma 9.1 *Given an MAS (9.1) that adheres to Assumptions 9.1 and 9.2, and considering a DoS attack model $\mathscr{S}_{\text{DoS}}$ with parameters T and T_{off} that are known, let's employ the pinning control law (9.7) and the ETM (9.6). For $\alpha_\varsigma > 0$, $\omega_\varsigma \in (0, 1)$, and $h \in (0, T_{\text{off}})$, suppose there are positive definite matrices H_ς, S_ς, Z_ς, W_ς, P_ς, R_ς, Q_ς, and matrices A_ς, U_ς, N_ς, X_ς, M_ς, and Y_ς, where $\varsigma \in \{1, 2\}$, that satisfy the subsequent inequalities*

$$\Delta_1 = \begin{bmatrix} \Delta_1^{11} & \Delta_1^{12} & \sqrt{h}\Psi_1^T(P_1 + Q_1) \\ * & \Delta_1^{22} & \boldsymbol{0} \\ * & * & -(P_1 + Q_1) \end{bmatrix} \tag{9.18}$$

$$\Delta_2 = \begin{bmatrix} \Delta_2^{11} & \Delta_2^{12} & \sqrt{h}\Psi_2^T(P_2 + Q_2) \\ * & \Delta_2^{22} & \boldsymbol{0} \\ * & * & -(P_2 + Q_2) \end{bmatrix} \tag{9.19}$$

where

$$\Delta_1^{11} = \Lambda_{11} + \Lambda_{12} + \Lambda_{12}^T$$

$$\Delta_2^{11} = \Lambda_{21} + \Lambda_{22} + \Lambda_{22}^T$$

$$\Delta_1^{22} = -e^{-2\alpha_1 h}\text{diag}\{Z_1, R_1, R_1, Q_1, P_1, P_1\}$$

$$\Delta_1^{12} = \begin{bmatrix} \sqrt{h}A_1 & \sqrt{h}M_1 & \sqrt{h}N_1 & \sqrt{h}U_1 & \sqrt{h}X_1 & \sqrt{h}Y_1 \end{bmatrix}$$

$$\Delta_2^{12} = \begin{bmatrix} \sqrt{h}A_2 & \sqrt{h}M_2 & \sqrt{h}N_2 & \sqrt{h}U_2 & \sqrt{h}X_2 & \sqrt{h}Y_2 \end{bmatrix}$$

$$\Delta_2^{22} = -\text{diag}\{Z_2, R_2, R_2, Q_2, P_2, P_2\}$$

$$\Psi_1 = [\rho I_N, \rho I_N, -\gamma\mathscr{L} - \mathscr{B}, -\gamma\mathscr{L} - \mathscr{B}, 0, 0, -\gamma\mathscr{L} - \mathscr{B}, -\gamma\mathscr{L} - \mathscr{B}]$$

$$\Psi_2 = [\rho I_N \quad \rho I_N \quad 0 \quad 0 \quad 0 \quad 0]$$

with

$$\Lambda_{11} = \begin{bmatrix} \Lambda_{11}^{11} & \Lambda_{11}^{12} & \overline{\mathscr{L}} & \overline{\mathscr{L}} & 0 & 0 & \overline{\mathscr{L}} & \overline{\mathscr{L}} \\ * & \Lambda_{11}^{22} & \overline{\mathscr{L}} & \overline{\mathscr{L}} & 0 & 0 & \overline{\mathscr{L}} & \overline{\mathscr{L}} \\ * & * & \Lambda_{11}^{33} & 0 & 0 & 0 & \Lambda_{11}^{37} & 0 \\ * & * & * & \Lambda_{11}^{44} & 0 & 0 & 0 & \Lambda_{11}^{48} \\ * & * & * & * & \chi S_1 & 0 & 0 & 0 \\ * & * & * & * & * & \chi H_1 & 0 & 0 \\ * & * & * & * & * & * & \Lambda_{11}^{77} & 0 \\ * & * & * & * & * & * & * & \Lambda_{11}^{88} \end{bmatrix}$$

$$\Lambda_{21} = \begin{bmatrix} \Lambda_{21}^{11} & (\psi - 2\alpha_2\psi)I_n & 0 & 0 & 0 & 0 \\ * & \Lambda_{21}^{22} & 0 & 0 & 0 & 0 \\ * & * & 0 & 0 & 0 & 0 \\ * & * & * & 0 & 0 & 0 \\ * & * & * & * & -S_2 & 0 \\ * & * & * & * & * & -H_2 \end{bmatrix}$$

$$\Lambda_{11}^{11} = 2\alpha_1\gamma(\mathscr{L} + \mathscr{L}^T) + 4\alpha_1\mathscr{B} + 4\epsilon I_N + S_1$$

$$\Lambda_{11}^{22} = 2\alpha_1 I_N + 4\epsilon I_N + H_1 + h(R_1 + Z_1)$$

$$\Lambda_{11}^{12} = 2\alpha_1 I_N + 2\gamma(\mathscr{L} + \mathscr{L}^T) + 4\mathscr{B}$$

$$\overline{\mathscr{L}} = -2(\gamma\mathscr{L} + \mathscr{B})$$

$$\Lambda_{11}^{33} = \Lambda_{11}^{37} = \omega_1 W_1, \quad \Lambda_{11}^{44} = \Lambda_{11}^{48} = \omega_2 W_2$$

$$\Lambda_{11}^{77} = (\omega_1 - 1)W_1, \quad \Lambda_{11}^{88} = (\omega_2 - 1)W_2$$

$$\Lambda_{12} = [A_1 + M_1, U_1 + X_1, -M_1 + N_1, -X_1 + Y_1, -A_1 - N_1, -U_1 - Y_1, \ 0, \ 0]$$

$$\chi = -e^{-2\alpha_1 h}$$

$$\Lambda_{21}^{11} = -2\alpha_2\psi\gamma(\mathscr{L} + \mathscr{L}^T) - 4\alpha_2\psi\mathscr{B} + 2\zeta\psi I_N + S_2$$

$$\zeta = \frac{2\rho_i + \rho_i}{2}$$

$$\Lambda_{21}^{22} = -2\alpha_2\psi I_N + 2\zeta\psi I_N + H_2 + h(R_2 + Z_2)$$

$$\Lambda_{22} = [A_2 + M_2, U_2 + X_2, -M_2 + N_2, -X_2 + Y_2, -A_2 - N_2, -U_2 - Y_2]$$

$$\epsilon = \max\{p, q + 1\}, \quad p = \max_{1 \le i \le N}\{\frac{3\rho_i + \rho_i}{2}\}, \quad q = \max_{1 \le i \le N}\{\frac{3\rho_i + \rho_i}{2}\}.$$

Then, the piecewise Lyapunov function (9.16) satisfies

$$V(t) \leq \begin{cases} e^{-2\alpha_1(t-\mathscr{C}_{1,n})} V_1(\mathscr{C}_{1,n}), & t \in [\mathscr{C}_{1,n}, \mathscr{C}_{2,n}) \\ e^{2\alpha_2(t-\mathscr{C}_{2,n})} V_2(\mathscr{C}_{2,n}), & t \in [\mathscr{C}_{2,n}, \mathscr{C}_{1,n+1}). \end{cases} \tag{9.20}$$

Proof According to the attack model (9.4), the proof procedure is bifurcated into two steps.

Step 1. When $t \in [\mathscr{C}_{1,n}, \mathscr{C}_{2,n})$, $\dot{V}_1(t)$ can be presented as

$$\begin{aligned} \dot{V}_1(t) \leq & -2\alpha_1 V_1(t) + 2\alpha_1[\tilde{x}^T, \tilde{v}^T]\Omega_1[\tilde{x}^T, \tilde{v}^T]^T - \Xi_1 - \Xi_2 - \Xi_3 + 2[\tilde{x}^T, \tilde{v}^T]\Omega_1[\dot{\tilde{x}}^T, \dot{\tilde{v}}^T]^T \\ & + \tilde{x}^T S_1 \tilde{x} + \tilde{v}^T H_1 \tilde{v} - \tilde{x}_h^T e^{-2\alpha_1 h} S_1 \tilde{x}_h - \tilde{v}_h^T e^{-2\alpha_1 h} H_1 \tilde{v}_h - \Xi_4 - \Xi_5 - \Xi_6 \\ & + h\dot{\tilde{x}}^T(t)(R_1 + Z_1)\dot{\tilde{x}}(t) + h\dot{\tilde{v}}^T(t)(P_1 + Q_1)\dot{\tilde{v}}(t) \\ & + 2\Theta_1^T(t)(A_1 r_1 + M_1 r_2 + N_1 r_3 + U_1 r_4 + X_1 r_5 + Y_1 r_6) \end{aligned}$$

where

$$\Theta_1(t) = [\tilde{x}^T(t), \tilde{v}^T(t), \tilde{x}_{k,n}^T(t), \tilde{v}_{k,n}^T(t), \tilde{x}_h^T(t), \tilde{v}_h^T(t), \varkappa_{k,n}^T(t), \kappa_{k,n}^T(t)]^T$$

$$\tilde{x}_h = \tilde{x}(t-h), \quad \tilde{v}_h = \tilde{v}(t-h)$$

$$\Xi_1 = \int_{t-h}^{t-\eta_{k,n}} \dot{\tilde{x}}(\tau)^T e^{-2\alpha_1 h} R_1 \dot{\tilde{x}}(\tau)\, d\tau, \quad \Xi_2 = \int_{t-\eta_{k,n}}^{t} \dot{\tilde{x}}(\tau)^T e^{-2\alpha_1 h} R_1 \dot{\tilde{x}}(\tau)\, d\tau$$

$$\Xi_3 = \int_{t-h}^{t} \dot{\tilde{x}}(\tau)^T e^{-2\alpha_1 h} Z_1 \dot{\tilde{x}}(\tau)\, d\tau, \quad \Xi_4 = \int_{t-h}^{t-\eta_{k,n}} \dot{\tilde{v}}(\tau)^T e^{-2\alpha_1 h} P_1 \dot{\tilde{v}}(\tau)\, d\tau$$

$$\Xi_5 = \int_{t-\eta_{k,n}}^{t} \dot{\tilde{v}}(\tau)^T e^{-2\alpha_1 h} P_1 \dot{\tilde{v}}(\tau)\, d\tau, \quad \Xi_6 = \int_{t-h}^{t} \dot{\tilde{v}}(\tau)^T e^{-2\alpha_1 h} Q_1 \dot{\tilde{v}}(\tau)\, d\tau$$

$$r_1 = \tilde{x}(t) - \tilde{x}_h(t) - \int_{t-h}^{t} \dot{\tilde{x}}(\tau)\, d\tau = 0, \quad r_2 = \tilde{x}(t) - \tilde{x}_{k,n}(t) - \int_{t-\eta_{k,n}}^{t} \dot{\tilde{x}}(\tau)\, d\tau = 0$$

$$r_3 = \tilde{x}_{k,n}(t) - \tilde{x}_h(t) - \int_{t-h}^{t-\eta_{k,n}} \dot{\tilde{x}}(\tau)\,d\tau = 0, \quad r_4 = \tilde{v}(t) - \tilde{v}_h(t) - \int_{t-h}^{t} \dot{\tilde{v}}(\tau)\,d\tau = 0$$

$$r_5 = \tilde{v}(t) - \tilde{v}_{k,n}(t) - \int_{t-\eta_{k,n}}^{t} \dot{\tilde{v}}(\tau)\,d\tau = 0, \quad r_6 = \tilde{v}_{k,n}(t) - \tilde{v}_h(t) - \int_{t-h}^{t-\eta_{k,n}} \dot{\tilde{v}}(\tau)\,d\tau = 0.$$

By employing a similar methodology as proposed in [9], it can be obtained that

$$\begin{aligned}\dot{V}_1(t) \le & -2\alpha_1 V_1(t) + \Theta_1^T(t)\big[\Delta_1^{11} + hA_1e^{2\alpha_1 h}Z_1^{-1}A_1^T + hM_1e^{2\alpha_1 h}R_1^{-1}M_1^T \\ & + hN_1e^{2\alpha_1 h}R_1^{-1}N_1^T + hU_1e^{2\alpha_1 h}Q_1^{-1}U_1^T + hX_1e^{2\alpha_1 h}P_1^{-1}X_1^T \\ & + hY_1e^{2\alpha_1 h}P_1^{-1}Y_1^T + h\Psi_1^T(P_1+Q_1)\Psi_1\big]\Theta_1(t). \end{aligned} \tag{9.21}$$

Based on Lemma 2.8, the condition $\Delta_1 < 0$ is equivalent to

$$\begin{aligned}&\Delta_1^{11} + hA_1e^{2\alpha_1 h}Z_1^{-1}A_1^T + hM_1e^{2\alpha_1 h}R_1^{-1}M_1^T + hN_1e^{2\alpha_1 h}R_1^{-1}N_1^T + hU_1e^{2\alpha_1 h}Q_1^{-1}U_1^T \\ &+ hX_1e^{2\alpha_1 h}P_1^{-1}X_1^T + hY_1e^{2\alpha_1 h}P_1^{-1}Y_1^T + h\Psi_1^T(P_1+Q_1)\Psi_1 < 0.\end{aligned}$$

This implies that the subsequent inequality is valid

$$\dot{V}(t) \le -2\alpha_1 V_1(t), \quad t \in [\mathscr{C}_{1,n}, \mathscr{C}_{2,n}). \tag{9.22}$$

Then, one can get

$$V(t) \le e^{-2\alpha_1(t-\mathscr{C}_{1,n})} V_1(\mathscr{C}_{1,n}). \tag{9.23}$$

Step 2. When $t \in [\mathscr{C}_{2,n}, \mathscr{C}_{1,n+1})$, it can be arrived that

$$\begin{aligned}\dot{V}(t) \le & 2\alpha_2 V_2(t) + \Theta_2^T(t)[\Delta_2^{11} + hA_2Z_2^{-1}A_2^T + hM_2R_2^{-1}M_2^T + hN_2R_2^{-1}N_2^T \\ & + hU_2Q_2^{-1}U_2^T + hX_2P_2^{-1}X_2^T + hY_2P_2^{-1}Y_2^T + h\Psi_2^T(P_2+Q_2)\Psi_2]\Theta_2(t)\end{aligned} \tag{9.24}$$

where

$$\Theta_2 = [\tilde{x}^T(t), \tilde{v}^T(t), \tilde{x}_{k,n}^T(t), \tilde{v}_{k,n}^T(t), \tilde{x}_h^T(t), \tilde{v}_h^T(t)]^T.$$

According to (9.19), one has

$$\dot{V}(t) \le 2\alpha_2 V_2(t), \quad t \in [\mathscr{C}_{2,n}, \mathscr{C}_{1,n+1}) \tag{9.25}$$

and

$$V(t) \le e^{2\alpha_2(t-\mathscr{C}_{2,n})} V_2(\mathscr{C}_{2,n}). \tag{9.26}$$

The proof is completed. □

Drawing from the results of Lemma 9.1, the conditions for the convergence of the state error system (9.14) will be specified.

Theorem 9.1 *Given an MAS (9.1) that adheres to Assumptions 9.1 and 9.2, and considering a DoS attack model $\mathscr{S}_{\mathrm{DoS}}$ with parameters T and T_{off} that are known, let's employ the pinning control law (9.7) and the ETM (9.6). Suppose that the parameters $\alpha_\varsigma > 0$, $\omega_\varsigma \in (0, 1)$, $\gamma > 0$, $\psi > 0$, $h \in (0, T_{\mathrm{off}})$, $b_i > 0$, and $\xi_\varsigma > 1$ satisfy the subsequent inequality*

$$\pi \triangleq 2\alpha_1 T_{\mathrm{off}} - 2\alpha_2(T - T_{\mathrm{off}}) - 2(\alpha_1 + \alpha_2)h - \ln \xi_1\xi_2 > 0. \tag{9.27}$$

In addition, there are positive definite matrices H_ς, S_ς, Z_ς, W_ς, P_ς, R_ς, Q_ς, and matrices A_ς, U_ς, N_ς, X_ς, M_ς, and Y_ς, where $\varsigma \in \{1, 2\}$, such that

$$\begin{bmatrix} \Delta_\varsigma^{11} & \Delta_\varsigma^{12} & \sqrt{h}\Psi_\varsigma^T(P_\varsigma + Q_\varsigma) \\ * & \Delta_\varsigma^{22} & \mathbf{0} \\ * & * & m_\varsigma^2(P_\varsigma + Q_\varsigma) - 2m_\varsigma I_N \end{bmatrix} < 0 \tag{9.28}$$

$$\Omega_1 \le \xi_2\Omega_2, \quad \Omega_2 \le e^{2(\alpha_1+\alpha_2)h}\xi_1\Omega_1 \tag{9.29}$$

$$\begin{aligned} &\begin{bmatrix} -\xi_{3-\varsigma}S_{3-\varsigma} & I_N \\ * & b_\varsigma^2 S_\varsigma 2b_\varsigma I_N \end{bmatrix} < 0, \quad \begin{bmatrix} -\xi_{3-\varsigma}H_{3-\varsigma} & I_N \\ * & c_\varsigma^2 H_\varsigma 2c_\varsigma I_N \end{bmatrix} < 0 \\ &\begin{bmatrix} -\xi_{3-\varsigma}R_{3-\varsigma} & I_N \\ * & o_\varsigma^2 R_\varsigma 2o_\varsigma I_N \end{bmatrix} < 0, \quad \begin{bmatrix} -\xi_{3-\varsigma}Z_{3-\varsigma} & I_N \\ * & g_\varsigma^2 Z_\varsigma 2g_\varsigma I_N \end{bmatrix} < 0 \\ &\begin{bmatrix} -\xi_{3-\varsigma}P_{3-\varsigma} & I_N \\ * & y_\varsigma^2 P_\varsigma 2y_\varsigma I_N \end{bmatrix} < 0, \quad \begin{bmatrix} -\xi_{3-\varsigma}Q_{3-\varsigma} & I_N \\ * & z_\varsigma^2 Q_\varsigma 2z_\varsigma I_N \end{bmatrix} < 0 \end{aligned} \tag{9.30}$$

where the definitions of Δ_ς^{11}, Δ_ς^{12}, Δ_ς^{22}, and Ψ_ς, $\varsigma \in \{1, 2\}$ can be found in Lemma 9.1; $b_\varsigma > 0$, $g_\varsigma > 0$, $y_\varsigma > 0$, $c_\varsigma > 0$, $m_\varsigma > 0$, $o_\varsigma > 0$, and $z_\varsigma > 0$ are constants. Consequently, the signals of the error system converge exponentially, and the parameter matrices W_1 and W_2 in the ETM can be determined by solving the LMIs. This implies that the MASs achieve cluster consensus despite the presence of DoS attacks.

Proof With respect to the Lyapunov function (9.16), when the Linear Matrix Inequalities (9.28) are satisfied, it follows from Lemma 9.1 that

For the selected piecewise Lyapunov function (9.16), if the LMIs (9.28) are satisfied, according to Lemma 9.1, it is easy to obtain that

$$V(t) \leq \begin{cases} e^{-2\alpha_1(t-\mathscr{C}_{1,n})}V_1(\mathscr{C}_{1,n}), & t \in [\mathscr{C}_{1,n}, \mathscr{C}_{2,n}) \\ e^{2\alpha_2(t-\mathscr{C}_{2,n})}V_2(\mathscr{C}_{2,n}), & t \in [\mathscr{C}_{2,n}, \mathscr{C}_{1,n+1}). \end{cases} \tag{9.31}$$

Based on (9.29) and (9.30), by using Lemma 2.8 and the relation

$$S_\varsigma^{-1} \geq 2b_\varsigma I_N - b_\varsigma^2 S_\varsigma$$

as in [12], the following conditions can be obtained

$$\begin{cases} \Omega_1 \leq \xi_2\Omega_2, & \Omega_2 \leq e^{2(\alpha_1+\alpha_2)h}\xi_1\Omega_1 \\ S_1 \leq \xi_2 S_2, & S_2 \leq \xi_1 S_1 \\ H_1 \leq \xi_2 H_2, & H_2 \leq \xi_1 H_1 \\ R_1 \leq \xi_2 R_2, & R_2 \leq \xi_1 R_1 \\ Z_1 \leq \xi_2 Z_2, & Z_2 \leq \xi_1 Z_1 \\ P_1 \leq \xi_2 P_2, & P_2 \leq \xi_1 P_1 \\ Q_1 \leq \xi_2 Q_2, & Q_2 \leq \xi_1 Q_1. \end{cases} \tag{9.32}$$

Thus, it gives

$$\begin{cases} \xi_2 V_2(\mathscr{C}_{1,n}^-) - V_1(\mathscr{C}_{1,n}) \geq 0 \\ e^{2(\alpha_1+\alpha_2)h}\xi_1 V_1(\mathscr{C}_{2,n}^-) - V_2(\mathscr{C}_{2,n}) \geq 0. \end{cases} \tag{9.33}$$

According to (9.31) and (9.33), one can get the following result when $t \in [\mathscr{C}_{1,n}, \mathscr{C}_{2,n})$:

$$\begin{aligned} V(t) \leq& \xi_2 e^{-2\alpha_1(t-\mathscr{C}_{1,n})}V_2(\mathscr{C}_{1,n}^-) \\ &\vdots \\ \leq& e^{-\pi n}V_1(\mathscr{C}_{1,0}) = e^{-\pi n}V_1(0). \end{aligned} \tag{9.34}$$

Notice that $t < \mathscr{C}_{2,n} = nT + T_{\text{off}}$, i.e., $n > \frac{t-T_{\text{off}}}{T}$, it follows from (9.34) that

$$V(t) \leq e^{\frac{\pi T_{\text{off}}}{T}} e^{-\frac{\pi}{T}t} V_1(0). \tag{9.35}$$

Similar to (9.35), if $t \in [\mathscr{C}_{2,n}, \mathscr{C}_{1,n+1})$, one can get

$$V(t) \leq \frac{V_1(0)}{\xi_2} e^{-\frac{\pi}{T}t}. \tag{9.36}$$

Defining

$$\mathscr{M} = \max\left\{e^{\frac{\pi T_{\text{off}}}{T}}, \frac{1}{\xi_2}\right\}$$

and combining (9.35) with (9.36), one can get

$$V(t) \leq \mathscr{M} e^{-\frac{\pi}{T}t} V_1(0) \tag{9.37}$$

Following that, assuming that condition (9.27) holds, it can be inferred that the system's error signals exhibit exponential stability, leading to the MAS reaching *cluster consensus* even under the influence of DoS attacks. The proof is completed. □

9.4 Simulation Examples

In this section, an MAS with a total of five follower agents and two leader agents that are divided into two distinct clusters $G_1 = \{1, 3, 4\}$ and $G_2 = \{2, 5\}$. The topology structure among these agents is directed and is illustrated in Fig. 9.1, where a directed spanning tree is present.

The dynamics of the leader agent are set as

$$\begin{aligned} f_{\sigma_i}(t, x_i, v_i) &= 1.5\cos 2.5t - \sin x_i - 0.25v_i, \quad i = 1, 3, 4 \\ f_{\sigma_i}(t, x_i, v_i) &= 0.5\sin 2t - 0.5v_i, \quad i = 2, 5. \end{aligned}$$

From the given conditions, we select $p = 1.625$, $q = 0.875$, and $\epsilon = 1.875$. Furthermore, for the DoS attack signal, we set the parameters $T = 2$ and $T_{\text{off}} = 1.7$. The values of $h = 0.01\,\alpha_1 = 0.18, \alpha_2 = 0.8, \xi_1 = 1.05, and \xi_2 = 1.005$ are selected to be in compliance with (9.27). Given $d = 20$, $\psi = 50$, $\gamma = 4$, $\omega_1 = \omega_2 = 0.02$, $b_\varsigma = c_\varsigma = o_\varsigma = 1.3$, $m_1 = 0.2$, $m_2 = 0.1$, and $g_\varsigma = y_\varsigma = z_\varsigma = 0.5$ with $\varsigma = \{1, 2\}$. By invoking Theorem 9.1, the matrices W_1 and W_2 are calculated as follows:

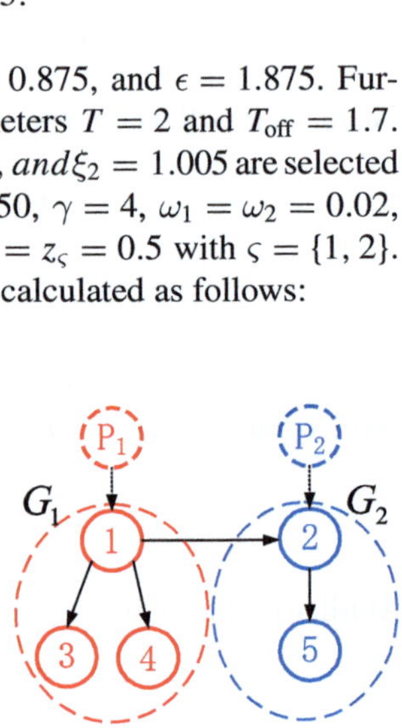

Fig. 9.1 Communication topology

$$W_1 = 10^3 \times \begin{bmatrix} 1.1607 & -0.0679 & -0.1314 & -0.1314 & 0.0268 \\ * & 1.2734 & -0.0348 & -0.0348 & -0.0722 \\ * & * & 0.1214 & -0.0396 & 0.0037 \\ * & * & * & 0.1214 & 0.0037 \\ * & * & * & * & 0.1099 \end{bmatrix},$$

$$W_2 = 10^3 \times \begin{bmatrix} 3.7407 & -0.3038 & -0.3928 & -0.3928 & 0.0745 \\ * & 4.0535 & -0.1332 & -0.1332 & -0.2215 \\ * & * & 0.5570 & -0.1825 & 0.0263 \\ * & * & * & 0.5570 & 0.0263 \\ * & * & * & * & 0.6020 \end{bmatrix}.$$

Figure 9.2 shows the system performance of the MAS (9.1) achieved through the application of the pinning control law (9.7) and the ETM (9.6). The consensus error $J(t)$ is defined as

$$J(t) = \frac{1}{N}\sqrt{\sum_{i=1}^{N}\left(x_i(t) - x_{\sigma_i}^r(t)\right)^2 + \sum_{i=1}^{N}\left(v_i(t) - v_{\sigma_i}^r(t)\right)^2}.$$

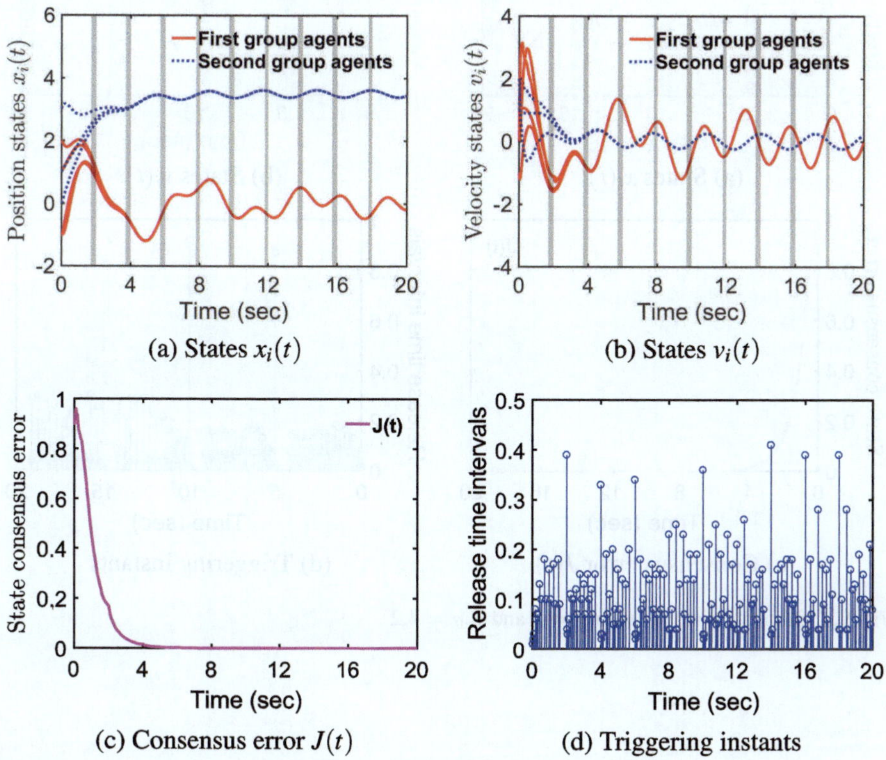

Fig. 9.2 Simulation results with $T = 2$ and $T_{\text{off}} = 1.7$

From Fig. 9.2, we can deduce the following conclusions: (1) the MAS achieves *cluster consensus*, facilitated by the pinning controller and the ETM, even in the presence of DoS attacks; and (2) the implemented ETM not only diminishes the frequency of controller updates but also mitigates the detrimental effects of DoS attacks on the system's performance.

In order to illustrate the effectiveness of the proposed control scheme under different attack durations, the following comparative simulations are performed. As depicted in Fig. 9.3, the secure *cluster consensus* of the MAS is maintained even under prolonged DoS attack durations, namely $T = 5$, $T_{\text{off}} = 4.2$, with the rest of the parameters staying the same. Figure 9.3 reveals that the performance in achieving *cluster consensus* is comparable to the scenario with shorter attack durations. Hence, it is evident that the designed pinning control scheme is applicable against DoS attacks of diverse durations.

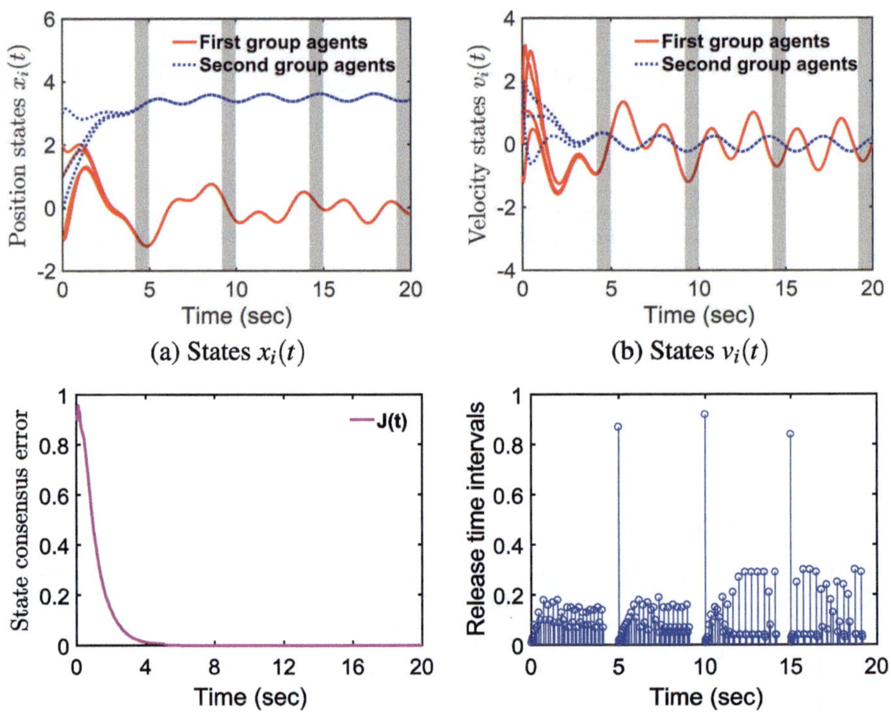

(a) States $x_i(t)$ (b) States $v_i(t)$

(c) Consensus error $J(t)$ (d) Triggering instants

Fig. 9.3 Simulation results with $T = 5$ and $T_{\text{off}} = 4.2$

9.5 Conclusion

This chapter focuses on the secure *cluster consensus* issue of nonlinear MASs that feature a directed topology graph, all while enduring periodic DoS attacks. A new control approach that employs event-triggered mechanism and pinning control technique is proposed. Additionally, the approach to determining the parameters of the event-triggered mechanism and the criteria for *cluster consensus* in MASs under the influence of periodic DoS attacks are presented in the form of LMI. Finally, the effectiveness of the *cluster consensus* strategy is demonstrated through simulation results. The exploration of *cluster consensus* control strategies for MASs with aperiodic DoS attacks is deemed significant and will be further investigated in the subsequent chapter.

References

1. F. Li, Z. Ma, Q. Duan, Clustering component synchronization in a class of unconnected networks via pinning control. Phys. A **525**, 394–401 (2019)
2. P.M. Liu, Z.Z Huang, X.G. Guo, Event-triggered secure group consensus of second-order multi-agent systems under periodic DoS attacks, in *2021 IEEE 10th Data Driven Control and Learning Systems Conference*, pp. 1007–1012 (2021)
3. X.G. Guo, H.J. Li, J.J. Zhao, W.W. Che, Static group synchronization of second-order multi-agent systems via pinning control. J. Frankl. Inst. **356**(9), 4842–4858 (2019)
4. L. Liu, X. Ding, W. Zhou, Prescribed-time cluster synchronization of uncertain complex dynamical networks with switching via pinning control. Neurocomputing **419**, 136–147 (2021)
5. L. Liu, W. Zhou, X. Li, Y. Sun, Dynamic event-triggered approach for cluster synchronization of complex dynamical networks with switching via pinning control. Neurocomputing **340**, 32–41 (2019)
6. X. Liu, T. Chen, Finite-time and fixed-time cluster synchronization with or without pinning control. IEEE Trans. Cybern. **48**(1), 240–252 (2019)
7. X.G. Guo, W.D. Xu, J.L. Wang, J.H. Park, Distributed neuroadaptive fault-tolerant sliding-mode control for 2-D plane vehicular platoon systems with spacing constraints and unknown direction faults. Automatica **129**, Art. No. 109675 (2021)
8. Y. Wan, G. Wen, X. Yu, T. Huang, Distributed consensus tracking of networked agent systems under denial-of-service attacks. IEEE Trans. Syst. Man Cybern. Syst. **51**(10), 6183–6196 (2021)
9. X. Chen, Y. Wang, S. Hu, Event-based robust stabilization of uncertain networked control systems under quantization and denial-of-service attacks. Inf. Sci. **459**, 369–386 (2018)
10. C. Deng, C. Wen, MAS-based distributed resilient control for a class of cyber-physical systems with communication delays under DoS attacks. IEEE Trans. Cybern. **51**(5), 2347–2358 (2021)
11. X.G. Guo, X. Fan, J.L. Wang, J.H. Park, Event-triggered switching-type fault detection and isolation for fuzzy control systems under DoS attacks. IEEE Trans. Fuzzy Syst. **29**(11), 3401–3414 (2021)
12. S. Hu, D. Yue, X. Xie, X. Chen, X. Yin, Resilient event-triggered controller synthesis of networked control systems under periodic DoS jamming attacks. IEEE Trans. Cybern. **49**(12), 4271–4281 (2019)
13. L. Ding, Q.L. Han, X.H. Ge, X.M. Zhang, An overview of recent advances in event-triggered consensus of multiagent systems. IEEE Trans. Cybern. **48**(4), 1110–1123 (2018)
14. X.G. Guo, X. Fan, C.K. Ahn, Adaptive event-triggered fault detection for interval type-2 T-S fuzzy systems with sensor saturation. IEEE Trans. Fuzzy Syst. **29**(8), 2310–2321 (2021)

15. A. Hu, J. Cao, M. Hu, L. Guo, Event-triggered group consensus for multi-agent systems subject to input saturation. J. Franklin Inst. **355**(15), 7384–7400 (2018)
16. H. Lu, Y. Hu, C. Guo, W. Zhou, Cluster synchronization for a class of complex dynamical network system with randomly occurring coupling delays via an improved event-triggered pinning control approach. J. Frankl. Inst. **357**(4), 2167–2184 (2020)
17. X.G. Guo, D.C. Tan, C.K. Ahn, J.L. Wang, Fully distributed adaptive fault-tolerant sliding-mode control for nonlinear leader-following multiagent systems with ANASs and IQCs. IEEE Trans. Cybern. **52**(5), 2763–2774 (2022)

Chapter 10
Fault-Tolerant Cluster Consensus of MASs Under Aperiodic DoS Attacks

Abstract This chapter details an event-triggered approach to achieving *cluster consensus* in heterogeneous nonlinear second-order multi-agent systems (MASs) that are susceptible to cyber attacks (aperiodic denial-of-service (DoS) attacks), actuator faults, and IQCs, all within a directed communication structure that has a directed spanning tree. The event-triggered adaptive *fault-tolerant* pinning control strategy is designed to reach *cluster consensus* through local communication, despite concurrent cyber attacks and actuator faults. This scheme operates without the need for a communication topology that maintains in-degree balance among different clusters. Moreover, the *fault-tolerant* controller only necessitates the estimation of one parameter per agent. Unlike the need for continuous overall information about MASs to set trigger instants as discussed in Chap. 9, this chapter develops an event-triggered mechanism (ETM) that dispenses with the need for periodic neighbor information sampling, thus conserving network resources. In addition, the *Zeno behavior* is excluded.

10.1 Introduction

Cluster consensus refers to a scenario in multi-agent systems (MASs) where agents are partitioned into separate groups for various tasks, and agents within the same group reach a consensus [1–4]. As discussed in Chap. 3, an increase in the number of agents in the system correlates with a rise in the volume of data that must be collected and transmitted. This can significantly increase the communication load, potentially leading to the collapse of wireless networks [5–9]. Therefore, to conserve communication and computational resources, event-triggered mechanisms (ETMs) are increasingly being used in place of continuous sampling [10–15]. Consequently, an increasing number of scholars are focusing on the theory and practical application of event-triggered *cluster consensus* mechanisms, which pose a greater challenge compared to the case of *complete consensus* case due to the concurrent deployment of ETMs within and across clusters. It is important to noted that the ETM designed

X.-G. Guo et al., *Anti-disturbance Intrusion- and Fault-Tolerant Control of Multi-agent Systems*, https://doi.org/10.1007/978-981-96-9614-7_10

in [16] operates in a centralized manner, whereas the one in [8] is asynchronous across clusters but synchronous within a single cluster, limiting their application scenarios. Additionally, the ETM presented in Chap. 9 necessitates the periodic sampling of overall information about the MASs to assess whether the triggering conditions are satisfied. Hence, it is both a significant endeavor to develop a distributed ETM that eliminates the need for periodic sampling of overall information gathering on MASs, thereby alleviating the communication load.

On the other hand, MASs operating in networked settings are susceptible to various type of *cyber attacks* [9, 17, 18], which can lead to significant performance degradation and potentially induce system instability [11, 19, 20]. In Chap. 9, the *cluster consensus* control strategy of MASs under periodic denial-of-service (DoS) attacks is studied. However, in real-world scenarios, DoS attacks are typically aperiodic. Thus, addressing the secure consensus challenge for MASs under aperiodic DoS attacks is an area of intense research interest (see [17, 21–23]). Despite this, current studies, such as [24] and [25], limit their scope to *complete consensus* only. Furthermore, in addition to *cyber attacks*, the physical layer's actuators may also encounter unforeseen faults, which can lead to a decline in system performance and even threaten the system's stability [20, 25–31]. Consequently, MASs also need to possess fault tolerance capabilities [32]. To date, there has been limited research on *fault-tolerant* consensus issues within the *cluster consensus* framework. While numerous effective *fault-tolerant* control methods have been devised to ensure the *complete consensus* of MASs [25, 31], these approaches are not directly applicable to addressing the *cluster consensus* challenge when confronted with the dual impacts of actuator faults and DoS attacks. Additionally, external *disturbances* and uncertainties are two primary factors that can impact system performance [33]. Particularly, the interactions among various agents mean that the external *disturbances* and uncertainties affecting one agent can have repercussions on the performance and stability of the entire system. Thus, it is desirable to develop effective strategies to mitigate the impact of *disturbances* and uncertainties. It is recognized from the previous chapter that integral quadratic constraints (IQCs) are more adept at characterizing the properties of external *disturbances* and uncertainties, as well as indicating their origins, compared to norm constraints [33]. However, the modeling of IQCs introduces new challenges, especially when there is strong coupling between actuator faults and IQCs. Consequently, it is both intriguing and theoretically significant to devise effective control schemes that can address the aforementioned trio of challenges, i.e., aperiodic DoS attacks, actuator faults, and IQCs, under the framework of *cluster consensus*.

In light of the aforementioned insights, the objective of this chapter is to develop an event-triggered adaptive *fault-tolerant cluster consensus* approach for heterogeneous second-order nonlinear MASs with a directed spanning tree, while dealing with aperiodic DoS attacks, actuator faults, and IQCs. Within this chapter, it is essential to answer the following three questions:

($Q1$) How to design a control strategy to enable the MAS to attain *fault-tolerant cluster consensus* under aperiodic DoS attacks?
($Q2$) How to establish an ETM to decrease the frequency of control law updates while also eliminating the need for periodic sampling of neighboring agents' information to assess if the triggering conditions are satisfied?
($Q3$) How to reduce the number of trigger instants while maintaining the performance of *cluster consensus*?

The main contributions of this chapter are outlined as follows:

(1) Aperiodic DoS attack: This chapter is dedicated to the investigation of event-triggered adaptive *fault-tolerant cluster consensus* in the presence of cyber attacks and actuator faults. Additionally, a novel lemma (Lemma 10.1) has been developed to function as a tool for assessing the finite-time upper bounds of the candidate piecewise Lyapunov functions under conditions of aperiodic DoS attacks.
(2) Novel distributed ETM: In contrast to the ETM of [22], this chapter introduces two nonnegative constants to determine and adjust the event-triggering times under DoS attacks. As a result, the proposed ETM offers greater flexibility and more efficiently conserves communication resources. Additionally, unlike the centralized ETM within the *cluster consensus* framework in Chap. 9 and the synchronous trigger mechanism within the same cluster in [8], the distributed ETM presented in this chapter under the *cluster consensus* framework lowers computational complexity. Moreover, it eliminates the need for periodic neighbor information sampling to check trigger conditions, which sets it apart from the ETM in [34] that reduces control input update frequency but not the communication load. Furthermore, this chapter ensures the avoidance of Zeno behavior.
(3) *Fault-tolerant* control strategy: In comparison with the fault-tolerant control strategies proposed in [20], the novel contributions of the fault-tolerant control strategy designed in this chapter are as follows: (a) The focus is shifted from achieving *complete consensus* to attaining *cluster consensus*; (b) Pinning control is utilized in place of a fully equipped control method, thereby reducing the control cost; (c) In contrast to [20], the strategy proposed in this chapter requires the estimation of only one parameter per agent, simplifying the structure and lowering the computational cost; (d) An ETM is implemented in place of a continuous sampling mechanism, which alleviates the communication load.

10.2 Problem Formulation and Preliminaries

10.2.1 Heterogeneous MAS Model

In this chapter, a system composed of N nonlinear heterogeneous second-order agents is considered, and the dynamics of each agent can be formulated as follows:

$$
\begin{aligned}
&\dot{x}_i(t) = v_i(t), \quad i = 1, 2, \cdots, N \\
&\dot{v}_i(t) = f_{n_i}(t, x_i(t), v_i(t)) + u_i^F(t) + \sum_{k=1}^{p_i} B_{ik} w_{ik}(t), \quad n_i = 1, 2, \cdots, s \\
&\xi_{ik}(t) = H_{ik} v_i(t) + G_{ik} u_i^F(t), \quad k = 1, 2, \cdots, p_i
\end{aligned} \tag{10.1}
$$

where $x_i(t) \in \mathbb{R}$, $v_i(t) \in \mathbb{R}$, and $u_i^F(t) \in \mathbb{R}$ represent the position state, speed state, and fault input signal, respectively, and the subscripts indicate that they belong to the ith vertex/agent; $w_{ik}(t) \in \mathbb{R}^{q_i}$ and $\xi_{ik}(t) \in \mathbb{R}^{n_i}$ are the IQC output and input, respectively; $f_{n_i}(t, x_i(t), v_i(t))$ represents the nonlinear dynamics of the agent itself and is a differentiable nonlinear bounded function; B_{ik}, H_{ik} and G_{ik} are parameter matrices with suitable dimensions. Note that agents in the same group (i.e., $n_i = p$) have the same homogeneous nonlinear dynamics $f_{n_i}(t, \cdot, \cdot)$.

In different clusters, the virtual leader of the agents is different. The leader dynamics of the cluster where the ith agent belongs is as follows:

$$
\begin{aligned}
&\dot{x}_{\sigma_i}^r(t) = v_{\sigma_i}^r(t) \\
&\dot{v}_{\sigma_i}^r(t) = f_{\sigma_i}(t, x_{\sigma_i}^r(t), v_{\sigma_i}^r(t)), \quad n_i = 1, 2, \cdots, s
\end{aligned} \tag{10.2}
$$

where $x_{\sigma_i}^r(t)$ and $v_{\sigma_i}^r(t)$ are the states of the virtual leader in the n_ith cluster; $f_{\sigma_i}(t, x_{\sigma_i}^r(t), v_{\sigma_i}^r(t))$ can be heterogeneous in different clusters.

10.2.2 Actuator Faults

In practice, the actuator may experience unpredictable faults. The actual input is given as

$$
u_i^F(t) = \begin{cases} u_i(t), & \text{fault-free} \\ \rho_i(t) u_i(t) + r_i(t), & \text{fault} \end{cases} \tag{10.3}
$$

in which $\rho_i(t)$ and $r_i(t)$ are fault scaling factors and bias faults, respectively; $u_i(t)$ is the fault-free control input.

Assumption 10.1 The fault scaling factor $\rho_i(t)$ has upper bound $\overline{\rho}_i$ and lower bound $\underline{\rho}_i$, and satisfies $0 < \underline{\rho}_i \le \rho_i(t) \le \overline{\rho}_i < \infty$; the bias fault $r_i(t)$ satisfies $|r_i(t)| \le \overline{r}_i$.

Remark 10.1 Assumptions akin to those in Assumption 10.1 are prevalent in the current literature [20]. However, this chapter distinguishes itself from [20] by allowing the fault scaling factor to potentially exceed 1, signifying the presence of excessive faults. These excessive faults are not uncommon, particularly in the circuits and sensors that may experience extreme electrical and thermal stress [29].

10.2.3 IQC Model

For MAS (10.1) and (10.2), the IQC output and input are assumed to satisfy the equation

$$w_{ik}(t) = \psi_{ik}(t, \xi_{ik}(\cdot)|_{t_0}^{t}) \tag{10.4}$$

where $\psi_{ik}(\cdot, \cdot)$ denotes nonlinear time-varying dynamic constraints. On this basis, IQCs are defined as follows.

Recalling Chap. 7, it gives

$$\begin{aligned}\sum_{k=1}^{p_i} \|B_{ik} w_{ik}(t)\| &\leq \sum_{k=1}^{p_i} \|B_{ik}\| \|\xi_{ik}(t)\| + \sum_{k=1}^{p_i} \|B_{ik}\| \delta_{wik} \\ &\leq \zeta_{1i} |u_i^F(t)| + \zeta_{2i} |v_i(t)| + \delta_{di}, \quad i \in \mathscr{V}\end{aligned} \tag{10.5}$$

where

$$\zeta_{1i} = \sum_{k=1}^{p_i} \|B_{ik}\| \|G_{ik}\| < \frac{\underline{\rho}_i}{\overline{\rho}_i}, \quad \zeta_{2i} = \sum_{k=1}^{p_i} \|B_{ik}\| \|H_{ik}\|, \quad \delta_{di} = \sum_{k=1}^{p_i} \|B_{ik}\| \delta_{wik} \tag{10.6}$$

with $\delta_{wik} > 0$ being an unknown bounded constant. In addition, according to (10.6), one has $\underline{\rho}_i - \zeta_{1i} \overline{\rho}_i > 0$.

10.2.4 Nonlinear Dynamics

To simplify the analysis, the following assumption is introduced.

Assumption 10.2 ([20]) For the continuous function $f(t, x_i(t), v_i(t))$ as in Lemma 2.10, there are unknown bounded constants $\alpha_i \geq 0$ and $\beta_i \geq 0$ that satisfy the inequalities when $(x_i(t), v_i(t)) \in S$

$$\left| \frac{\partial f(t, x_i(t), v_i(t))}{\partial x_i} \right| \leq \alpha_i, \quad \left| \frac{\partial f(t, x_i(t), v_i(t))}{\partial v_i} \right| \leq \beta_i. \tag{10.7}$$

Based on Assumption 10.2 and Lemma 2.10, we have

$$|f(t, x_i(t), v_i(t)) - f(t, x_i(t_0), v_i(t_0))| \leq \alpha_i |x_i(t) - x_i(t_0)| + \beta_i |v_i(t) - v_i(t_0)|. \tag{10.8}$$

Remark 10.2 Assumption 10.2, which is commonly adopted in the related research (see [20] and the reference therein), posits that the partial derivatives of the nonlinear function $f(t, x_i(t), v_i(t))$ are bounded. By integrating this assumption with Lemma 2.10, the Lipschitz condition (10.8) is derived. Furthermore, Assumption 10.2 offers a guideline for the determination of the constants α_i and β_i.

10.2.5 Cyber Attack: Aperiodic DoS Attack Model

DoS attacks not only disrupt the communication channel but also inflict damage on communication infrastructure, hindering information transmission. The primary objective of a DoS attack is to obstruct the channel, rendering the system incapable of transmitting and exchanging information. This chapter considers a scenario where both communication and control channels are compromised, i.e., the agent's controller fails, and the flow of information among agents is disrupted. Additionally, it is assumed that following a DoS assault, the MAS is capable of re-establishing its original communication state via a recovery mechanism.

Even though DoS attacks can be highly damaging to systems, attackers must take time to regain energy post-attack due to their limited energy resources. As a result, the timeline can be segmented into periods with and without DoS attacks. In the absence of an attack, communication can proceed uninterrupted, but when an attack is underway, communication is obstructed. The mth attack interval is defined as $D_m = [t_m, t_m + \tau_m), m = 1, 2, 3, \cdots$, where t_m represents the time instant when the mth attack starts, and τ_m represents the time duration of the mth attack. On $[t_1, t_2)$, the set of DoS attack time periods is denoted as

$$\Xi_d(t_1, t_2) = \bigcup_{m \in \mathbb{N}} D_m \cap [t_1, t_2).$$

The complement of $\Xi_d(t_1, t_2)$ is the set of time periods during which no attack occurs, and is denoted as

$$\Xi_h(t_1, t_2) = [t_1, t_2) \setminus \Xi_d(t_1, t_2).$$

During the time period $\Xi_d(t_1, t_2)$, due to the existence of the recovery mechanism, the MAS periodically attempts to communicate within the interval $\pi > 0$ and $t_{m+1} - t_m > \pi$. It can be seen from Fig. 10.1 that the MAS cannot resume communication immediately after the DoS attack. So the effective time period of the mth DoS attack can be expressed as $\overline{D}_m = [t_m, t_m + \overline{\tau}_m)$, where $\overline{\tau}_m = (N(t_m, t_m + \tau_m) + 1)\pi$ with $N(t_m, t_m + \tau_m)$ representing the number of communication attempts during $[t_m, t_m + \tau_m)$. The new DoS attack time period set and its complement can be expressed as

$$\overline{\Xi}_d(t_1, t_2) = \bigcup_{m \in \mathbb{N}} \overline{D}_m \cap [t_1, t_2]$$
$$\overline{\Xi}_h(t_1, t_2) = [t_1, t_2] \setminus \overline{\Xi}_d(t_1, t_2).$$

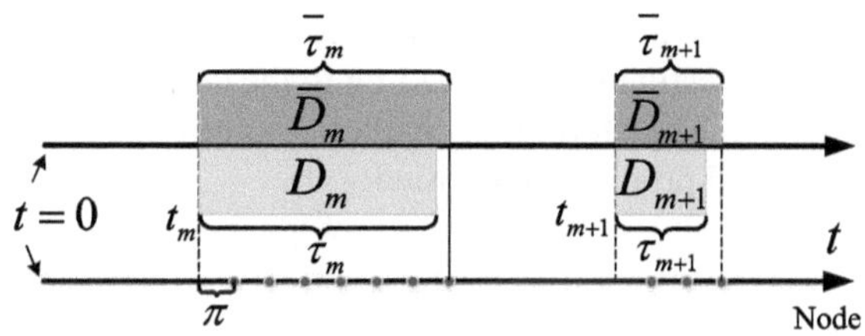

Fig. 10.1 Schematic diagram of the DoS attack sequence. The red circles indicate periodic communication attempts

In addition to considering the duration of DoS attacks, we also characterize them by their frequency. The frequency of DoS attacks is defined as follows.

Definition 10.1 Let $N_a(t_1, t_2)$ represent the number of DoS attacks in the interval $[t_1, t_2)$, then the frequency of attacks can be expressed as

$$F_a(t_1, t_2) = \frac{N_a(t_1, t_2)}{t_2 - t_1}. \tag{10.9}$$

Assumption 10.3 Let $|\Xi_d(t_1, t_2)|$ represent the duration of the DoS attack in the interval $[t_1, t_2)$, that is, $|\Xi_d(t_1, t_2)|$ represents the total length of the time period in set $\Xi_d(t_1, t_2)$. Then, there are constants $T_0 \geq 0$, $F_0 \geq 0, 0 < 1/T_1 < 1$, and $0 < 1/F_1 < 1$ such that

$$|\Xi_d(t_1, t_2)| \leq T_0 + \frac{t_2 - t_1}{T_1} \tag{10.10}$$

$$N_a(t_1, t_2) \leq F_0 + \frac{t_2 - t_1}{F_1} \tag{10.11}$$

where $1/T_1$ and $1/F_1$ are attack strength and attack frequency, respectively.

Remark 10.3 The Assumptions regarding the duration and frequency of DoS attacks as stated in Assumption 10.3 are standard in the related research, such as [9, 17]. The parameters T_0 and F_0 are employed to restrict the occurrence of DoS attacks at the initial time. The duration of an attack is limited by $1/T_1$, which can be interpreted as the proportion of the attack duration relative to the overall time period. Similarly, $1/F_1$ denotes the upper limit for the frequency of attacks.

Remark 10.4 Cyber attackers typically avoid conducting attacks in a sporadic or periodic manner, as these do not align with their objectives. The goal is to launch intensive attacks over a relatively short period to effectively disrupt the target after penetrating its defenses. Consequently, this chapter adopts a more realistic approach by considering aperiodic DoS attacks, focusing on the frequency and duration of these attacks, in contrast to the periodic DoS model and probabilistic DoS model found in [35, 36].

Remark 10.5 Unlike packet loss, a DoS attack can persist for an extended period. Packet loss is quantified as an integer number, with the count of successive packet losses typically being low. Moreover, the detection of DoS attacks has been extensively investigated [37], leading to the development of numerous effective detection methods. Therefore, this chapter assumes that DoS attacks can consistently be precisely and promptly identified.

10.2.6 Control Objective and Preliminaries

Definition 10.2 For any initial state, if the speed and position state of each agent in the system satisfies (10.12), then the heterogeneous MAS is said to achieve *cluster consensus* in finite-time.

$$\begin{aligned}
&\lim_{t\to t_f} \|x_i(t) - x^r_{\sigma_i}(t)\| = 0, \quad i \in \mathscr{V}, \quad t \in [0, t_f) \\
&\lim_{t\to t_f} \|v_i(t) - v^r_{\sigma_i}(t)\| = 0, \quad n_i \in \{1, 2, \cdots, s\}.
\end{aligned} \tag{10.12}$$

According to Definition 10.2, the aim of this chapter is to propose an event-triggered adaptive *fault-tolerant* pinning control scheme. This scheme is intended to ensure that the heterogeneous second-order MAS (10.1) and (10.2) with a directed graph that includes a directed spanning tree under aperiodic DoS attacks can reach *cluster consensus* within a *finite time* frame.

To attain the control objective, the following lemmas and assumptions are essential.

Assumption 10.4 There is a directed spanning tree in the directed communication topology of MAS (10.1) and (10.2).

Serving as the foundation for the derivation of the primary results in this chapter, the subsequent technical lemma analyses the finite-time upper bounds of the candidate piecewise Lyapunov functions under aperiodic DoS attacks.

Lemma 10.1 *Consider MAS (10.1) and (10.2) with the DoS attack sequence as shown in Fig. 10.1 and satisfying Assumption 10.3. If the system-related piecewise Lyapunov function $V(t)$ satisfies*

(i) if t belongs to a period of no DoS attack, namely $t \in \overline{\Xi}_h$,

$$V(t) = V_1(t),\ \dot{V}(t) \leq -\rho_1 V_1(t) + \rho_2 \tag{10.13}$$

(ii) if t belongs to a period of DoS attack, namely $t \in \overline{\Xi}_d$,

$$V(t) = V_2(t),\ \dot{V}(t) \leq \rho_3 V_2(t) + \rho_4 \tag{10.14}$$

where ρ_1, ρ_2, ρ_3, and ρ_4 are positive constants, and the constants T_1 and F_1 defined in Assumption 10.3 satisfy the following inequalities:

$$\frac{1}{T_1} < \frac{\rho_1 - \eta^*}{\rho_1 + \rho_3} \tag{10.15}$$

$$\frac{1}{F_1} < \frac{\eta^*}{2\ln\kappa + (\rho_1 + \rho_3)\pi} \tag{10.16}$$

where $0 < \eta^* < \rho_1$, $\pi > 0$ is the time interval between attempts to resume communication, $\kappa \geq 1$ is a constant satisfying

$$\begin{cases} \kappa V_2((t_m + \overline{\tau}_m)^-) - V_1(t_m + \overline{\tau}_m) \geq 0 \\ \kappa V_1(t_{m+1}^-) - V_2(t_{m+1}) \geq 0 \end{cases} \tag{10.17}$$

where $m \in \mathbb{N}$. Then, the signals of $V(t)$ are bounded in finite-time.

Proof First, in order to facilitate expression and understanding, use $t \in [t_{2s}, t_{2s+1})$ and $t \in [t_{2s+1}, t_{2s+2})$ with $s \in \mathbb{N}$ to represent $t \in \overline{\Xi}_h$ and $t \in \overline{\Xi}_d$, respectively; and define the following expression:

$$\epsilon(t) = \begin{cases} -\rho_1, & t \in [t_{2s}, t_{2s+1}) \\ \rho_3, & t \in [t_{2s+1}, t_{2s+2}) \end{cases} \tag{10.18}$$

$$\iota(t) = \begin{cases} \rho_2, & t \in [t_{2s}, t_{2s+1}) \\ \rho_4, & t \in [t_{2s+1}, t_{2s+2}). \end{cases} \tag{10.19}$$

Next, by integrating both sides of (10.13) and (10.14), the following inequality can be obtained:

$$V(t) \leq \begin{cases} e^{\epsilon(t_{2s})(t-t_{2s})} V_1(t_{2s}) + \int_{t_{2s}}^{t} e^{\epsilon(t_{2s})(t-z)} \iota(z)\, dz, & t \in [t_{2s}, t_{2s+1}) \\ e^{\epsilon(t_{2s+1})(t-t_{2s+1})} V_2(t_{2s+1}) + \int_{t_{2s+1}}^{t} e^{\epsilon(t_{2s+1})(t-z)} \iota(z)\, dz, & t \in [t_{2s+1}, t_{2s+2}). \end{cases} \tag{10.20}$$

Then, according to the definition of κ in (10.15), the following expression can be derived:

$$\begin{cases} \kappa V_2(t_{2s}^-) - V_1(t_{2s}) \geq 0 \\ \kappa V_1(t_{2s+1}^-) - V_2(t_{2s+1}) \geq 0. \end{cases} \tag{10.21}$$

Following [9], when $t \in [t_{2s}, t_{2s+1})$, combining (10.20) with (10.21), and through the iterative method, one can obtain

$$
\begin{aligned}
V(t) &\leq \kappa e^{\epsilon(t_{2s})(t-t_{2s})} V_2(t_{2s}^-) + \int_{t_{2s}}^{t} e^{\epsilon(t_{2s})(t-z)} \iota(z)\, dz \\
&\ \ \vdots \\
&\leq \kappa^{2s} e^{\epsilon(t_{2s})|\overline{\Xi}_h(t_0,t)|+\epsilon(t_{2s-1})|\overline{\Xi}_d(t_0,t)|} V(t_0) \\
&\quad + \kappa^{2s} \int_{t_0}^{t_1} e^{\Gamma_{2s}(t,2)+\epsilon(t_0)(t_1-z)} \iota(z)\, dz + \kappa^{2s} \int_{t_1}^{t_2} e^{\Gamma_{2s}(t,3)+\epsilon(t_1)(t_2-z)} \iota(z)\, dz \\
&\quad + \cdots + \kappa^2 \int_{t_{2s-2}}^{t_{2s-1}} e^{\Gamma_{2s}(t,2s)+\epsilon(t_{2s-2})(t_{2s-1}-z)} \iota(z)\, dz \\
&\quad + \kappa^2 \int_{t_{2s-1}}^{t_{2s}} e^{\Gamma_{2s}(t,2s+1)+\epsilon(t_{2s-1})(t_{2s}-z)} \iota(z)\, dz + \int_{t_{2s}}^{t} e^{\epsilon(t_{2s})(t-z)} \iota(z)\, dz
\end{aligned} \tag{10.22}
$$

where

$$
\Gamma_{2s}(t, p) = \epsilon(t_{2s})(t - t_{2s}) + \sum_{q=p}^{2s} \epsilon(t_{q-1})(t_q - t_{q-1}).
$$

Similarly, when $t \in [t_{2s+1}, t_{2s+2})$, one can get the following expression:

$$
\begin{aligned}
V(t) &\leq \kappa^{2(s+1)} e^{\epsilon(t_{2s})|\overline{\Xi}_h(t_0,t)|+\epsilon(t_{2s+1})|\overline{\Xi}_d(t_0,t)|} V(t_0) \\
&\quad + \kappa^{2(s+1)} \int_{t_0}^{t_1} e^{\Gamma_{2s+1}(t,2)+\epsilon(t_0)(t_1-z)} \iota(z)\, dz + \kappa^{2(s+1)} \int_{t_1}^{t_2} e^{\Gamma_{2s+1}(t,3)+\epsilon(t_1)(t_2-z)} \iota(z)\, dz \\
&\quad + \cdots + \kappa^4 \int_{t_{2s-1}}^{t_{2s}} e^{\Gamma_{2s+1}(t,2s+1)+\epsilon(t_{2s-1})(t_{2s}-z)} \iota(z)\, dz \\
&\quad + \kappa^2 \int_{t_{2s}}^{t_{2s+1}} e^{\Gamma_{2s+1}(t,2s+2)+\epsilon(t_{2s})(t_{2s+1}-z)} \iota(z)\, dz + \kappa^2 \int_{t_{2s+1}}^{t} e^{\epsilon(t_{2s+1})(t-z)} \iota(z)\, dz
\end{aligned} \tag{10.23}
$$

where

$$
\Gamma_{2s+1}(t, p) = \epsilon(t_{2s+1})(t - t_{2s+1}) + \sum_{q=p}^{2s+1} \epsilon(t_{q-1})(t_q - t_{q-1}).
$$

According to Definition 10.1, it can be found that when $t \in [t_{2s}, t_{2s+1})$, the number of DoS attacks $N_a(t_0, t) = s$, and when $t \in [t_{2s+1}, t_{2s+2})$, the number of attacks $N_a(t_0, t) = s + 1$. Therefore, it is deduced by combining inequalities (10.22) with (10.23) that

$$V(t) \leq \kappa^{2N_a(t_0,t)} e^{-\rho_1|\overline{\Xi}_h(t_0,t)|+\rho_3|\overline{\Xi}_d(t_0,t)|} V(t_0) + \int_{t_0}^{t} \kappa^{2N_a(z,t)} e^{-\rho_1|\overline{\Xi}_h(z,t)|+\rho_3|\overline{\Xi}_d(z,t)|} \iota \, dz \tag{10.24}$$

where $\iota = \max\{\rho_2, \rho_4\}$.

It can be known from Fig. 10.1 that

$$|\overline{\Xi}_h(t_0, t)| = t - t_0 - |\overline{\Xi}_d(t_0, t)|$$
$$|\overline{\Xi}_d(t_0, t)| \leq |\Xi_d(t_0, t)| + N_a(t_0, t)\pi.$$

Then, by utilizing (10.24) and Assumption 10.3 under the conditions (10.15) and (10.16), one can yield

$$V(t) \leq c e^{-\varkappa(t-t_0)} V(t_0) + \frac{\iota c}{\varkappa} \tag{10.25}$$

where

$$c = e^{(\rho_+\rho_3)T_0+[(\rho_+\rho_3)\pi+\ln\kappa]F_0}$$
$$\varkappa = \rho_1 - (\rho_+\rho_3)\frac{1}{T_1} - \eta^* > 0.$$

Then, there exists $0 \leq t_f < \infty$ such that $V(t_f) = V_{\max}$, where $V_{\max} > 0$ is a sufficiently large constant. In addition, it can be obtained from (10.25) that

$$t_f = \frac{\ln \frac{V(t_f) - \frac{\iota c}{\varkappa}}{cV(t_0)}}{-\varkappa} + t_0 = \frac{\ln \frac{V_{\max} - \frac{\iota c}{\varkappa}}{cV(t_0)}}{-\varkappa} + t_0 \tag{10.26}$$

which implies that $V(t) \leq V_{\max}$ if $t > t_f$. This shows that the signals in $V(t)$ are bounded in finite-time. The proof is completed. □

Remark 10.6 Lemma 10.1 implies that by constraining both the frequency and duration of DoS attacks, it is possible to guarantee that the signals of the Lyapunov function $V(t)$ remain bounded throughout the entire time span. Furthermore, the maximum allowable duration of a DoS attack can be ascertained through Assumption 10.3 in conjunction with condition (10.15).

10.3 Event-Triggered Distributed Fault-Tolerant Pinning Control

In this section, a novel event-triggered distributed *fault-tolerant* pinning control strategy is formulated. The schematic diagram of the strategy is illustrated in Fig. 10.2. In addition, the analysis of *cluster consensus* under the designed control strategy is carried out.

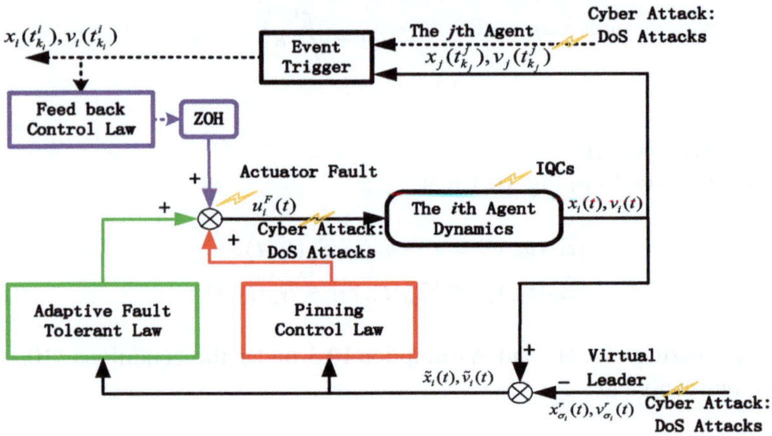

Fig. 10.2 Schematic of the designed event-triggered distributed fault-tolerant pinning controller

10.3.1 Event-Triggered Fault-Tolerant Pinning Controller

In the designed strategy, *fault-tolerant cluster consensus* is targeted through the implementation of pinning control and *fault-tolerant* control techniques. Moreover, an ETM is deployed to economize on communication resources. The designed control strategy is as follows:

$$\begin{aligned} u_i(t) &= \begin{cases} u_{1i}(t) + u_{2i}(t) + u_{3i}(t), & t \in \overline{\Xi}_h \cap [t^i_{k_i}, t^i_{k_i+1}) \\ 0, \quad t \in \overline{\Xi}_d \end{cases} \\ u_{1i}(t) &= -\varphi \sum_{j=1}^{N} l_{ij} x_j(t^j_{k_j}) - \varphi \sum_{j=1}^{N} l_{ij} v_j(t^j_{k_j}) \\ u_{2i}(t) &= \begin{cases} -b_i \tilde{x}_i(t) - b_i \tilde{v}_i(t) + \mathscr{P}_i(t), & i \in \tilde{V}_{\sigma_i} \\ -b_i \tilde{x}_i(t) - b_i \tilde{v}_i(t), \quad \mathscr{D}_i = 0 \\ 0, \text{otherwise} \end{cases} \\ u_{3i}(t) &= -\text{sgn}(s_i(t)) \Psi_i(t) \end{aligned} \tag{10.27}$$

where

$$s_i(t) = \tilde{x}_i(t) + \tilde{v}_i(t), \quad \tilde{x}_i(t) = x_i(t) - x^r_{\sigma_i}(t), \quad \tilde{v}_i(t) = v_i(t) - v^r_{\sigma_i}(t).$$

In addition, φ and $b_i > 0$ are the coupling coefficient and the pinning gains of the pinned agents, respectively; $t^j_{k_j}$ is the k_jth triggered time instance for the jth agent, and $k_j = \arg\min_m\{t - t_m^j | t > t_m^j, m \in \mathbb{N}\}$ denotes the sequential number of the most recent trigger for the jth agent, $\mathscr{P}_i(t)$ is defined as

$$\mathscr{P}_i(t) = \varphi \sum_{j=1}^{N} l_{ij} x_{y_j}^r(t) + \varphi \sum_{j=1}^{N} l_{ij} v_{y_j}^r(t) \tag{10.28}$$

and $\Psi_i(t)$ is designed as

$$\Psi_i(t) = \frac{\zeta_{2i}|v_i(t)| + \hat{\varpi}_{iw}}{\underline{\rho}_i - \zeta_{1i}\overline{\rho}_i} \tag{10.29}$$

$$\dot{\hat{\varpi}}_{iw}(t) = \begin{cases} -\vartheta_i \hat{\varpi}_{iw}(t) + z_i(t)|s_i(t)|, & t \in \overline{\Xi}_h \\ 0, \quad t \in \overline{\Xi}_d \end{cases} \tag{10.30}$$

where $z_i(t)$ is a positive bounded function and satisfies

$$\lim_{t\to\infty} \int_{t_0}^{t} z_i(\tau)\, d\tau \le \overline{z}_i < +\infty$$

with $\overline{z}_i > 0$ being an unknown constant; ζ_{1i} and ζ_{2i} are defined in (10.5); $\hat{\varpi}_{iw}$ is an estimate of ϖ_{iw}, where ϖ_{iw} is an unknown constant and will be defined in (10.50) below. The sequence of triggering instants of the jth agent is also expressed as $\{t_{k_j}^j\}$. For the ith agent, the next triggering instant is designed by

$$t_{k_i+1}^i = \inf\{t > t_{k_i}^i \ge t_0 | \mathscr{H}_i(t) \ge 0\} \tag{10.31}$$

in which

$$\begin{aligned} \mathscr{H}_i(t) &= \|e_{x_i}(t)\|^2 + \|e_{v_i}(t)\|^2 - \Phi_i(t, t_{k_j}^j) - \mu_i e^{-\theta_i(t-t_0)} \\ \Phi_i(t, t_{k_j}^j) &= \varepsilon_1 \left[\left(\sum_{j=1}^{N} l_{ij}\left(- x_j(t_{k_j}^j) + x_{n_j}^r(t) \right) \right)^2 + \left(\sum_{j=1}^{N} l_{ij}\left(- v_j(t_{k_j}^j) + v_{n_j}^r(t) \right) \right)^2 \right] \\ &\quad + \varepsilon_2 \left(\tilde{x}_i^2(t) + \tilde{v}_i^2(t) \right) \end{aligned} \tag{10.32}$$

where

$$e_{x_i}(t) = x_i(t_{k_i}^i) - x_i(t), \quad e_{v_i}(t) = v_i(t_{k_i}^i) - v_i(t).$$

Furthermore, μ_i, θ_i, ε_1, and ε_2 are positive constants.

Remark 10.7 As shown in Fig. 10.2, the control signal (10.27) consists of three distinct parts: the feedback control part, the pinning control part, and the adaptive *fault-tolerant* part. Unlike the approach in [13], our proposed control strategy does not necessitate the in-degree balance in the communication topology among different clusters, enhancing its versatility. Additionally, the feedback control part $u_{1i}(t)$ is formulated to enable the MASs to achieve *cluster consensus*.

Remark 10.8 The pinning control law $u_{2i}(t)$ within the control framework is designed to target specific agents for pinning, rather than pinning every agent in the system. Echoing the principles in [2], there are two principal types of agents that warrant pinning: the first type includes agents capable of receiving information from agents of other clusters, i.e., $i \in \tilde{V}_{\sigma_i}$. The second type encompasses agents, specifically the root node in the communication graph, that do not receive information from any other agents, i.e., $\mathscr{D}_i = 0$. For agents in the first type, a non-zero $\mathscr{P}_i(t)$ is required to counteract the influence of agents from other clusters; for the others, $\mathscr{P}_i(t) = 0$ applies. Contrasting with the fully equipped control approach detailed in [20], the pinning control method presented in this chapter conserves physical resources and computational load. Moreover, in comparison to the random node selection for pinning as in [38], the method in this chapter diminishes the element of randomness. These benefits become particularly pronounced as the number of agents grows.

Remark 10.9 The controller u_{3i} is utilized to counteract the adverse effects of faults and IQCs. Unlike the *fault-tolerant* method in [20] that necessitates the estimation of three parameters, the *fault-tolerant* controller proposed in this chapter requires the estimation of only one parameter. This significantly lightens the computational load and complexity, and conserves resources. Furthermore, barring the onset and cessation of DoS attacks, the *fault-tolerant* controller $u_{3i}(t)$ is free from chattering issues, given that $s_i(t)$ and $\Psi_i(t)$ are continuous and exhibit no abrupt changes.

Remark 10.10 The ETM proposed in this chapter indicates that the trigger conditions (10.32) require only the latest state data $x_j(t_{k_j}^j)$ and $v_j(t_{k_j}^j)$ of neighboring agents to assess whether the conditions for triggering are met, instead of the continuous information $x_j(t)$ as required in the trigger conditions found in [5] and [7]. Although the results in [5, 7] reduce the controller update frequency, they do not alleviate the communication burden. Moreover, in contrast to the synchronous triggering within clusters [8], the asynchronous and distributed ETM designed in this chapter reduces computational complexity.

Remark 10.11 It should be noted that the event-triggering condition (10.32) solely necessitates the state information of the agent and its neighbors. Moreover, the changes of the parameters of the conditions adhere to specific rules (see Remark 10.13). Consequently, the ETM can be readily implemented in practical systems.

10.3.2 Cluster Consensus Analysis

The subsequent theorem establishes that the MAS can achieve *cluster consensus*, with the state error system's signals being bounded in *finite time*.

Theorem 10.1 *Consider MAS (10.1) and (10.2) satisfying Assumptions 10.1, 10.2 and 10.4 and DoS attacks satisfying Assumption 10.3. Under the controller (10.27) with event-triggering conditions (10.32) and adaptive law (10.29) and (10.30), the system can achieve the control objective (10.12), if there exists* $\eta^* \in (0, \rho_1)$ *such that the conditions (10.15) and (10.16) in Lemma 10.1 and the following conditions are satisfied for a given* $\rho_1 > 0$*:*

$$\lambda_{\max}(\tilde{L}_l) < -\frac{\sigma + \varphi\chi}{\varphi} \tag{10.33}$$

$$\lambda_{\max}(\tilde{L}_l) + \chi < 0 \tag{10.34}$$

$$b_i > \frac{\lambda_{\max}(B - CM_l^{-1}C^T)}{\tilde{\rho}_i}, \; i \in \tilde{V}_{\sigma_i} \text{ or } \mathscr{D}_i = 0 \tag{10.35}$$

where

$$\tilde{L} = \frac{(-\tilde{\rho}\mathscr{L} - \mathscr{L}^T\tilde{\rho})}{2}, \quad \tilde{\rho} = \operatorname{diag}\{\tilde{\rho}_1, \tilde{\rho}_2, \cdots, \tilde{\rho}_N\}, \quad \tilde{\rho}_i = \underline{\rho}_i - \zeta_{1i}\overline{\rho}_i$$
$$\sigma = \max\{m, n+1\}, \quad m = \max_{1\le i\le N}\left\{\frac{3\alpha_i + \beta_i}{2}\right\}, \quad n = \max_{1\le i\le N}\left\{\frac{3\beta_i + \alpha_i}{2}\right\}. \tag{10.36}$$

In addition, $\tilde{L}_l$ *is a matrix composed of the last* $N - l$ *row-columns of* $\tilde{L}$*;* α_i *and* β_i *are defined in (10.8); B, C and* M_l *are proposed in Lemma 2.11;* χ *is defined in (10.49). The parameters defined by Lemma 10.1 in this theorem are expressed as*

$$\kappa = \max\left\{\frac{\lambda_{\max}(\Omega_1)}{\lambda_{\min}(\Omega_2)}, \frac{\lambda_{\max}(\Omega_2)}{\lambda_{\min}(\Omega_1)}\right\} \ge 1$$

where Ω_1 *is defined in (10.44),* Ω_2 *is defined in (10.60), and* ρ_1 *and* ρ_3 *are defined in (10.58) and (10.62), respectively.*

Proof Choose the following candidate piecewise Lyapunov function:

$$V(t) = \begin{cases} V_1(t), & t \in \overline{\Xi}_h(t_0, t) \\ V_2(t), & t \in \overline{\Xi}_d(t_0, t). \end{cases} \tag{10.37}$$

The proof is carried out in three steps. The initial two steps provide the decay estimates for $V_1(t)$ and $V_2(t)$, respectively. The third step demonstrates that all error signals within the MAS are uniformly bounded in *finite time*.

Step 1. When there is no DoS attack, namely $t \in \overline{\Xi}_h$.

In this case, the state error system can be defined as

$$\begin{aligned}
&\dot{\tilde{x}}_i(t) = \tilde{v}_i(t)\\
&\dot{\tilde{v}}_i(t) = \rho_l(t)u_i(t) + r_i(t) + \sum_{k=1}^{p_i} B_{ik}w_{ik}(t) + f_{\sigma_i}(t, x_i(t), v_i(t)) - f_{\sigma_i}(t, x^r_{\sigma_i}(t), v^r_{\sigma_i}(t)).
\end{aligned} \tag{10.38}$$

Note that when all agents satisfy $\mathscr{H}_i(t) < 0$ in (10.31), and one has

$$\begin{aligned}
\|e(t)\|^2 \triangleq& \sum_{i=1}^{N}(\|e_{x_i}(t)\|^2 + \|e_{v_i}(t)\|^2)\\
\leq&\varepsilon_1\big[\mathscr{L}\big(e_x(t) + \tilde{x}(t)\big)\big]^T\big[\mathscr{L}\big(e_x(t) + \tilde{x}(t)\big)\big] + \varepsilon_2\big(\tilde{x}^T(t)\tilde{x}(t)+\tilde{v}^T(t)\tilde{v}(t)\big)\\
&+\varepsilon_1\big[\mathscr{L}\big(e_v(t) + \tilde{v}(t)\big)\big]^T\big[\mathscr{L}\big(e_v(t) + \tilde{v}(t)\big)\big]+\sum_{i=1}^{N}\mu_i e^{-\theta_i(t-t_0)}\\
\leq&\varepsilon_1[e^T(t)(I_2\otimes\mathscr{L}^T\mathscr{L})e(t) + \varsigma^T(t)(I_2\otimes\mathscr{L}^T\mathscr{L})\varsigma(t)\\
&+2e^T(t)(I_2\otimes\mathscr{L}^T\mathscr{L})\varsigma(t)]+\varepsilon_2\varsigma^T(t)(I_2\otimes I_N)\varsigma(t) + \sum_{i=1}^{N}\mu_i e^{-\theta_i(t-t_0)}
\end{aligned} \tag{10.39}$$

where e_{x_i} and e_{v_i} are defined as in (10.32), and

$$\begin{aligned}
e(t) &= [e_x^T(t), e_v^T(t)]^T, \quad \varsigma(t) = [\tilde{x}^T(t), \tilde{v}^T(t)]^T\\
e_x^T(t) &= [e_{x_1}^T(t), e_{x_2}^T(t), \cdots, e_{x_N}^T(t)], \quad e_v^T(t) = [e_{v_1}^T(t), e_{v_2}^T(t), \cdots, e_{v_N}^T(t)]\\
\tilde{x}(t) &= [\tilde{x}_1^T(t), \tilde{x}_2^T(t), \cdots, \tilde{x}_N^T(t)]^T, \quad \tilde{v}(t) = [\tilde{v}_1^T(t), \tilde{v}_2^T(t), \cdots, \tilde{v}_N^T(t)]^T.
\end{aligned}$$

Using *Young's inequality* in Lemma 2.7 for any positive constant a, the formula (10.39) is transformed into

$$\begin{aligned}
\|e(t)\|^2 \leq& \varepsilon_1[e^T(t)(I_2\otimes\mathscr{L}^T\mathscr{L})e(t) + \varsigma^T(t)(I_2\otimes\mathscr{L}^T\mathscr{L})\varsigma(t)\\
&+ ae^T(t)e(t)+\frac{1}{a}\varsigma^T(t)(I_2\otimes\mathscr{L}^T\mathscr{L}\mathscr{L}^T\mathscr{L})\varsigma(t)]\\
&+ \varepsilon_2\varsigma^T(t)(I_2\otimes I_N)\varsigma(t) + \sum_{i=1}^{N}\mu_i e^{-\theta_i(t-t_0)}.
\end{aligned} \tag{10.40}$$

By simplifying, one has

$$\|e(t)\| \leq \eta\|\varsigma(t)\| + \gamma\sqrt{\sum_{i=1}^{N}\frac{\mu_i}{\varepsilon_1}e^{-\theta_i(t-t_0)}} \tag{10.41}$$

where η and λ are defined as

$$\eta = \sqrt{\frac{\lambda_{\max}(\mathscr{L}^T\mathscr{L} + \frac{\varepsilon_2}{\varepsilon_1}I_N + \frac{1}{a}\mathscr{L}^T\mathscr{L}\mathscr{L}^T\mathscr{L})}{\lambda_{\min}[(\frac{1}{\varepsilon_1} - a)I_N - \mathscr{L}^T\mathscr{L}]}} \tag{10.42}$$

$$\gamma = \sqrt{\frac{1}{\lambda_{\min}[(\frac{1}{\varepsilon_1} - a)I_N - \mathscr{L}^T\mathscr{L}]}}. \tag{10.43}$$

Then, choose the following Lyapunov function:

$$V_1(t) = \frac{1}{2}\varsigma^T(t)\Omega_1\varsigma(t) + \frac{1}{2}\sum_{i=1}^{N}\overline{z}_i^{-1}\tilde{\varpi}_{iw}^2 \tag{10.44}$$

where $\bar{z}_i$ is defined as in (10.29),

$$\begin{aligned}
\tilde{\rho} &= \text{diag}\{\tilde{\rho}_1, \tilde{\rho}_2, \cdots, \tilde{\rho}_N\}, \quad \tilde{\rho}_i = (\underline{\rho}_i - \zeta_{1i}\overline{\rho}_i) \\
\mathscr{B} &= \text{diag}\{b_1, b_2, \cdots, b_l, 0, \cdots, 0\}, \quad b_i > 0 \\
\tilde{\varpi}_{iw} &= \hat{\varpi}_{iw} - \varpi_{iw}, \quad \dot{\tilde{\varpi}}_{iw} = \dot{\hat{\varpi}}_{iw} \\
\Omega_1 &= \begin{bmatrix} \varphi(\tilde{\rho}\mathscr{L} + \mathscr{L}^T\tilde{\rho}) + 2\tilde{\rho}\mathscr{B} & I_N \\ I_N & I_N \end{bmatrix}.
\end{aligned}$$

In addition, based on Lemmas 2.8 and 2.1, when φ and b_i take appropriate values such that

$$\varphi(\tilde{\rho}\mathscr{L} + \mathscr{L}^T\tilde{\rho}) + 2\tilde{\rho}D > 0$$

the matrix Ω_1 is positive definite matrix.

According to the state error system (10.38), the derivative of $V_1(t)$ is

$$\begin{aligned}
\dot{V}_1(t) &= \tilde{x}^T[\varphi(\tilde{\rho}\mathscr{L} + \mathscr{L}^T\tilde{\rho}) + 2\tilde{\rho}\mathscr{B}]\tilde{v} + \tilde{v}^T\dot{\tilde{v}} + \tilde{v}^T\tilde{v} + \tilde{x}^T\dot{\tilde{v}} + \sum_{i=1}^{N}\overline{z}_i^{-1}\dot{\tilde{\varpi}}_{iw}\tilde{\varpi}_{iw} \\
&= \tilde{x}^T[\varphi(\tilde{\rho}\mathscr{L} + \mathscr{L}^T\tilde{\rho}) + 2\tilde{\rho}\mathscr{B}]\tilde{v} + \tilde{v}^T\tilde{v} + \sum_{i=1}^{N}(\tilde{x}_i + \tilde{v}_i)\rho_i(t)u_i \\
&\quad + \sum_{i=1}^{N}(\tilde{x}_i + \tilde{v}_i)(f_{\sigma_i}(t, x_i, v_i) - f_{\sigma_i}(t, x_{\sigma_i}^r, v_{\sigma_i}^r)) + \sum_{i=1}^{N}(\tilde{x}_i + \tilde{v}_i)r_i(t) \\
&\quad + \sum_{i=1}^{N}(\tilde{x}_i + \tilde{v}_i)\sum_{k=1}^{p_i}B_{i,k}w_{i,k}(t) + \sum_{i=1}^{N}\overline{z}_i^{-1}\dot{\tilde{\varpi}}_{iw}\tilde{\varpi}_{iw}.
\end{aligned} \tag{10.45}$$

Recalling $s_i = \tilde{x}_i + \tilde{v}_i$, and according to (10.5), one can obtain

$$\begin{aligned}\dot{V}_1(t) \le\, & \tilde{x}^T[\varphi(\tilde{\rho}\mathscr{L} + \mathscr{L}^T\tilde{\rho}) + 2\tilde{\rho}\mathscr{B}]\tilde{v} + \tilde{v}^T\tilde{v} + \sum_{i=1}^{N} s_i\rho_i(t)u_i \\ & + \sum_{i=1}^{N} s_i(f_{\sigma_i}(t, x_i, v_i) - f_{\sigma_i}(t, x^r_{\sigma_i}, v^r_{\sigma_i})) + \sum_{i=1}^{N} \zeta_{1i}\overline{\rho}_i|u_i||s_i| \\ & + \sum_{i=1}^{N}[\zeta_{2i}|v_i(t)| + (1+\zeta_{1i})\overline{r}_i + \delta_{di}]|s_i| + \sum_{i=1}^{N}\overline{z}_i^{-1}\dot{\tilde{\varpi}}_{iw}\tilde{\varpi}_{iw}. \end{aligned} \tag{10.46}$$

In order to simplify the calculation, define $\tilde{u}_1$ as

$$\begin{aligned}\tilde{u}_1 \triangleq\, & \sum_{i=1}^{N}[s_i\rho_i(t)u_i + \zeta_{1i}\overline{\rho}_i|u_i||s_i| + \overline{z}_i^{-1}\dot{\tilde{\varpi}}_{iw}\tilde{\varpi}_{iw}] \\ & + \sum_{i=1}^{N}[\zeta_{2i}|v_i(t)| + (1+\zeta_{1i})\overline{r}_i + \delta_{di}]|s_i| \\ = & \sum_{i=1}^{N}[s_i\rho_i(u_{1i} + u_{2i}) + \zeta_{1i}\overline{\rho}_i|u_{1i} + u_{2i}||s_i|] \\ & + \sum_{i=1}^{N}[s_i\rho_i(t)u_{3i} + \zeta_{1i}\overline{\rho}_i|u_{3i}||s_i| + \overline{z}_i^{-1}\dot{\tilde{\varpi}}_{iw}\tilde{\varpi}_{iw}] \\ & + \sum_{i=1}^{N}[\zeta_{2i}|v_i(t)| + (1+\zeta_{1i})\overline{r}_i + \delta_{di}]|s_i| \\ = & \,\tilde{u}_{11} + \tilde{u}_{12} \end{aligned} \tag{10.47}$$

where

$$\tilde{u}_{11} \triangleq \sum_{i=1}^{N}[s_i\rho_i(u_{1i} + u_{2i}) + \zeta_{1i}\overline{\rho}_i|u_{1i} + u_{2i}||s_i|]$$

and $\tilde{u}_{12}$ is the part of (10.47) except $\tilde{u}_{11}$. Then, one has

$$
\begin{aligned}
\tilde{u}_{11} \le & -\sum_{i=1}^{N} s_i \rho_i \varphi \sum_{j=1}^{N} l_{ij}[x_j(t_{k_j}^j) + v_j(t_{k_j}^j)] - \sum_{i=1}^{N} s_i \rho_i (b_i s_i - \mathscr{P}_i) \\
& + \sum_{i=1}^{N} \zeta_{1i} \overline{\rho}_i |\varphi \sum_{j=1}^{N} l_{ij}[x_j(t_{k_j}^j) + v_j(t_{k_j}^j)]||s_i| + \sum_{i=1}^{N} \zeta_{1i} \overline{\rho}_i |b_i s_i + \mathscr{P}_i||s_i| \\
= & -\sum_{i=1}^{N} (\underline{\rho}_i - \zeta_{1i}\overline{\rho}_i) \left| s_i \varphi \sum_{j=1}^{N} l_{ij} \left(x_j(t_{k_j}^j) + v_j(t_{k_j}^j) - x_{n_j}^r(t) - v_{n_j}^r(t) \right) + b_i s_i^2 \right| \\
= & -\sum_{i=1}^{N} (\underline{\rho}_i - \zeta_{1i}\overline{\rho}_i) \left| \varphi \sum_{j=1}^{N} s_i l_{ij} s_j + b_i s_i^2 \right| - \sum_{i=1}^{N} (\underline{\rho}_i - \zeta_{1i}\overline{\rho}_i) \left| s_i \varphi \sum_{j=1}^{N} l_{ij} \left(e_{x_j}(t) + e_{v_j}(t) \right) \right|
\end{aligned} \tag{10.48}
$$

where

$$
\begin{aligned}
& -\sum_{i=1}^{N} (\underline{\rho}_i - \zeta_{1i}\overline{\rho}_i) \left| s_i \varphi \sum_{j=1}^{N} l_{ij} \left(e_{x_j}(t) + e_{v_j}(t) \right) \right| \\
& = -\varphi \left| \begin{bmatrix} \tilde{x}_i(t) \\ \tilde{v}_i(t) \end{bmatrix}^T \begin{bmatrix} \tilde{\rho}\mathscr{L} & \tilde{\rho}\mathscr{L} \\ \tilde{\rho}\mathscr{L} & \tilde{\rho}\mathscr{L} \end{bmatrix} \begin{bmatrix} e_x(t) \\ e_v(t) \end{bmatrix} \right| \\
& \le 2\varphi \|\tilde{\rho}\mathscr{L}\| \|e(t)\| \|\varsigma(t)\| \\
& \le \varphi \chi \|\varsigma(t)\|^2 + \varphi \|\tilde{\rho}\mathscr{L}\| \gamma \sum_{i=1}^{N} \frac{\mu_i}{\varepsilon_1} e^{-\theta_i(t-t_0)}
\end{aligned} \tag{10.49}
$$

with $\chi = 2\|\tilde{\rho}\mathscr{L}\|(\eta + \gamma/2)$. In addition,

$$
\begin{aligned}
\tilde{u}_{12} \le & \sum_{i=1}^{N} [s_i \rho_i(t) u_{3i} + \zeta_{1i}\overline{\rho}_i |u_{3i}||s_i| + \overline{z}_i^{-1} \dot{\tilde{\varpi}}_{iw} \tilde{\varpi}_{iw}] \\
& + \sum_{i=1}^{N} [\zeta_{2i}|v_i(t)| + (1+\zeta_{1i})\overline{r}_i + \delta_{di}]|s_i|
\end{aligned} \tag{10.50}
$$

where $\varpi_{iw} := (1+\zeta_{1i})\overline{r}_i + \delta_{di}$. If $s_i(t) = 0$, MAS achieves *cluster consensus*, then $u_{3i}(t) = 0$, otherwise

$$
u_{3i}(t) = -\frac{s_i(t)(\zeta_{2i}|v_i(t)| + \hat{\varpi}_{iw})}{|s_i(t)|(\underline{\rho}_i - \zeta_{1i}\overline{\rho}_i)}.
$$

Hence the inequality (10.50) can be written as follows:

$$
\tilde{u}_{12} \le -\sum_{i=1}^{N} (\tilde{\varpi}_{iw}|s_i| - \overline{z}_i^{-1} \dot{\tilde{\varpi}}_{iw} \tilde{\varpi}_{iw}). \tag{10.51}
$$

Combining (10.48) with (10.51), it follows from (10.47) that

$$\tilde{u}_1 \leq -\sum_{i=1}^{N}(\tilde{\varpi}_{iw}|s_i| - \overline{z}_i^{-1}\dot{\tilde{\varpi}}_{iw}\tilde{\varpi}_{iw}) + \varphi\chi\varsigma^T(t)\varsigma(t)$$
$$-\sum_{i=1}^{N}(\underline{\rho}_i - \zeta_{1i}\overline{\rho}_i)\left|\varphi\sum_{j=1}^{N}s_i l_{ij}s_j + b_i s_i^2\right| + \varphi\|\rho\mathscr{L}\|\gamma\sum_{i=1}^{N}\frac{\mu_i}{\varepsilon_1}e^{-\theta_i(t-t_0)}. \quad (10.52)$$

Based on Assumption 10.2, one can get

$$\sum_{i=1}^{N}\tilde{x}_i^T\left(f_{\sigma_i}(t,x_i,v_i) - f_{\sigma_i}(t,x_{\sigma_i}^r,v_{\sigma_i}^r)\right) \leq \sum_{i=1}^{N}\tilde{x}_i^T(\alpha_i|\tilde{x}_i| + \beta_i|\tilde{v}_i|)$$
$$\leq \sum_{i=1}^{N}\left(\frac{2\alpha_i+\beta_i}{2}\tilde{x}_i^T\tilde{x}_i + \frac{\beta_i}{2}\tilde{v}_i^T\tilde{v}_i\right) \quad (10.53)$$

and

$$\sum_{i=1}^{N}\tilde{v}_i^T\left(f_{\sigma_i}(t,x_i,v_i) - f_{\sigma_i}(t,x_{\sigma_i}^r,v_{\sigma_i}^r)\right) \leq \sum_{i=1}^{N}\left(\frac{2\beta_i+\alpha_i}{2}\tilde{v}_i^T\tilde{v}_i + \frac{\alpha_i}{2}\tilde{x}_i^T\tilde{x}_i\right). \quad (10.54)$$

It can be obtained from (10.53) and (10.54) that

$$\sum_{i=1}^{N}(\tilde{x}_i+\tilde{v}_i)^T(f_{\sigma_i}(t,x_i,v_i) - f_{\sigma_i}(t,x_{\sigma_i}^r,v_{\sigma_i}^r))$$
$$\leq \sum_{i=1}^{N}\left(\frac{3\alpha_i+\beta_i}{2}\tilde{x}_i^T\tilde{x}_i + \frac{3\beta_i+\alpha_i}{2}\tilde{v}_i^T\tilde{v}_i\right)$$
$$\leq \max_{1\leq i\leq N}\left\{\frac{3\alpha_i+\beta_i}{2}\right\}\tilde{x}^T\tilde{x} + \max_{1\leq i\leq N}\left\{\frac{3\beta_i+\alpha_i}{2}\right\}\tilde{v}^T\tilde{v}$$
$$= m\tilde{x}^T\tilde{x} + n\tilde{v}^T\tilde{v} \quad (10.55)$$

where m and n are defined in Theorem 10.1.

Upon applying (10.52) and (10.55) into (10.46), it gives

$$\dot{V}_1(t) \leq \tilde{x}^T\left(\Lambda I_N - \varphi\tilde{\rho}\mathscr{L} - \tilde{\rho}\mathscr{B}\right)\tilde{x} + \tilde{v}^T\left(\Lambda I_N - \varphi\tilde{\rho}\mathscr{L} - \tilde{\rho}\mathscr{B}\right)\tilde{v}$$
$$-\sum_{i=1}^{N}(\tilde{\varpi}_{iw}|s_i| - \overline{z}_i^{-1}\dot{\tilde{\varpi}}_{iw}\tilde{\varpi}_{iw}) + \varphi\|\tilde{\rho}\mathscr{L}\|\gamma\sum_{i=1}^{N}\frac{\mu_i}{\varepsilon_1}e^{-\theta_i(t-t_0)} \quad (10.56)$$

where $\dot{\tilde{\varpi}}_{iw}(t) = \dot{\hat{\varpi}}_{iw}(t) > 0$, which means that the adaptive law $\tilde{\varpi}_{iw}$ is nonnegative if $\tilde{\varpi}_{iw}(0) \geq 0$; and $\Lambda = \sigma + \varphi\chi$ with σ and χ being defined in Theorem 10.1.

It is deduced by the definition of $\tilde{L}$ in (10.36) that

$$\dot{V}_1(t) \leq \tilde{x}^T\left(\Lambda I_N + \varphi\tilde{L} - \tilde{\rho}\mathscr{B}\right)\tilde{x} + \tilde{v}^T\left(\Lambda I_N + \varphi\tilde{L} - \tilde{\rho}\mathscr{B}\right)\tilde{v}$$
$$- \sum_{i=1}^{N} \vartheta_i \overline{z}_i^{-1} \tilde{\varpi}_{iw}^2 + \varphi\|\tilde{\rho}\mathscr{L}\|\gamma \sum_{i=1}^{N} \frac{\mu_i}{\varepsilon_1} e^{-\theta_i(t-t_0)}. \tag{10.57}$$

In light of Lemma 2.11, when the conditions (10.33), (10.34), and (10.35) in Theorem 10.1 are satisfied, $\Lambda I_N + \varphi\tilde{L} - \tilde{\rho}\mathscr{B}$ is negative definite. Therefore, the following inequality holds:

$$\begin{aligned}\dot{V}_1(t) &\leq \lambda_{\min}(\Theta)\varsigma^T(t)\varsigma(t) - \sum_{i=1}^{N} \vartheta_i \overline{z}_i^{-1} \tilde{\varpi}_{iw}^2 + \rho_2 \\ &\leq -\rho_1 V_1(t) + \rho_2 \end{aligned} \tag{10.58}$$

where

$$\rho_1 = \min\left\{\frac{-\lambda_{\min}(\Theta)}{\lambda_{\max}(\Omega_1)}, 2\vartheta\right\} > 0, \quad \vartheta = \min_{1\leq i\leq N}\{\vartheta_i\}$$
$$\rho_2 = \varphi\|\tilde{\rho}L\|\gamma \sum_{i=1}^{N} \frac{\overline{\mu}}{\varepsilon_1}, \quad \overline{\mu} = \max_{1\leq i\leq N}\{\mu_i\}$$
$$\Theta = \begin{bmatrix} \Lambda I_N + \varphi\tilde{L} - \tilde{\rho}\mathscr{B} & 0 \\ 0 & \Lambda I_N + \varphi\tilde{L} - \tilde{\rho}\mathscr{B} \end{bmatrix}.$$

Note that when $\Lambda I_N + \varphi\tilde{L} - \tilde{\rho}\mathscr{B}$ is negative definite, Ω_1 must be positive definite.

Step 2. When there are DoS attacks, namely $t \in \overline{\Xi}_d$.

When the channels for communication and control are impeded, the control input is nullified, that is, $u_i(t) = 0$, and the error system can be written as

$$\begin{aligned}\dot{\tilde{x}}_i(t) &= \tilde{v}_i(t), \\ \dot{\tilde{v}}_i(t) &= \sum_{k=1}^{p_i} B_{ik}w_{ik}(t) + f_{\sigma_i}(t, x_i(t), v_i(t)) - f_{\sigma_i}(t, x^r_{\sigma_i}(t), v^r_{\sigma_i}(t)). \end{aligned} \tag{10.59}$$

Then, choose a Lyapunov function as

$$V_2(t) = \frac{1}{2}\varsigma^T(t)\Omega_2\varsigma(t) + \frac{\lambda_{\min}(\Omega_2)}{2\lambda_{\max}(\Omega_1)} \sum_{i=1}^{N} \overline{z}_i^{-1}\tilde{\varpi}_{iw}^2 \tag{10.60}$$

where

$$\Omega_2 = \begin{bmatrix} I_N & 0 \\ 0 & I_N \end{bmatrix}.$$

Since $\dot{\tilde{\varpi}}_{iw}(t) = \dot{\hat{\varpi}}_{iw}(t) = 0$ when $t \in \overline{\Xi}_d$, the time derivative of $V_2(t)$ is derived as:

$$\begin{aligned} \dot{V}_2(t) &= \tilde{x}^T \tilde{v} + \tilde{v}^T \dot{\tilde{v}} + \frac{\lambda_{\min}(\Omega_2)}{\lambda_{\max}(\Omega_1)} \sum_{i=1}^{N} \overline{z}_i^{-1} \dot{\tilde{\varpi}}_{iw} \tilde{\varpi}_{iw} \\ &\leq \tilde{x}^T \tilde{v} + \sum_{i=1}^{N} \tilde{v}_i (\alpha_i \tilde{x}_i + \beta_i \tilde{v}_i) + \sum_{i=1}^{N} \tilde{v}_i (\zeta_{2i} |v_i| + \delta_{di}) \\ &\leq \tilde{x}^T \tilde{v} + \sum_{i=1}^{N} \left(\frac{2\beta_i + \alpha_i}{2} \tilde{v}^T \tilde{v} + \frac{\alpha_i}{2} \tilde{x}^T \tilde{x} \right) + \sum_{i=1}^{N} \tilde{v}_i (\zeta_{2i} |v_i| + \delta_{di}). \end{aligned} \tag{10.61}$$

According to the actual situation, it can be assumed that the speed of the agent has an upper bound $\overline{v}$, that is, $\overline{v} \geq \sup_{t \geq t_0} \{v_i\}$, $1 \leq i \leq N$. Then, it follows from (10.61) that

$$\begin{aligned} \dot{V}_2(t) &\leq \tilde{x}^T \tilde{v} + \phi \sum_{i=1}^{N} (\tilde{v}^T \tilde{v} + \tilde{x}^T \tilde{x}) + \sum_{i=1}^{N} 2\overline{v}(\zeta_{2i} \overline{v} + \overline{\delta}_d) \\ &\leq \varsigma^T(t) R \varsigma(t) + \rho_4 \\ &\leq \rho_3 V_2(t) + \rho_4 \end{aligned} \tag{10.62}$$

where

$$\begin{aligned} \phi &= \max_{1 \leq i \leq N} \left\{ \frac{2\beta_i + \alpha_i}{2} \right\}, \quad \overline{\delta} = \max_{1 \leq i \leq N} \{\delta_{di}\} \\ \rho_3 &= 2 \frac{\lambda_{\max}(R)}{\lambda_{\min}(\Omega_2)}, \quad \rho_4 = \sum_{i=1}^{N} 2\overline{v}(\zeta_{2i} \overline{v} + \overline{\delta}_d) \\ R &= \begin{bmatrix} \phi I_N & 0 \\ I_N & \phi I_N \end{bmatrix}. \end{aligned}$$

Step 3. Prove that the signals of the MAS are cluster uniformly bounded in finite-time.
Based on the above analysis, the following expression can be obtained

$$\dot{V}(t) \leq \begin{cases} -\rho_1 V_1(t) + \rho_2, & t \in \overline{\Xi}_h \\ \rho_3 V_2(t) + \rho_4, & t \in \overline{\Xi}_d. \end{cases} \tag{10.63}$$

With Lemma 10.1 being applied, it is easily deduced that all signals of $V(t)$ are bounded in *finite time*, i.e., the signals of the error system are finite-time bounded. In other words, it can be stated that all signals of the MAS are uniformly bounded in *finite time*. The proof is completed. □

Remark 10.12 In the recent years, a multitude of significant findings, such as [20, 33], have been dedicated to the adaptive *fault-tolerant* control challenges in MASs with IQCs and actuator faults. Distinguished from the pursuit of *complete consensus* in [33] and [20], this chapter tackles the more complex and significant objective of achieving *cluster consensus* for MASs under *cyber attacks* and actuator faults. This is accomplished by integrating event-triggered mechanisms with pinning control methods.

The occurrence of *Zeno behavior*must be precluded in the ETM of Theorem 10.1. This issue is specifically analyzed by Theorem 10.2.

Theorem 10.2 *For MAS (10.1) and (10.2), consider the event-triggered scheme (10.31) and controller (10.27). If the conditions in Theorem 10.1 are satisfied, then Zeno behavior is excluded. Mathematically, it satisfies* $t^i_{k_i+1} - t^i_{k_i} > 0$ *for the ith agent.*

Proof In the interval $[t^i_{k_i}, t^i_{k_i+1})$, if the conditions in (10.32) for event-triggering are not satisfied, one can get

$$\|e_{x_i}(t)\|^2 + \|e_{v_i}(t)\|^2 \leq \Phi_i(t, t^j_{k_j}) + \mu_i e^{-\theta_i(t-t_0)}. \tag{10.64}$$

One of its sufficient conditions is

$$\|e_{x_i}(t)\|^2 + \|e_{v_i}(t)\|^2 + \|e^r_{xi}(t)\|^2 + \|e^r_{vi}(t)\|^2 \leq \Phi_i(t, t^j_{k_j}) + \mu_i e^{-\theta_i(t-t_0)} \tag{10.65}$$

where

$$e^r_{xi}(t) = x^r_{\sigma_i}(t^i_{k_i}) - x^r_{\sigma_i}(t), \quad e^r_{vi}(t) = v^r_{\sigma_i}(t^i_{k_i}) - v^r_{\sigma_i}(t).$$

At this point, it can be said that when *Zeno behavior* is excluded under condition (10.65), then *Zeno behavior* under condition (10.64) is also excluded.

Next, taking the right-hand derivative of

$$e^*(t) = \|e_{x_i}(t)\| + \|e_{v_i}(t)\| + \|e^r_{xi}(t)\| + \|e^r_{vi}(t)\|$$

results in the following:

$$
\begin{aligned}
D^+(e^*(t)) &\le \|\dot{e}_{x_i}(t)\| + \|\dot{e}_{v_i}(t)\| + \|\dot{e}^r_{xi}(t)\| + \|\dot{e}^r_{vi}(t)\| \\
&\le \|e_{v_i}(t)\| + \|v_i(t^i_{k_i})\| + \left\| f_{\sigma_i}(t, x_i, v_i) + u_i^F(t) + \sum_{k=1}^{p_i} B_{ik} w_{ik} \right\| \\
&\quad + \|v^r_{\sigma_i}(t)\| + \|f_{\sigma_i}(t, x^r_{\sigma_i}, v^r_{\sigma_i})\| \\
&\le \|e_{v_i}(t)\| + \|e^r_{v_i}(t)\| + \|v_i(t^i_{k_i})\| + \|v^r_{\sigma_i}(t)\| + 2\overline{f} + (1+\zeta_{1i})\|u_{1i}(t)\| \\
&\quad + (1+\zeta_{1i})\overline{\rho}_i b_i e^*(t) + (1+\zeta_{1i})\overline{\rho}_i b_i \left\| x_i(t^i_{k_i}) - x^r_{\sigma_i}(t^i_{k_i}) + v_i(t^i_{k_i}) \right. \\
&\quad \left. - v^r_{\sigma_i}(t^i_{k_i}) \right\| + (1+\zeta_{1i})\overline{\rho}_i b + c \\
&\le \left[1 + (1+\zeta_{1i})\overline{\rho}_i b_i\right] e^*(t) + \mathscr{F}\left(t^i_{k_i}, t^k_{k_j}\right) \\
&= \Upsilon e^*(t) + \mathscr{F}\left(t^i_{k_i}, t^j_{k_j}\right)
\end{aligned}
\tag{10.66}
$$

where $\overline{f}$ is the upper bound of $f_{\sigma_i}(t, x_i, v_i)$.

$$
b = \|\mathscr{L}\|(\eta + \overline{\varsigma} + \gamma \sum_{i=1}^{N} \frac{\overline{\mu}}{\varepsilon_1}), \quad \overline{\mu} = \max_{1\le i\le N} \{\mu_i\}
$$

$$
\overline{\varsigma} = \sup_{t\ge t_0} \|\varsigma(t)\|, \quad c = \left(\frac{\overline{\rho}_i - \zeta_{1i}\underline{\rho}_i}{\underline{\rho}_i - \zeta_{1i}\overline{\rho}_i} - 1 \right) (\overline{v} + \overline{\varpi}_{iw})
$$

with $\overline{\varpi}_{iw}$ being the upper bound of $\tilde{\varpi}_{iw}$. In addition, Υ and $\mathscr{F}\left(t^i_{k_i}, t^j_{k_j}\right)$ are defined as follows:

$$
\begin{aligned}
\Upsilon &= 1 + (1+\zeta_{1i})\overline{\rho}_i b_i \\
\mathscr{F}\left(t^i_{k_i}, t^j_{k_j}\right) &= \|v_i(t^i_{k_i})\| + \|v^r_{\sigma_i}(t^i_{k_i})\| + 2\overline{f} + (1+\zeta_{1i})\overline{\rho}_i b + c \\
&\quad + (1+\zeta_{1i})\|u_{1i}(t)\| + (1+\zeta_{1i})\overline{\rho}_i b_i \left\| x_i(t^i_{k_i}) \right. \\
&\quad \left. - x^r_{\sigma_i}(t^i_{k_i}) + v_i(t^i_{k_i}) - v^r_{\sigma_i}(t^i_{k_i}) \right\|.
\end{aligned}
$$

Noting that Theorem 10.1 establishes the stability of the closed-loop system, so $\overline{f}$, $\overline{\varpi}_{iw}$ and $\overline{\varsigma}$ in (10.66) exist and are finite. According to the event-triggered scheme (10.31) and (10.32), $e_{x_i}(t^i_{k_i}) = e^r_{xi}(t^i_{k_i}) = e_{v_i}(t^i_{k_i}) = e^r_{v_i}(t^i_{k_i}) = 0$ can be obtained, then the following expression holds:

$$
e^*(t) \le \frac{\mathscr{F}\left(t^i_{k_i}, t^j_{k_j}\right)}{\Upsilon} \left(e^{\Upsilon(t - t^i_{k_i})} - 1 \right).
\tag{10.67}
$$

In addition, when the event-triggering condition is next satisfied, one has

$$
\|e_{x_i}(t^i_{k_i+1})\|^2 + \|e_{v_i}(t^i_{k_i+1})\|^2 \ge \Phi_i\left(t^i_{k_i+1}, t^i_{k_i}, t^j_{k_j}\right) + \mu_i e^{-\theta_i(t^i_{k_i+1} - t_0)}.
\tag{10.68}
$$

It is deduced by considering (10.67) and (10.68) that

$$\begin{aligned}
&\mu_i e^{-\theta_i(t^i_{k_i+1}-t_0)}\\
&\leq \Phi_i\left(t^i_{k_i+1}, t^i_{k_i}, t^j_{k_j}\right) + \mu_i e^{-\theta_i(t^i_{k_i+1}-t_0)}\\
&\leq \left\|e_{x_i}(t^i_{k_i+1})\right\|^2 + \left\|e_{v_i}(t^i_{k_i+1})\right\|^2 + \left\|e^r_{xi}(t^i_{k_i+1})\right\|^2 + \left\|e^r_{vi}(t^i_{k_i+1})\right\|^2\\
&\leq \left(\left\|e_{x_i}(t^i_{k_i+1})\right\| + \left\|e_{v_i}(t^i_{k_i+1})\right\| + \left\|e^r_{xi}(t^i_{k_i+1})\right\| + \left\|e^r_{vi}(t^i_{k_i+1})\right\|\right)^2\\
&\leq \left[\frac{\mathscr{F}\left(t^i_{k_i}, t^j_{k_j}\right)}{\Upsilon}\left(e^{\Upsilon(t^i_{k_i+1}-t^i_{k_i})} - 1\right)\right]^2.
\end{aligned} \tag{10.69}$$

The next step is using the method of contradiction to prove that *Zeno behavior* is excluded. Suppose $\lim_{k_i\to\infty} t^i_{k_i} = t^i_* < \infty$, then there exists k_i such that $t^i_{k_i+1} - t^i_{k_i} \to 0$, and from (10.69), the following can be derived:

$$\mu_i e^{-\theta_i(t^i_{k_i+1}-t^i_{k_i})} \leq e^{\theta_i(t^i_*-t_0)}\left[\frac{\mathscr{F}\left(t^i_{k_i}, t^j_{k_j}\right)}{\Upsilon}\left(e^{\Upsilon(t^i_{k_i+1}-t^i_{k_i})} - 1\right)\right]^2 \tag{10.70}$$

which means $\mu_i \leq 0$. This contradicts the fact that $\mu_i > 0$. Therefore, the hypothesis does not hold, and it is easy to get that

$$t^i_{k_i+1} - t^i_{k_i} > 0, \quad \lim_{k_i\to\infty} t^i_{k_i} = t^i_* = \infty.$$

Thus, it has been proved that *Zeno behavior* will not occur under the condition (10.65). Furthermore, the event-triggered scheme (10.31) and (10.32) will not cause *Zeno behavior* to occur. The proof is completed. □

Remark 10.13 The ETM proposed in this chapter takes inspiration from [22]. However, given the influence of actuator faults, IQCs, and DoS attacks, the position error $\tilde{x}_i(t)$ and velocity error $\tilde{v}_i(t)$ approach zero but do not reach it. This situation significantly narrows the parameter selection for the ETM of [22], which nearly degenerates the ETM to a continuous-time triggering scheme. In contrast to [22], this chapter introduces two positive constants μ_i and θ_i to ascertain and modulate the trigger moments during DoS attacks, thereby enhancing the flexibility of the proposed ETM.

10.4 Simulation Examples

In this section, a simulation example is provided to validate the efficacy of the proposed control scheme. The communication graph of the MAS is depicted in Fig. 8.2. The graph reveals that the system comprises five followers and two virtual leaders. These followers are split into two distinct clusters, with the agents in each cluster exhibiting heterogeneous dynamics. The interactions among the agents are directed, and the *directed graph* includes a single directed spanning tree, fulfilling the conditions of Assumption 10.4. The dynamics of the agents in the two clusters are expressed as

$$f_{n_1}(t, x_i(t), v_i(t)) = -\sin(x_i) - 0.25v_i + 6\cos(2.5t), \quad i \in \{1, 3, 4\}$$
$$f_{n_2}(t, x_i(t), v_i(t)) = 5\sin(2t) - 0.5v_i, \quad i \in \{2, 5\}.$$

Based on Assumption 10.2 and the definition of σ, m, and n in Theorem 10.1, it is easy to obtain that $m = 1.625$, $n = 0.875$ and $\sigma = 1.875$.

The gain matrices of IQCs and IQC output w_{i1} in the MAS are reproduced [20] as

$$B_{i1} = \begin{bmatrix} -0.5 & 0.5i \end{bmatrix}, \quad H_{i1} = \begin{bmatrix} -1 \\ i \end{bmatrix}, \quad G_{i1} = \begin{bmatrix} -0.2 \\ 0.2i \end{bmatrix}$$
$$w_{i1} = \sin(\xi_{i1})$$
$$\zeta_{1i} = 0.1\sqrt{1+i^2}, \quad \zeta_{2i} = 0.5\sqrt{1+i^2}, \quad i \in \{1, 2, \cdots, 5\}$$

and the fault scaling factor $\rho_i(t)$ and bias fault $r_i(t)$ are given as

$$\begin{cases} \rho_1(t) = 1.5 + 0.1\cos(0.1t) \\ \rho_2(t) = 0.8 - 0.1\sin(t) \\ \rho_3(t) = 0.8 + 0.2\sin(1.2t), \quad 5 \le t < 30 \\ \rho_4(t) = 0.8 + 0.1\sin(1.2t), \quad i \in \{1, 2, \cdots, 5\} \\ \rho_5(t) = 0.8 - 0.1\cos(0.5t) \\ r_i(t) = 0.5\sin(0.1i \times t). \end{cases}$$

The parameters of the event-triggered scheme and the adaptive law are selected as $\varepsilon_1 = \varepsilon_2 = \vartheta_i = 0.00001$, $\mu_i = 0.8$, $\theta_i = 0.008$; $z_i(t) = 0.5e^{-0.3t}$, and $\hat{\varpi}_{iw}(0) = 3$.

In accordance with the control strategy (10.27), since $\mathscr{D}_1 = 0$ and $i = 2 \in \tilde{V}_2$, the first and second agents are seted as pinning nodes. Consequently, the coupling coefficient and pinning gains that satisfy the criteria given in Theorem 10.1 are selected to be $\varphi = 12$, $d_1 = d_2 = 35$, and $d_3 = d_4 = d_5 = 0$. Furthermore, the initial states for the virtual leaders and followers are detailed in Table 10.1.

Table 10.1 The initial state values for agents

Agent	P_1	P_2	1	2	3	4	5
Position $x_i(0)$	1	1	–1	5	–3	4	2
Speed $v_i(0)$	1	1	–2	4	3	–4	6

The parameters for the DoS attacks under consideration are specified as: $T_0 = 3$, $F_0 = 4$, and $\eta^* = 0.393$. In this example, the length of the DoS attack is $|\Xi_d(0, 30)| = 3.79$, and the total number of attacks is $N_a(0, 30) = 5$, both of which comply with the respective limits of 4.294 and 5.320 as obtained from, resp., the conditions (10.15) and (10.16) in Lemma 10.1. The simulation diagram features grey bars that denote the specific times of the DoS attack.

Under the finite-time *cluster consensus* controller (10.27) with adaptive update laws (10.29) and the ETM (10.31) in Theorem 10.1, the secure *cluster consensus* results under both *cyber attacks* and actuator faults are shown in Fig. 10.3. The consensus errors are defined as

$$\text{Position Error} = \frac{1}{N}\sqrt{\sum_{i=1}^{N} \|x_i(t) - x^r_{\sigma_i}(t)\|^2}$$

$$\text{Velocity Error} = \frac{1}{N}\sqrt{\sum_{i=1}^{N} \|v_i(t) - v^r_{\sigma_i}(t)\|^2}.$$

As seen in Fig. 10.3d, the MAS is capable of achieving bounded *cluster consensus* despite the presence of actuator faults, IQCs, and DoS attacks. This illustrates the efficacy of the designed control strategy.

To highlight the effects of actuator faults, IQCs, and *cyber attacks* on *cluster consensus*, the control strategy designed in [3] is implemented on the same MAS under identical *cyber attacks* and actuator faults, with the simulation results presented in Fig. 10.4. Observations from Fig. 10.4 reveal that *cyber attacks*, IQCs, and actuator faults lead to the instability of the MAS. This underscores the effectiveness of the strategy designed in this chapter in mitigating the impacts of *cyber attacks* and actuator faults.

For the purpose of comparison, Fig. 10.5 displays the comparison between the proposed *fault-tolerant* controller in Theorem 10.1 and the method in [39] where $J(t)$ is designed as follows:

$$J(t) = \frac{1}{N}\sqrt{\sum_{i=1}^{N} \left(x_i(t) - x^r_{\sigma_i}(t)\right)^2 + \sum_{i=1}^{N} \left(v_i(t) - v^r_{\sigma_i}(t)\right)^2}.$$

It should be noted that regardless of the presence or absence of *cyber attacks*, the *fault-tolerant* control strategy presented in this chapter outperforms the control strategy proposed in [39]. Furthermore, by comparing Fig. 10.5a with b, it can be concluded that *cyber attacks* can impair or even annihilate the system's *cluster consensus* performance.

Furthermore, when comparing the designed ETM (10.31) with the one presented in [22] under similar system performance, it can be inferred from Fig. 10.6 and Table 10.2 that the communication load is significantly lessened by employing the

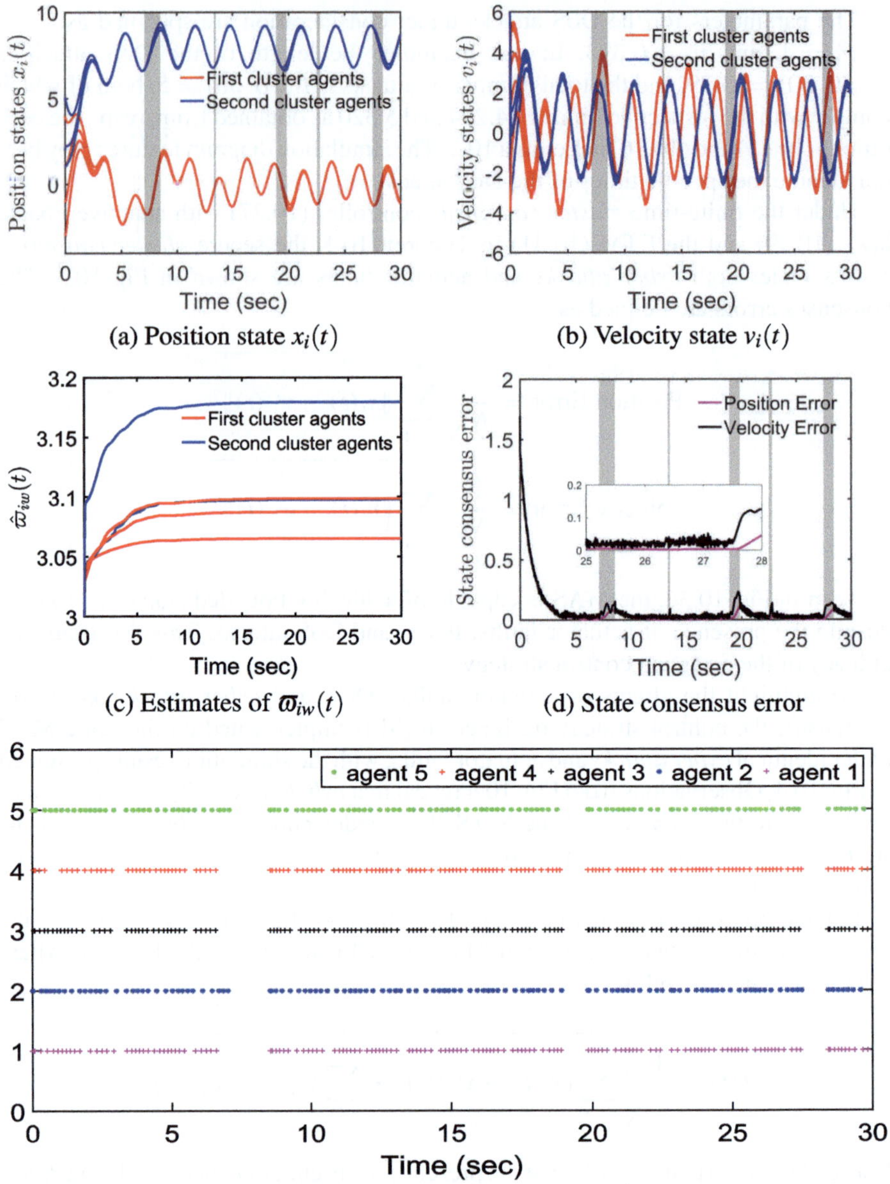

(a) Position state $x_i(t)$

(b) Velocity state $v_i(t)$

(c) Estimates of $\varpi_{iw}(t)$

(d) State consensus error

(e) Trigger instant for agents

Fig. 10.3 Simulation results based on the controller (10.27) in Theorem 10.1 under cyber attacks and actuator faults

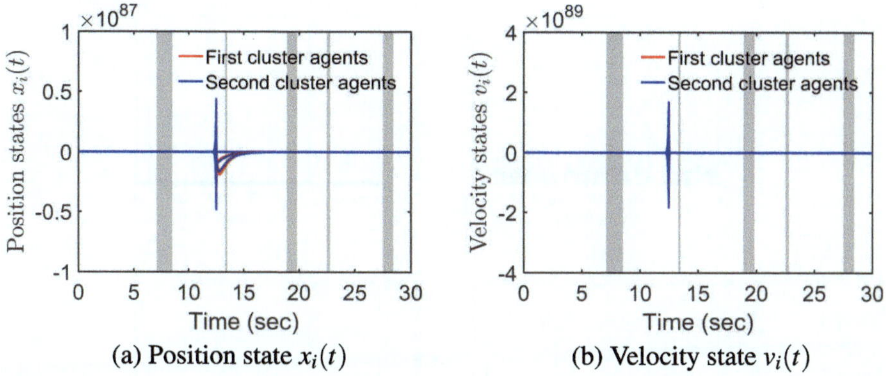

Fig. 10.4 Simulation results based on the controller of [3] under cyber attacks and actuator faults

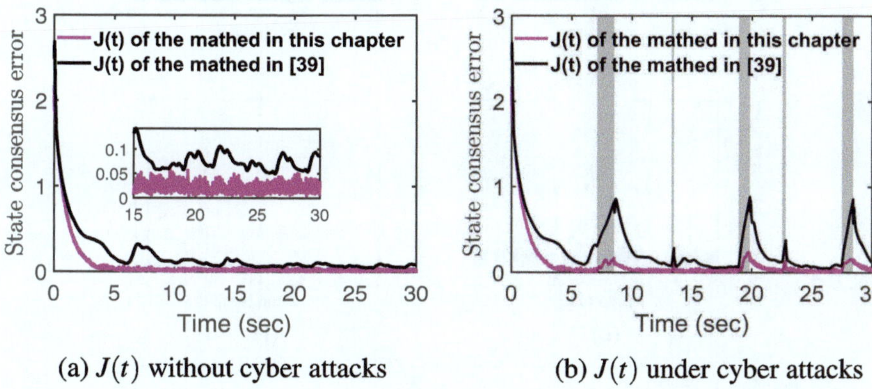

Fig. 10.5 Comparison between proposed fault-tolerant controller in Theorem 10.1 and the method in [39]

ETM (10.31). Additionally, as observed from Table 10.3, the frequency of triggered instants for each agent is contingent upon the values of the parameters μ_i, θ_i, ε_1, and ε_2. It is evident from Table 10.3 that the ETM proposed in (10.31) reverts to the one of [22] under the conditions $\mu_i = 0$ and $\theta_i = 1$. Moreover, it is noticeable that the lower the value of the parameter θ_i, the fewer the trigger numbers when μ_i, ε_1, and ε_2 remain constant. In a similar vein, the higher the value of the parameter μ_i, the fewer the trigger numbers when θ_i, ε_1, and ε_2 are held fixed. Thus, the judicious selection of parameters is pivotal for reducing the consumption of communication bandwidth.

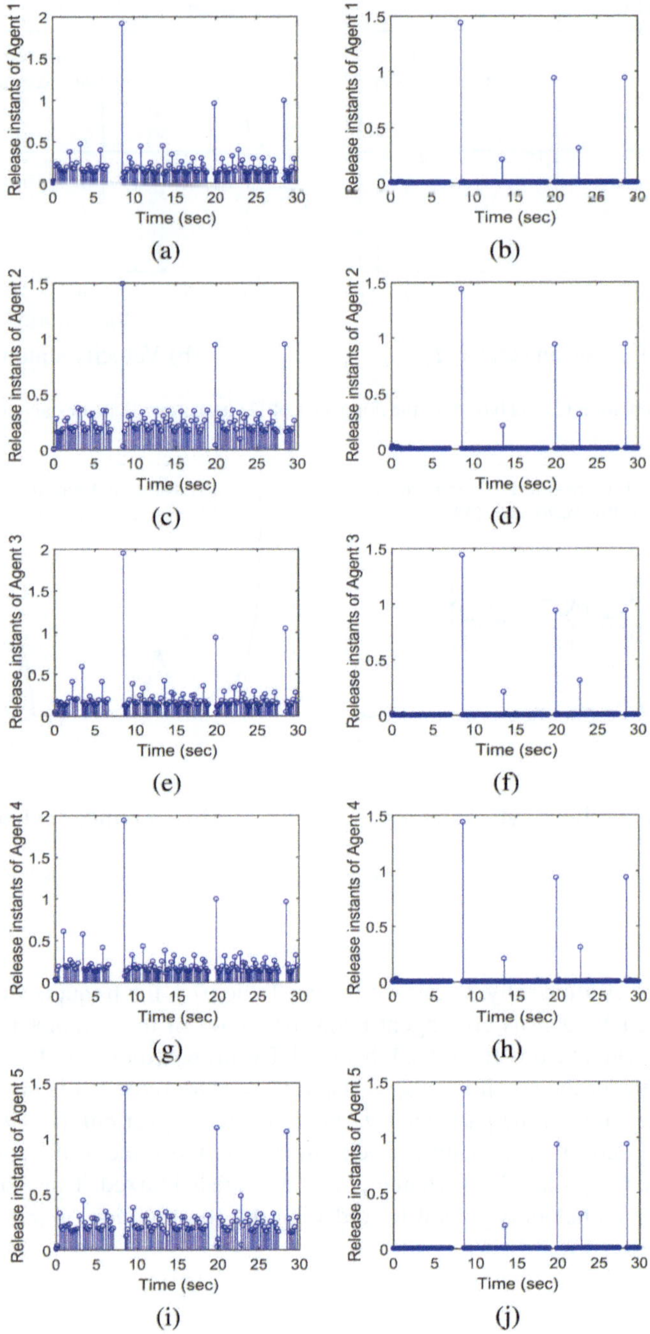

Fig. 10.6 Release instants and release time intervals with the proposed ETM in (10.31) (left) and the one proposed by [22] (right) under similar consensus performance

Table 10.2 Number of triggers under different triggered mechanisms

Method	Agents				
	1	2	3	4	5
ET[a] in this chapter	153	124	151	149	127
ET[a] in [22] with fault[b]	5235	5211	5237	5203	5237
ET[a] in [22] without fault[b]	3150	3648	3634	3470	3534

[a] ET is a shorthand for event-triggered
[b] The fault here includes actuator failure, IQCs and DoS attacks

Table 10.3 The triggers number of θ_i and μ_i in different situations under DoS attacks, where $\varepsilon_1 = \varepsilon_2 = 0.00001$

μ_i	θ_i	Triggers number of agents				
		1	2	3	4	5
0	1	5235	5211	5237	5203	5237
0.3	1	4201	4294	4519	4515	4660
0.3	0.2	1127	1079	1921	1842	2428
0.5	0.2	925	830	1536	1505	2052
0.5	0.05	264	210	263	256	211
0.8	0.008	153	124	151	149	127

10.5 Conclusion

This chapter designs an innovative event-triggered adaptive *fault-tolerant* pinning control scheme, aimed at guiding heterogeneous second-order nonlinear MASs to achieve finite-time bounded *cluster consensus* under a directed spanning tree structure. This achievement is effective even in the presence of *cyber attacks*, actuator faults, and IQCs. The proposed ETM mitigates communication strain by eliminating the need for ongoing neighbor state updates, and the pinning control approach curtails computational demands. Additionally, a sufficient condition is established to determine the appropriate coupling coefficients and pinning gains. For DoS attacks, this chapter provides maximum thresholds for attack duration and frequency to maintain cluster uniform boundedness of all MAS signals in finite time. The effectiveness and advantages of the control strategy are substantiated by simulations. Upcoming research will delve into *cluster consensus* under multiple *cyber attacks* and the development of adaptive *fault-tolerant* strategies for scenarios with unknown fault scaling factor bounds, which will be introduced in subsequent chapters.

References

1. L.V. Gambuzza, M. Frasca, A criterion for stability of cluster synchronization in networks with external equitable partitions. Automatica **100**, 212–218 (2019)
2. T.H. Lee, Q. Ma, S.Y. Xu, J.H. Park, Pinning control for cluster synchronisation of complex dynamical networks with semi-Markovian jump topology. Int. J. Control **88**(6), 1223–1235 (2015)
3. Q. Ma, Z. Wang, G.Y. Miao, Second-order group consensus for multi-agent systems via pinning leader-following approach. J. Frankl. Inst. **351**(3), 1288–1300 (2014)
4. V.T. Pham, N. Messai, N. Manamanni, Impulsive observer-based control in clustered networks of linear multi-agent systems. IEEE Trans. Netw. Sci. Eng. **7**(3), 1840–1851 (2020)
5. X.W. Li, Y. Tang, H.R. Karimi, Consensus of multi-agent systems via fully distributed event-triggered control. Automatica **116**, Art. No. 108898 (2020)
6. J.L. Wang, Q. Wang, H.N. Wu, T.W. Huang, Finite-time consensus and finite-time H_∞ consensus of multi-agent systems under directed topology. IEEE Trans. Netw. Sci. Eng. **7**(3), 1619–1632 (2020)
7. W. Xu, D.W.C. Ho, J. Zhong, B. Chen, Event/self-triggered control for leader-following consensus over unreliable network with dos attacks. IEEE Trans. Neural Netw. Learn. Syst. **30**(10), 3137–3149 (2019)
8. W.Y. Xu, D.W.C. Ho, Clustered event-triggered consensus analysis: an impulsive framework. IEEE Trans. Ind. Electron. **63**(11), 7133–7143 (2016)
9. Y. Yang, Y. Li, D. Yue, Y.C. Tian, X. Ding, Distributed secure consensus control with event-triggering for multiagent systems under DoS attacks. IEEE Trans. Cybern. **51**(6), 2916–2928 (2021)
10. X.G. Guo, X. Fan, C.K. Ahn, Adaptive event-triggered fault detection for interval type-2 T-S fuzzy systems with sensor saturation. IEEE Trans. Fuzzy Syst. **29**(8), 2310–2321 (2021)
11. X.G. Guo, X. Fan, J.L. Wang, J.H. Park, Event-triggered switching-type fault detection and isolation for fuzzy control systems under DoS attacks. IEEE Trans. Fuzzy Syst. **29**(11), 3401–3414 (2021)
12. J.L. Liu, T.T. Yin, J. Cao, D. Yue, H.R. Karimi, Security control for T-S fuzzy systems with adaptive event-triggered mechanism and multiple cyber-attacks. IEEE Trans. Syst. Man Cybern. Syst. **51**(10), 6544–6554 (2020)
13. H. Lu, Y. Hu, C. Guo, W. Zhou, Cluster synchronization for a class of complex dynamical network system with randomly occurring coupling delays via an improved event-triggered pinning control approach. J. Frankl. Inst. **357**(4), 2167–2184 (2020)
14. L.Q. Wang, J.X. Dong, C.J. Xi, Event-triggered adaptive consensus for fuzzy output-constrained multi-agent systems with observers. J. Frankl. Inst. **357**(1), 82–105 (2020)
15. M. Zhao, C. Peng, W.L. He, Y. Song, Event-triggered communication for leader-following consensus of second-order multiagent systems. IEEE Trans. Cybern. **48**(6), 1888–1897 (2018)
16. L. Liu, W. Zhou, X. Li, Y. Sun, Dynamic event-triggered approach for cluster synchronization of complex dynamical networks with switching via pinning control. Neurocomputing **340**, 32–41 (2019)
17. Y. Wan, G. Wen, X. Yu, T. Huang, Distributed consensus tracking of networked agent systems under denial-of-service attacks. IEEE Trans. Syst. Man Cybern. Syst. **51**(10), 6183–6196 (2021)
18. D. Zhang, G. Feng, A new switched system approach to leader-follower consensus of heterogeneous linear multiagent systems with DoS attack. IEEE Trans. Syst. Man Cybern. Syst. **51**(2), 1258–1266 (2019)
19. Z. Gu, C.K. Ahn, D. Yue, X.P. Xie, Event-triggered H_∞ filtering for T-S fuzzy-model-based nonlinear networked systems with multisensors against DoS attacks. IEEE Trans. Cybern. **52**(6), 5311–5321 (2020)
20. L. Zhao, G.H. Yang, Adaptive fault-tolerant control for nonlinear multi-agent systems with DoS attacks. Inf. Sci. **526**, 39–53 (2020)

21. C. Deng, C.Y. Wen, Distributed resilient observer-based fault-tolerant control for heterogeneous multiagent systems under actuator faults and DoS attacks. IEEE Trans. Control Netw. Syst. **7**(3), 1308–1318 (2020)
22. T. Dong, Y.L. Gong, Leader-following secure consensus for second-order multi-agent systems with nonlinear dynamics and event-triggered control strategy under DoS attack. Neurocomputing **416**, 95–102 (2020)
23. Z. Feng, G. Hu, Secure cooperative event-triggered control of linear multiagent systems under DoS attacks. IEEE Trans. Control Syst. Technol. **28**(3), 741–752 (2020)
24. J. Liu, T. Yin, D. Yue, H.R. Karimi, J, Cao, Event-based secure leader-following consensus control for multiagent systems with multiple cyber attacks. IEEE Trans. Cybern. **51**(1), 162–173 (2021)
25. X. Wang, G.H. Yang, Fault-tolerant consensus tracking control for linear multiagent systems under switching directed network. IEEE Trans. Cybern. **50**(5), 1921–1930 (2020)
26. J.C.L. Chan, T.H. Lee, C.P. Tan, A sliding mode observer for robust fault reconstruction in a class of nonlinear non-infinitely observable descriptor systems. Nonlinear Dyn. **101**(2), 1023–1036 (2020)
27. J.J. Fu, Y.Z. Lv, G.H. Wen, X.H. Yu, T.W. Huang, Velocity and input constrained coordination of second-order multi-agent systems with relative output information. IEEE Trans. Netw. Sci. Eng. **7**(3), 1925–1938 (2020)
28. T.H. Lee, C.P. Lim, S. Nahavandi, R.G. Roberts, Observer-based $\mathscr{H}_\infty$ fault-tolerant control for linear systems with sensor and actuator faults. IEEE Syst. J. **13**(2), 1981–1990 (2019)
29. E. Ribeiro, A.J.M. Cardoso, C. Boccaletti, Fault-tolerant strategy for a photovoltaic DC-DC converter. IEEE Trans. Power Electron. **28**(6), 3008–3018 (2013)
30. Y. Song, Y. Wang, C. Wen, Adaptive fault-tolerant PI tracking control with guaranteed transient and steady-state performance. IEEE Trans. Autom. Control **62**(1), 481–487 (2017)
31. W. Zou, C.K. Ahn, Z. Xiang, Fuzzy-approximation-based distributed fault-tolerant consensus for heterogeneous switched nonlinear multiagent systems. IEEE Trans. Fuzzy Syst. **29**(10), 2916–2925 (2021)
32. C. Ke, C.D. Li, Q.Q. Zhang, Distributed adaptive fault-tolerant consensus of nonlinear multi-agent systems via state-constraint impulsive protocols with time-delay. IEEE Trans. Netw. Sci. Eng. **7**(4), 3112–3121 (2020)
33. X.G. Guo, D.C. Tan, C.K. Ahn, J.L. Wang, Fully distributed adaptive fault-tolerant sliding-mode control for nonlinear leader-following multiagent systems with ANASs and IQCs. IEEE Trans. Cybern. **52**(5), 2763–2774 (2022)
34. A. Hu, J. Cao, M. Hu, L. Guo, Event-triggered group consensus for multi-agent systems subject to input saturation. J. Frankl. Inst. **355**(15), 7384–7400 (2018)
35. S. Liu, S. Li, B. Xu, Event-triggered resilient control for cyber-physical system under denial-of-service attacks. Int. J. Control **93**(8), 1907–1919 (2020)
36. Y. Xu, M. Fang, P. Shi, Z.G. Wu, Event-based secure consensus of mutiagent systems against DoS attacks. IEEE Trans. Cybern. **50**(8), 3468–3476 (2020)
37. G. Carl, G. Kesidis, R.R. Brooks, S. Rai, Denial-of-service attack-detection techniques. IEEE Internet Comput. **10**(1), 82–89 (2006)
38. X. Liu, T. Chen, Finite-time and fixed-time cluster synchronization with or without pinning control. IEEE Trans. Cybern. **48**(1), 240–252 (2019)
39. Y.J. Wang, Y.D. Song, F.L. Lewis, Robust adaptive fault-tolerant control of multiagent systems with uncertain nonidentical dynamics and undetectable actuation failures. IEEE Trans. Ind. Electron. **62**(6), 3978–3988 (2015)

Chapter 11
Intrusion- and Fault-Tolerant Cluster Consensus of MASs with TVNUD Faults

Abstract This chapter explores the achievement of *cluster consensus* for heterogeneous multi-agent systems (MASs) with time-varying unknown nonidentical direction (TVUND) faults under stochastic false data injection attacks (FDIAs). A *fault-tolerant* method is introduced, which is based on the *Nussbaum function* and does not require the knowledge of the fault directions and the upper and lower bounds of the fault scaling factors. To address the challenges posed by the summation effect of multiple *Nussbaum functions*, a new set of *Nussbaum functions* is designed. Additionally, a hybrid event-triggered mechanism (ETM) is proposed, which incorporates both adaptive and memory capabilities, to lessen the communication load by modulating the trigger threshold based on both the consensus error and the trajectory dynamics of agent.

11.1 Introduction

In recent years, the need for systems to perform multiple tasks concurrently has sparked a surge in interest in the concept of *cluster consensus* [1, 2]. The challenges posed by open network environments and the nature of distributed control have highlighted the importance of considering cyber threats and communication bandwidth constraints when designing control systems. To mitigate the burden on communication network, the event-triggered mechanism (ETM) has become a popular choice [3–6]. There are three primary ways to establish the threshold for ETM: the fixed threshold method [6], the adaptive threshold method [4], and the memory-based threshold method [3]. The fixed threshold method is essentially a simplified version of the adaptive threshold method. The adaptive threshold method in [4] adjusts the threshold based on consensus error. However, the consensus performance of the multi-agent system (MAS) deteriorates as the trajectory approaches its peaks or troughs. To address this, the ETM proposed in [3] employs a memory threshold method, which dynamically adjusts the threshold based on the agent's own motion trajectory, ensuring performance even at peaks or troughs. In practice, both threshold-setting approaches mentioned above are essential for the ETM to guarantee *cluster consensus* in MASs. However, simply combining these methods does not suffice to

X.-G. Guo et al., *Anti-disturbance Intrusion- and Fault-Tolerant Control of Multi-agent Systems*, https://doi.org/10.1007/978-981-96-9614-7_11

achieve the desired result. Moreover, the stable operation of MASs is increasingly jeopardized by the threat of *cyber attacks* [1, 7–9]. Unlike stochastic *disturbances* addressed in [10], stochastic false data injection attacks (FDIAs) target the consensus by corrupting measurements and control signals with falsified data [9, 11]. In discrete systems, FDIAs following a Bernoulli distribution are particularly prevalent [7]. However, continuous systems are not immune, as discussed in [8], which analyzes bounded consensus in continuous MASs under continuous FDIAs. However, the FDIA that obeys the Bernoulli distribution is more destructive. Therefore, under stochastic FDIAs, designing an ETM with adaptive and memory capabilities is crucial to alleviate communication burdens and ensure robust *cluster consensus*.

During operation, an actuator may experience faults that can degrade system performance or even lead to instability [12–14]. The loss of efficiency faults and the bias faults are two prevalent types of actuator faults that have been effectively addressed in numerous studies on *fault-tolerant* control challenges [11, 12, 15]. Nevertheless, actuators can also be subject to excessive faults [16] and reverse faults [17], which render the aforementioned fault tolerance strategies ineffective. Notably, in the chemical and industrial processes, the directions of faults may vary over time and remain unknown [18–20], causing the fault scale factor to extend beyond the traditional range of (0, 1]. The *Nussbaum function* technique is recognized as a potent strategy for addressing issues of unknown control directions in multi-agent systems (MASs) [21]. However, in [22], it is assumed that the unknown control directions for all agents must be the same. Additionally, other research posits that the control gain matrix of the agents is either unknown but negative, positive, constant [23], or partially unknown [24]. These assumptions are not suitable for MASs that encounter time-varying unknown direction faults, given that the occurrence of faults is unpredictable and their directions are typically time-varying, unknown, and nonidentical in different agents. To tackle nonidentical control directions, new Nussbaum functions have been devised in [25, 26] to overcome the challenges posed by the summation of multiple Nussbaum functions. Despite its effectiveness, the construction of the *Nussbaum function* in [26] involves a rather intricate monotonically increasing sequence, and the frequency modulation of the *Nussbaum function* is comparatively inflexible. Additionally, the *Nussbaum function* developed in [25] necessitates restrictive parameter design to ensure the boundedness of certain scalars, which in turn amplifies the conservatism of the function. Consequently, it holds both theoretical and practical importance to introduce a *Nussbaum function* that features a straightforward design process, flexible frequency adjustments, and no constraints on parameter design. Such a function would be instrumental in addressing time-varying unknown nonidentical direction (TVUND) faults.

Motivated by the preceding discourse, this chapter explores the *cluster consensus* of heterogeneous MASs that are subjected to TVUND faults, uncertainty, and stochastic FDIAs. The main contributions of this chapter are as follows:

(1) TVUND faults: Actuator faults characterized by TVUNDs are considered in the cluster consensus control of MASs under FDIAs, which were not considered in Chap. 9. This fault model is more comprehensive than those in [1, 12, 13]

and Chap. 10. The simultaneous presence of TVUND faults and FDIAs renders existing *fault-tolerant* control strategies ineffective. To the best of our knowledge, this is the first study to propose a *cluster consensus* control scheme for MASs that accounts for TVUND faults under FDIAs.

(2) New Nussbaum functions: In this chapter, we introduce a series of innovative Nussbaum functions to assess the complex interactions between multiple *Nussbaum functions*. It's significant to point out that, as detailed in Remark 11.4, by doing away with rigid parameter constraints and making the design process more straightforward, the *Nussbaum functions* developed here are more convenient and less conservative than those in [25, 26]. Furthermore, different from the compensation-based *fault-tolerant* control method in Chap. 10 are suited for resolving issues of loss of efficiency fault, the Nussbaum function-based fault-tolerant control strategies in this chapter are equipped to handle not just loss of efficiency faults but also reverse faults.

(3) The ETM proposed in this chapter allows for threshold adjustments based on both the consensus error and the trajectory of individual agent. Therefore, the proposed ETM enhances the *cluster consensus* performance, particularly at peaks and troughs of the trajectory of individual agent. This is a great improvement over ETMs that possess a single characteristic, i.e., either adaptability or memory, as seen in [3, 4] and Chap. 10.

11.2 Problem Formulation

11.2.1 Heterogeneous Second-Order MAS Model

Consider a heterogeneous MAS composed of N follower agents. The state equation of the ith agent is described by

$$\begin{aligned} \dot{x}_i(t) &= v_i(t), \quad \sigma_i = 1, 2, \cdots, s \\ \dot{v}_i(t) &= f_{\sigma_i}(t, x_i, v_i) + \hat{\vartheta}_i(t) + u_i^F(t) + w_i(t, x_i, v_i), \quad i = 1, 2, \cdots, N \end{aligned} \tag{11.1}$$

where $x_i(t) \in \mathbb{R}$ and $v_i(t) \in \mathbb{R}$ are the position and velocity states, respectively. The function $f_{\sigma_i}(t, x_i, v_i)$ represents the dynamic acceleration of the agent itself, and the subscript σ_i indicates that the agents are heterogeneous. Let $\hat{\vartheta}_i(t)$ denote the deviation of the FDIAs on the ith agent; $u_i^F(t)$ and $w_i(t, x_i, v_i)$ are the fault input and unknown uncertain term, respectively.

There exist virtual leaders in different clusters, and they are characterized by

$$\begin{aligned} \dot{x}_{\sigma_i}^r(t) &= v_{\sigma_i}^r(t) \\ \dot{v}_{\sigma_i}^r(t) &= f_{\sigma_i}(t, x_{\sigma_i}^r, v_{\sigma_i}^r) \end{aligned} \tag{11.2}$$

where $\sigma_i = 1, 2, \cdots, s$, $x^r_{\sigma_i}(t) \in \mathbb{R}$ and $v^r_{\sigma_i}(t) \in \mathbb{R}$ are the position and velocity states of the σ_ith virtual leader, respectively.

The actuator of the agent may suffer from actuator faults. In this chapter, the fault model is considered as follows:

$$u_i^F(t) = \begin{cases} u_i(t), & \text{fault-free} \\ \rho_i(t)u_i(t) + r_i(t), & \text{fault} \end{cases} \tag{11.3}$$

where $u_i(t)$ is the control input and $\rho_i(t)$ and $r_i(t)$ denote the time-varying fault scaling factors with unknown signs and magnitudes and the time-varying bias fault, respectively.

Assumption 11.1 The directed topological graph of MAS (11.1) and (11.2) contains a directed spanning tree.

Assumption 11.2 For each agent, the fault scaling factor and bias fault are time-varying, unknown, and satisfy

$$|r_i(t)| \leq \overline{r}_i, \quad t \geq 0$$
$$0 < \underline{\rho}_i \leq |\rho_i(t)| \leq \overline{\rho}_i < \infty$$

where $\underline{\rho}_i$, $\overline{\rho}_i$, and $\overline{r}_i$ are unknown positive constants.

Remark 11.1 An assumption similar to Assumption 11.1 is frequently adopted in the field of *fault-tolerant* control studies [13, 15]. However, in contrast to the faults in [13, 15], the multiplicative fault factor discussed in this chapter is capable of depicting a wider array of fault cases. These encompass efficiency loss faults $0 < \rho_i(t) < 1$, excessive faults $\rho_i(t) > 1$, and reverse faults $\rho_i(t) < 0$. Moreover, the values for $\underline{\rho}_i$ and $\overline{\rho}_i$ are not mandate to be known in advance.

In this chapter, FDIAs are considered, and the attack model is represented as

$$\hat{\vartheta}_i(t) = \sum_{j \in N_i^*} \gamma_{ij}(t)\vartheta_j(t) \tag{11.4}$$

where $\vartheta_j(t) \in \mathbb{R}$ represents the false data generated by the attacker and $\gamma_{ij}(t)$ is an independent stochastic variable that obeys the Bernoulli distribution and is defined as

$$\gamma_{ij}(t) = \begin{cases} 1, & \text{successful attack on channel } (j, i) \text{ at instant t} \\ 0, & \text{otherwise} \end{cases} \tag{11.5}$$

where the probability of a successful attack is described by

$$\text{prob}\{\gamma_{ij}(t) = 1\} = p_{ij}, \quad \text{prob}\{\gamma_{ij}(t) = 0\} = 1 - p_{ij}. \tag{11.6}$$

Let $\hat{\gamma}(t) = [(\gamma_{ij}(t) - 1)^2]_{N\times N}$ and $\Upsilon = \mathbb{E}(\hat{\gamma}(t))$ with

$$\Upsilon = \begin{bmatrix} 0 & p_{12} & \cdots & p_{1N} \\ p_{21} & 0 & \cdots & p_{2N} \\ \vdots & \vdots & \ddots & \vdots \\ p_{N1} & p_{N2} & \cdots & 0 \end{bmatrix}. \tag{11.7}$$

In addition, attackers are usually limited by energy, so the attack signal $\vartheta_j(t)$ satisfies

$$\|\vartheta\|^2 \leq \overline{\vartheta} \tag{11.8}$$

where $\vartheta = [\vartheta_1, \vartheta_2, \cdots, \vartheta_N]^T$ with $\overline{\vartheta} > 0$ being a constant. Moreover, a similar energy-limited constraint can be found in [9] and the references therein.

Remark 11.2 Many excellent results about FDI attacks are based on discrete systems [7, 27]. This is because analysis is easier by introducing the probability-based FDI attacks into discrete systems. However, continuous systems are inevitably attacked by false data injection from malicious attackers. In studies in which attacks are introduced into continuous systems, such as in [8, 28], the attack model is similar to that of disturbances, and the attack acts continuously on the communication channel. However, it is more realistic to introduce probability-based FDI attacks into a continuous MAS. In addition, a similar energy-limited constraint (11.8) can be found in [9] and the references therein.

For the dynamic acceleration $f_{\sigma_i}(t, x_i, v_i)$ and uncertain term $w_i(t, x_i, v_i)$ in (11.1), the following assumptions hold:

Assumption 11.3 For $\forall\ i \in \mathscr{V}, \sigma_i \in \{1, 2, \cdots, s\}$, the dynamic acceleration $f_{\sigma_i}(t, x_i, v_i)$ satisfies the following Lipschitz condition:

$$f_{\sigma_i}(t, x_i, v_i) - f_{\sigma_i}(t, x^r_{\sigma_i}, v^r_{\sigma_i}) \leq \alpha_i|\tilde{x}_i| + \beta_i|\tilde{v}_i|,\ t \geq 0 \tag{11.9}$$

where α_i and β_i are positive constants and $\tilde{x}_i = x_i - x^r_{\sigma_i}$ and $\tilde{v}_i = v_i - v^r_{\sigma_i}$ are the tracking errors.

Assumption 11.4 For the uncertain term $w_i(t, x_i, v_i), \forall i \in \mathscr{V}$, in (11.1), there exists an unknown constant $\overline{p}_i$ and a nonlinear function $\overline{\psi}_i(x_i, v_i)$ such that

$$|w_i(t, x_i, v_i)| \leq \overline{p}_i\overline{\psi}_i(x_i, v_i).$$

Remark 11.3 Condition (11.9) in Assumption 11.2 is commonly adopted in the literature, including [1, 11]. The term $w_i(t)$ in the system model (11.1) represents various uncertain factors, such as bounded time-varying disturbances, bounded parameters, and parametrized nonlinear terms with unknown parameters but known structures, for which Assumption 11.3 is typically satisfied. Furthermore, Assumption 11.3 is frequently encountered in single-agent systems [13].

11.2.2 Control Objective

According to Definition 2.2, MAS (11.1) and (11.2) achieves bounded *cluster consensus* when the tracking error $\tilde{x}_i(t)$ and $\tilde{v}_i(t)$ of the systems are such that

$$\mathbb{E}\{\|\varsigma(t)\|\} \le \ell$$

where ℓ is the error bound represented by a constant.

$$\varsigma(t) = [\tilde{x}^T(t), \tilde{v}^T(t)]^T, \quad \tilde{x}(t) = [\tilde{x}_1^T(t), \tilde{x}_2^T(t), \cdots, \tilde{x}_N^T(t)]^T,$$
$$\tilde{v}(t) = [\tilde{v}_1^T(t), \tilde{v}_2^T(t), \cdots, \tilde{v}_N^T(t)]^T.$$

Thus, the objective of this chapter can be described as the following: for MAS (11.1) and (11.2), bounded *cluster consensus* can be guaranteed via the proposed *fault-tolerant* controller.

11.3 Event-Triggered Fault-Tolerant Control

In this section, for the TVUND faults, a group of Nussbaum functions with different frequencies are proposed. There exist intervals of the variable κ that make the Nussbaum functions reinforce each other. Next, an adaptive memory event-triggered *fault-tolerant* controller is proposed for the MAS to achieve cluster consensus. Finally, a cluster consensus analysis is conducted.

11.3.1 Nussbaum Function

When multiple Nussbaum functions, such as $x^2\cos(x)$, $e^{x^2}\cos(x)$, and $\cosh(x)\sin(x)$, simultaneously appear in an inequality, the interaction among them is not clear. This poses a challenge for analyzing the relationships between Nussbaum functions. Inspired by [26], this chapter designs a set of novel Nussbaum functions as follows:

$$\mathcal{N}_i(\kappa) = 2\varepsilon_{1i}\kappa\sinh(\varepsilon_{1i}\kappa^2)\cos(\varepsilon_{2i}^i\kappa) - \varepsilon_{2i}^i\cosh(\varepsilon_{1i}\kappa^2)\sin(\varepsilon_{2i}^i\kappa), \tag{11.10}$$

where ε_{1i} and ε_{2i} are positive constants that need to be designed. Direct calculation gives

$$H_i(\kappa) \triangleq \int_0^{\kappa} \mathcal{N}_i(\tau)\, d\tau = \cosh(\varepsilon_{1i}\kappa^2)\cos(\varepsilon_{2i}^i\kappa). \tag{11.11}$$

Then, based on (11.10) and (11.11), the following lemmas are given.

Lemma 11.1 *The properties of the Nussbaum function $\mathcal{N}_i(\kappa)$ are listed as follows:*

(a) $\mathcal{N}_i(\kappa)$ is an odd function of κ;

(b) $H_i(\kappa) \triangleq \int_0^{\kappa} \mathcal{N}_i(\tau)\, d\tau$ is an even function of κ;

(c) The Nussbaum function $\mathcal{N}_i(\kappa)$ is such that

$$\lim_{\kappa\to\infty} \sup \frac{1}{\kappa}\int_0^{\kappa} \mathcal{N}_i(\tau)\, d\tau = +\infty$$

$$\lim_{\kappa\to\infty} \inf \frac{1}{\kappa}\int_0^{\kappa} \mathcal{N}_i(\tau)\, d\tau = -\infty. \tag{11.12}$$

Proof The above properties are explicit and the proof process is omitted here. □

Lemma 11.2 *For $\forall i \in \{1, 2, \cdots, N\}$, when $\varepsilon_{2i} > 1$, there exists $\kappa \in [\underline{\hat{a}}_{iq}, \overline{\hat{a}}_{iq}]$ ($q \in \{1, 2, \cdots, \infty\}$) such that*

$$\mathrm{sign}(\rho_i)\cos(\varepsilon_{2i}^i\kappa) < -0.5.$$

In addition, when $\kappa \in [\underline{\hat{a}}_{Nq}, \overline{\hat{a}}_{Nq}]$, the following inequality holds:

$$\sum_{i=1}^{N} \mathrm{sign}(\rho_i)\cos(\varepsilon_{2i}^i\kappa) < -\frac{N}{2} \tag{11.13}$$

where

$$\begin{cases} \underline{\hat{a}}_{iq} = \dfrac{2q\pi+\frac{2\pi}{3}}{\varepsilon_{2i}^i}, \\ \overline{\hat{a}}_{iq} = \dfrac{2q\pi+\frac{4\pi}{3}}{\varepsilon_{2i}^i}, \end{cases} \quad \mathrm{sign}(\rho_i) = 1$$

or

$$\begin{cases} \underline{\hat{a}}_{iq} = \dfrac{2q\pi+\frac{5\pi}{3}}{\varepsilon_{2i}^i}, \\ \overline{\hat{a}}_{iq} = \dfrac{2q\pi+\frac{7\pi}{3}}{\varepsilon_{2i}^i}, \end{cases} \quad \mathrm{sign}(\rho_i) = -1.$$

In addition, the length of the interval $[\underline{\hat{a}}_{Nq}, \overline{\hat{a}}_{Nq}]$ *is* $\hat{l} = \frac{2\pi}{3\varepsilon_{2i}^N}$.

Proof According to Lemma 11.1, the case of $\kappa > 0$ is considered in the following proof procedure, whereas the case of $\kappa < 0$ is similar to the case of $\kappa > 0$ and is omitted here. Assuming that $\text{sign}(\rho_i) = 1$ $(i \in \{1, 2, \cdots, N-1\})$, one can obtain that $\text{sign}(\rho_i)\cos(\varepsilon_{2i}^i \kappa) < 0$ when $\kappa \in [(2n_1\pi + \pi/2)/\varepsilon_{2i}^i, (2n_1\pi + 3\pi/2)/\varepsilon_{2i}^i]$ $(n_1 \in \mathbb{N})$. For the $(i+1)$th agent, the analysis is divided into two cases.

Case 1: $\text{sign}(\rho_{i+1})\cos(\varepsilon_{2i}^{i+1}\kappa) < 0$ when $\kappa \in \left[\frac{2n_2\pi+\frac{\pi}{2}}{\varepsilon_{2i}^{i+1}}, \frac{2n_2\pi+\frac{3\pi}{2}}{\varepsilon_{2i}^{i+1}}\right]$ $(n_2 \in \mathbb{N})$, if $\text{sign}(\rho_{i+1}) = 1$.

Case 2: $\text{sign}(\rho_{i+1})\cos(\varepsilon_{2i}^{i+1}\kappa) < 0$ when $\kappa \in \left[\frac{2n_2\pi+\frac{3\pi}{2}}{\varepsilon_{2i}^{i+1}}, \frac{2n_2\pi+\frac{5\pi}{2}}{\varepsilon_{2i}^{i+1}}\right]$ $(n_2 \in \mathbb{N})$, if $\text{sign}(\rho_{i+1}) = -1$.

Let $n_2 = n_1\varepsilon_{2i} + n_3$ $(n_3 \in \mathbb{N})$. When $\varepsilon_{2i} > 1$, there exists n_3 such that

$$\begin{cases} 2n_3\pi + \frac{\pi}{2} > \frac{\varepsilon_{2i}\pi}{2} \\ 2n_3\pi + \frac{3\pi}{2} < \frac{3\varepsilon_{2i}\pi}{2} \end{cases} \text{or} \quad \begin{cases} 2n_3\pi + \frac{3\pi}{2} > \frac{\varepsilon_{2i}\pi}{2} \\ 2n_3\pi + \frac{5\pi}{2} < \frac{3\varepsilon_{2i}\pi}{2}. \end{cases}$$

Then, one has

$$\begin{cases} \frac{2n_2\pi+\frac{\pi}{2}}{\varepsilon_{2i}^{i+1}} > \frac{2n_1\pi+\frac{\pi}{2}}{\varepsilon_{2i}^{i}} \\ \frac{2n_1\pi+\frac{3\pi}{2}}{\varepsilon_{2i}^{i}} < \frac{2n_2\pi+\frac{3\pi}{2}}{\varepsilon_{2i}^{i+1}} \end{cases} \text{or} \quad \begin{cases} \frac{2n_2\pi+\frac{3\pi}{2}}{\varepsilon_{2i}^{i+1}} > \frac{2n_1\pi+\frac{\pi}{2}}{\varepsilon_{2i}^{i}} \\ \frac{2n_1\pi+\frac{5\pi}{2}}{\varepsilon_{2i}^{i}} < \frac{2n_2\pi+\frac{3\pi}{2}}{\varepsilon_{2i}^{i+1}} \end{cases}$$

and the following conclusions can be drawn.

For $\forall\ i \in \{1, 2, \cdots, N\}$, when $\varepsilon_{2i} > 1$, there exists $\kappa \in [\underline{\hat{a}}_{iq}, \overline{\hat{a}}_{iq}]$ $(q \in \{1, 2, \cdots, \infty\})$ such that

$$\text{sign}(\rho_i)\cos(\varepsilon_{2i}^i\kappa) < -0.5$$

and when $\kappa \in [\underline{\hat{a}}_{(i+1)q}, \overline{\hat{a}}_{(i+1)q}]$ such that

$$\text{sign}(\rho_i)\cos(\varepsilon_{2i}^i\kappa) + \text{sign}(\rho_{i+1})\cos(\varepsilon_{2i}^{i+1}\kappa) < -1$$

where $\underline{\hat{a}}_{iq}$ and $\overline{\hat{a}}_{iq}$ are defined in (11.13). By using the idea of recursion, it can be obtained that the following inequality holds

$$\sum_{i=1}^{N} \text{sign}(\rho_i)\cos(\varepsilon_{2i}^i\kappa) < -\frac{N}{2}$$

when $\kappa \in [\underline{\hat{a}}_{Nq}, \overline{\hat{a}}_{Nq}]$ $(q \in \{1, 2, \cdots, \infty\})$, and the length of the interval $[\underline{\hat{a}}_{Nq}, \overline{\hat{a}}_{Nq}]$ is $\hat{l} = \frac{2\pi}{3\varepsilon_2^N}$.

When $\text{sign}(\rho_i) = -1$, similar to $\text{sign}(\rho_i) = 1$, the same conclusion can be derived. The proof is completed. □

By utilizing Lemmas 11.1 and 11.2, the following technical theorem analyzes the boundedness of some scalars and the interaction of multiple *Nussbaum functions* $\mathscr{N}_i(\kappa)$ that simultaneously appear in an inequality and superpose each other.

Theorem 11.1 *For any non-negative function $V(t) \geq 0$, smooth function $\kappa_i(t)(i \in \mathscr{V})$, and a group of Nussbaum function*

$$\mathscr{N}_i(\kappa_i) = 2\varepsilon_{1i}\kappa_i \sinh(\varepsilon_{1i}\kappa_i^2)\cos(\varepsilon_{2i}^i\kappa_i) - \varepsilon_{2i}^i \cosh(\varepsilon_{1i}\kappa_i^2)\sin(\varepsilon_{2i}^i\kappa_i)$$

if the following inequality is satisfied:

$$\mathbb{E}\{V(t)\} \leq \mathbb{E}\left\{\sum_{i=1}^{N}\int_0^t d_i\big(\rho_i(\tau)\mathscr{N}_i(\kappa_i) - 1\big)\dot{\kappa}_i(\tau)\,d\tau\right\} + \phi \tag{11.14}$$

where d_i and ϕ are positive constants and $\rho_i(\tau)$ is a nonzero variable. Then, $\mathbb{E}\{\kappa_i(t)\}$, $\mathbb{E}\{\mathscr{N}_i(t)\}$, $\mathbb{E}\{V(t)\}$, and $\mathbb{E}\{\int_0^t d_i\big(\rho_i(\tau)\mathscr{N}_i(\kappa_i) - 1\big)\dot{\kappa}_i(\tau)\,d\tau\}$ are bounded.

Proof The process of proof uses the idea of finding contradictions. Assume $\kappa_1(t), \kappa_2(t), \cdots, \kappa_l(t)$ are unbounded and $\kappa_{l+1}(t), \cdots, \kappa_N(t)$ are bounded. By Lemma 11.2, for $\forall i \in \mathscr{V}$, there exists $\kappa_i(t) \in [\underline{\hat{a}}_{Nq}, \hat{a}_{Nq}]$ $(q \in \{1, 2, \cdots, \infty\})$ such that

$$\operatorname{sign}(\rho_i)\cos(\varepsilon_{2i}^i\kappa_i(t)) < -0.5 \tag{11.15}$$

$$\sum_{i=1}^{N}\operatorname{sign}(\rho_i)\cos(\varepsilon_{2i}^i\kappa_i(t)) < -\frac{N}{2} \tag{11.16}$$

are satisfied. From (11.14), we have

$$\mathbb{E}\{V(t)\} \leq \mathbb{E}\left\{\sum_{i=1}^{N}\int_0^t \rho_i(\tau)d_i\mathscr{N}_i(\kappa_i)\dot{\kappa}_i(\tau)\,d\tau - \Gamma\right\} + \phi_1 \tag{11.17}$$

where

$$\phi_1 = \phi + \sum_{i=1}^{N}\kappa_i(0), \quad \Gamma = \sum_{i=1}^{N}d_i\kappa_i(t).$$

The variable $\kappa_i(t)$ of different agents has different trajectories due to different fault types and other factors. Inspired by [25], the proof is divided into two cases according to the distribution of $\kappa_i(t)$ for different agents, $i = 1, 2, \cdots, l$.

Case 1: For any $i \in \{1, 2, \cdots, l\}$, $\kappa_i(t) \in [\underline{\hat{a}}_{Nq}, \overline{\hat{a}}_{Nq}]$ $(q \in \{1, 2, \cdots, \infty\})$.

$$\begin{aligned}
\mathbb{E}\{V(t)\} &\leq \mathbb{E}\Bigg\{\sum_{i=1}^{N}\int_{0}^{t}\rho_i(\tau)d_i\mathcal{N}_i(\kappa_i)\dot{\kappa}_i(\tau)\,d\tau - \sum_{i=1}^{N}d_i\kappa_i(t)\Bigg\} + \phi_1 \\
&\leq \mathbb{E}\Bigg\{\sum_{i=1}^{l}\underline{\rho}_i d_i \mathrm{sign}(\rho_i)H_i(\kappa_i) - \sum_{i=1}^{N}d_i\kappa_i(t) \\
&\quad + \sum_{i=l+1}^{N}\int_{0}^{t}\rho_i(\tau)d_i\mathcal{N}_i(\kappa_i)\dot{\kappa}_i(\tau)\,d\tau\Bigg\} + \phi_1 \\
&\leq \mathbb{E}\Bigg\{-\frac{1}{2}\sum_{i=1}^{l}\underline{\rho}_i d_i \cosh(\varepsilon_{1i}\kappa_i^2) - \sum_{i=1}^{N}d_i\kappa_i(t)\Bigg\} + \phi_2
\end{aligned} \tag{11.18}$$

where

$$\phi_2 = \phi_1 + \phi_3, \quad \phi_3 = \sum_{i=l+1}^{N}\int_{0}^{t}\rho_i(\tau)d_i\mathcal{N}_i(\kappa_i)\dot{\kappa}_i(\tau)\,d\tau$$

which is bounded due to the boundedness of $\kappa_i(t)$ $(i = l+1, \cdots, N)$. When $q \to \infty$, $\kappa_i(t)$ $(i = 1, 2, \cdots, l)$ is unbounded and $\cosh(\varepsilon_{1i}\kappa_i^2(t))$ is also unbounded. Then, one has $\mathbb{E}\{V(t)\} \leq -\infty$. This is contradictory to the fact that $V(t)$ is a non-negative function.

Case 2: Only part of $\kappa_i(t) \in [\underline{\hat{a}}_{Nq}, \overline{\hat{a}}_{Nq}]$ $(q \in \{1, 2, \cdots, \infty\})$.

Define

$$\kappa_m(t) = \max\{\kappa_1(t), \kappa_2(t), \cdots, \kappa_l(t)\}.$$

Because $\kappa_m(t)$ is continuous and unbounded, there must exist a monotonous time sequence $\{t_q\}$ such that $\kappa_m(t)$ satisfies

$$\kappa_m(t_q) = \underline{\hat{a}}_{Nq} + \bar{l}_q$$

where $\bar{l}_q > 0$ is a constant satisfying $\bar{l}_q < \hat{l}$ $(q = 1, 2, \cdots, \infty)$. Then, according to the distribution of $\kappa_i(t_q)$, the set $\{\kappa_i(t_q)\}$ $(i = 1, 2, \cdots, l)$ is divided into the following two sets.

$$\begin{aligned}
\Delta_1 &= \{i \mid 0 \leq \kappa_i(t_q) < \underline{\hat{a}}_{Nq}\} \\
\Delta_2 &= \{i \mid \underline{\hat{a}}_{Nq} \leq \kappa_i(t_q) \leq \underline{\hat{a}}_{Nq} + \bar{l}_q\}.
\end{aligned}$$

Let c_1 and c_2 denote the number of terms in Δ_1 and Δ_2, respectively. Then, for $i \in \Delta_2$, one can obtain

$$\sum_{i\in\Delta_2}\int_0^{t_q}\rho_i(\tau)d_i\mathcal{N}_i(\kappa_i)\dot{\kappa}_i(\tau)\,d\tau \le -\frac{1}{2}b_{\min}\underline{\rho}_{\min}\cosh(\varepsilon_{1i}(\underline{\hat{a}}_{Nq}+\bar{l}_q)^2).$$

Next, for $i\in\Delta_1$, one can obtain

$$\begin{aligned}\sum_{i\in\Delta_1}\int_0^{t_q}\rho_i(\tau)d_i\mathcal{N}_i(\kappa_i)\dot{\kappa}_i(\tau)\,d\tau &\le c_1 b_{\max}\overline{\rho}_{\max}\cosh(\varepsilon_{1i}\underline{\hat{a}}_{Nq}^2)\\ &\le Nb_{\max}\overline{\rho}_{\max}\cosh(\varepsilon_{1i}\underline{\hat{a}}_{Nq}^2).\end{aligned}$$

Recalling (11.14), it can be proved that

$$\begin{aligned}\mathbb{E}\{V(t)\} &\le \mathbb{E}\left\{\sum_{i=1}^{N}\int_0^{t_q}\rho_i(\tau)d_i\mathcal{N}_i(\kappa_i)\dot{\kappa}_i(\tau)\,d\tau-\sum_{i=1}^{N}d_i\kappa_i(t)\right\}+\phi_1\\ &\le \mathbb{E}\left\{\sum_{i=1}^{l}\int_0^{t_q}\rho_i(\tau)d_i\mathcal{N}_i(\kappa_i)\dot{\kappa}_i(\tau)\,d\tau-\sum_{i=1}^{N}d_i\kappa_i(t)\right\}+\phi_2\\ &\le \mathbb{E}\{(-\frac{1}{2}b_{\min}\underline{\rho}_{\min}\cosh(\varepsilon_{1i}(\underline{\hat{a}}_{Nq}+\bar{l}_q)^2)\\ &\quad+Nb_{\max}\overline{\rho}_{\max}\cosh(\varepsilon_{1i}\underline{\hat{a}}_{Nq}^2))-\Gamma\}+\phi_2\end{aligned}\tag{11.19}$$

where Γ and ϕ_2 are defined in (11.17) and (11.18), respectively. When $q\to\infty$, $\underline{\hat{a}}_{N\infty}\to\infty$. Thus, inequality (11.19) can be rewritten as

$$\mathbb{E}\{V(t)\}\le\mathbb{E}\left\{\left(-\frac{1}{2}b_{\min}\underline{\rho}_{\min}e^{\hbar_1}+\hbar_2\right)e^{\varepsilon_{1i}\underline{\hat{a}}_{N\infty}^2}-\Gamma\right\}+\phi_2\tag{11.20}$$

where $\hbar_1=\varepsilon_{1i}(2\underline{\hat{a}}_{N\infty}\bar{l}_\infty+\bar{l}_\infty^2)$; $\hbar_2=Nb_{\max}\overline{\rho}_{\max}$ is a finite constant. Because the exponential term $e^{\varepsilon_{1i}(2\underline{\hat{a}}_{N\infty}\bar{l}_\infty+\bar{l}_\infty^2)}$ is much larger than $\hbar$ as $\underline{\hat{a}}_{N\infty}\to\infty$, it is clear from (11.20) that

$$-\frac{1}{2}b_{\min}\underline{\rho}_{\min}e^{\varepsilon_{1i}(2\underline{\hat{a}}_{N\infty}\bar{l}_\infty+\bar{l}_\infty^2)}+\hbar<0.$$

Thus, it can also be concluded that $\mathbb{E}\{V(t)\}\le-\infty$. This is contradictory to the fact that $V(t)$ is a non-negative function. Therefore, $\kappa_i(t)$ $(i=1,2,\cdots,N)$ is bounded, and according to (11.10) and (11.14), $\mathbb{E}\{\mathcal{N}_i(t)\}$, $\mathbb{E}\{\int_0^t d_i(\rho_i(\tau)\mathcal{N}_i(\kappa_i)-1)\dot{\kappa}_i(\tau)\,d\tau\}$, and $\mathbb{E}\{V(t)\}$ are also bounded. The proof is completed. □

Remark 11.4 Unlike the *Nussbaum function* introduced in [26], where the design involves constructing a complex monotone increasing sequence based on an exponential term, this chapter simplifies the process by leveraging the properties of trigonometric functions. Specifically, a series of intervals $[\underline{\hat{a}}_{Nq}, \bar{\hat{a}}_{Nq}](q \in \{1, 2, \cdots, \infty\})$ for $\kappa_i(t)$ is identified, streamlining the design. Additionally, in [26], the frequency adjustment of the *Nussbaum function* relies on the exponential term, which poses challenges in high-frequency scenarios due to the steep slope, making parameter tuning difficult. In addition, the *Nussbaum function* developed in [25] requires parameter constraints to ensure scalar boundedness. In contrast, the *Nussbaum function* designed in this chapter eliminates such constraints, thereby reducing conservatism and enhancing flexibility in parameter design.

11.3.2 Design of Event-Triggered Fault-Tolerant Controller

The *Nussbaum function* method and adaptive *fault-tolerant* approach are employed to address the challenges posed by TVUND faults. To further alleviate computational load and conserve system resources, the pinning control and ETM are integrated. The schematic diagram of the designed control framework is shown in Fig. 11.1.

The proposed controller is given as follows:

$$\begin{aligned} u_i(t) &= \mathscr{N}_i(\kappa_i)\overline{\kappa}_i(t), \quad \dot{\kappa}_i(t) = d_i s_i(t)\overline{\kappa}_i(t) \\ \overline{\kappa}_i(t) &= -c\sum_{j=1}^{N} l_{ij}x_j(t_{k_j}^j) - c\sum_{j=1}^{N} l_{ij}v_j(t_{k_j}^j) + \mathscr{P}_i(t) \end{aligned} \tag{11.21}$$

where d_i and c are positive constants.

$$s_i(t) = \tilde{x}_i(t) + \tilde{v}_i(t), \quad \tilde{x}_i(t) = x_i(t) - x_{\sigma_i}^r(t), \quad \tilde{v}_i = v_i(t) - v_{\sigma_i}^r(t).$$

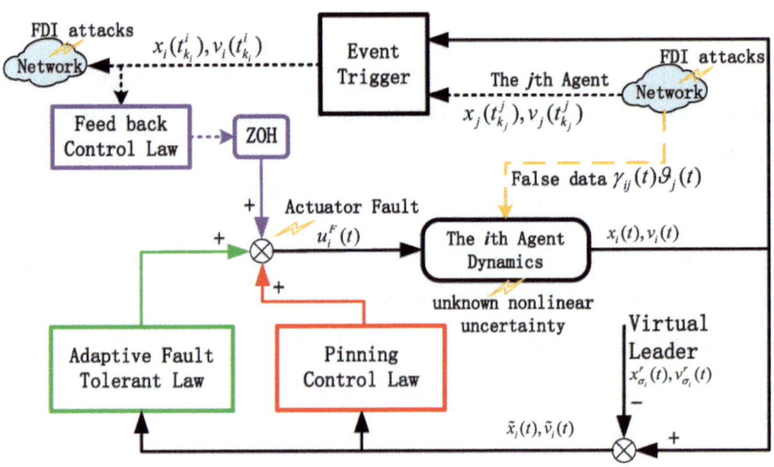

Fig. 11.1 Schematic of the proposed controller

Furthermore, $x_j(t^j_{k_j})$ are the trigger instants of the jth agent. $\mathscr{P}_i(t) = \mathscr{P}_{1i}(t) + \mathscr{P}_{2i}(t)$ with

$$\mathscr{P}_{1i}(t) = \begin{cases} -b_i s_i(t) + \mathscr{P}_{3i}(t), & i \in \tilde{\mathscr{V}}_{\sigma_i} \\ -b_i s_i(t), & \mathscr{D}_i = 0 \\ 0, & \text{otherwise} \end{cases} \tag{11.22}$$

$$\mathscr{P}_{2i}(t) = -\frac{\hat{\eta}_i s_i (1 + \overline{\psi}_i(x_i, v_i))^2}{\sqrt{s_i^2 (1 + \overline{\psi}_i(x_i, v_i))^2 + \zeta_i^2}} \tag{11.23}$$

$$\mathscr{P}_{3i}(t) = c \sum_{j=1}^{N} l_{ij} x^r_{\sigma_j}(t) + c \sum_{j=1}^{N} l_{ij} v^r_{\sigma_j}(t) \tag{11.24}$$

where $b_i > 0$ is the pinning gain and $\overline{\psi}_i(x_i, v_i)$ is defined in Assumption 11.3.

$$\zeta_i(t) = h_{i1} e^{-h_{i2} t}$$

with $h_{i1} > 0$ and $h_{i2} > 0$ being constants, $\hat{\eta}_i$ being the estimation of η_i, and η_i is defined in (11.33). In addition, the adaptive update law is given by

$$\dot{\hat{\eta}}_i = -\rho_i \hat{\eta}_i + \frac{\iota_i s_i^2 (1 + \overline{\psi}_i(x_i, v_i))^2}{\sqrt{s_i^2 (1 + \overline{\psi}_i(x_i, v_i))^2 + \zeta_i^2}} \tag{11.25}$$

where $\rho_i > 0$ and $\iota_i > 0$ are constants. The existence of ζ_i^2 ensures the nonsingularity of the fractional terms.

The hybrid ETM is expressed as

$$t^i_{k_i+1} = \inf\{t | t > t^i_{k_i} + \overline{\tau}, g_i(t) \geq 0, k \in \mathbb{Z}_{>0}\} \tag{11.26}$$

$$g_i(t) = \|e_{v_i}(t)\|^2 + \|e_{x_i}(t)\|^2 - \theta_i(t)\mu_i(t) \tag{11.27}$$

where $\overline{\tau}$ is a constant that needs to be designed and indicates the minimal transmission interval.

$$e_{v_i}(t) = v_i(t^i_{k_i}) - v_i(t), \quad e_{x_i}(t) = x_i(t^i_{k_i}) - x_i(t) \tag{11.28}$$

$$\dot{\theta}_i(t) = -\varepsilon_1 \theta_i^2(t)(\|\tilde{x}_i(t)\|^2 + \|\tilde{v}_i(t)\|^2), \ \theta_i(0) \in (0, 1] \tag{11.29}$$

$$\mu_i(t) = \overline{\mu}_i - (\overline{\mu}_i - \underline{\mu}_i) e^{-\varepsilon_2 \left| x_i(t) - \frac{1}{\varepsilon_3} \sum_{\varepsilon_4=1}^{\varepsilon_3-1} x_i(t^i_{k_i-\varepsilon_4}) \right|^2}. \tag{11.30}$$

Remark 11.5 In existing research on *cluster consensus* control [29], the in-degree balance condition is often imposed to simplify both the controller design and the analysis process. However, this condition restricts the applicability of the control scheme. Unlike the approach in [29], the control strategy (11.21) removes this limitation by introducing $\mathscr{P}_{3i}(t)$, thereby reducing the conservatism and broadening the applicability of the proposed method.

Remark 11.6 The ETM (11.26) proposed in this chapter integrates the features of both the adaptive ETM [4] and the memory ETM [3]. The adaptive ETM dynamically adjusts the triggering instant based on the consensus error [4], while the memory ETM modifies the triggering instant according to the agents motion trajectory [3]. As shown in (11.27), the number of triggers depends on $\theta_i(t)$ and $\mu_i(t)$. Considering $\mu_i(t)$ in (11.29), when an agents motion trajectory approaches peaks or troughs, both $\mu_i(t)$ and $x_i(t) - x_i(t^i_{k_i})$ take smaller values. This implies that the triggering condition (11.26) is more easily satisfied than when $\mu_i(t)$ is constant, leading to an increased number of triggers and enhanced consensus performance at trajectory extremes. In other cases, $\mu_i(t)$ and $x_i(t) - x_i(t^i_{k_i})$ take larger values, reducing the number of triggers. When the leaders trajectory is constant (without peaks or troughs) and the system achieves *cluster consensus*, the ETM naturally degenerates into a fixed-threshold ETM, maintaining normal operation. Furthermore, as indicated by (11.26), the triggering condition $g_i(t) \geq 0$ is evaluated only after a minimum transmission interval $\overline{\tau}$ has passed since the last event, effectively preventing the occurrence of *Zeno behavior* [5].

11.3.3 Cluster Consensus Analysis

Here, the stability theorem for the proposed *fault-tolerant cluster consensus* strategy is presented under the schematic shown in Fig. 11.1.

Theorem 11.2 *Under Assumptions 11.1–11.4, MAS (11.1)–(11.3) with unknown nonlinear uncertainty and TVUND faults under FDIAs are considered. The event-triggered fault-tolerant controller (11.21) can ensure that the MAS achieves cluster consensus if the following conditions are satisfied:*

$$\lambda_{\max}(\tilde{L}_l) < \min\left\{-\frac{\chi}{c}, -1 - \theta\|\mathscr{L}\|\right\} \tag{11.31}$$

$$b_i > \lambda_{\max}(B - CM_i^{-1}C^T) \tag{11.32}$$

where

$$\chi = m_3 + 1 + c\theta\|\mathscr{L}\|, \quad \theta = \max_{1\leq i\leq N}\{\theta_i(0)\}, \quad \tilde{L} = (-\mathscr{L} - \mathscr{L}^T)/2.$$

Furthermore, m_3 is defined in (11.45) and $\tilde{L}_l$ is a submatrix of $\tilde{L}$, which does not include the first l row-columns of $\tilde{L}$. M_l, C, and B are appropriate dimension matrices defined in Lemma 2.11.

Proof Choose a Lyapunov function candidate as

$$V(t) = \begin{bmatrix} \tilde{x} \\ \tilde{v} \end{bmatrix}^T \Omega \begin{bmatrix} \tilde{x} \\ \tilde{v} \end{bmatrix} + \sum_{i=1}^{N} \frac{1}{2\iota_i} \tilde{\eta}_i^2 \tag{11.33}$$

where

$$\Omega = \begin{bmatrix} c(\mathscr{L} + \mathscr{L}^T) + 2D & I_N \\ I_N & I_N \end{bmatrix}$$

$$\tilde{\eta}_i = \hat{\eta}_i - \eta_i, \quad \eta_i = \max \left\{ \overline{p}_i, \sup_{t \geq 0}(r_i(t) + \sum_{j \in N_i^*} \vartheta_j(t)) \right\}$$

and ι_i is defined in (11.25). In addition, in view of Lemmas 2.1 and 2.8, it can be proved that Ω is a positive definite matrix by choosing the appropriate value for c and b_i.

Substituting (11.1) into the infinitesimal operator of $V(t)$ gives

$$\begin{aligned} \mathscr{I}V(t) &= \tilde{x}^T[c(L + L^T) + 2D]\tilde{v} + \tilde{v}^T\tilde{v} + (\tilde{x}^T + \tilde{v}^T)\dot{\tilde{v}} + \sum_{i=1}^{N} \frac{1}{\iota_i} \tilde{\eta}_i \dot{\tilde{\eta}}_i \\ &\leq \tilde{x}^T[c(\mathscr{L} + \mathscr{L}^T) + 2D]\tilde{v} + \tilde{v}^T\tilde{v} \\ &\quad + \sum_{i=1}^{N} s_i \left(f_{\sigma_i}(t, x_i^r, v_i^r) - f_{\sigma_i}(t, x_{\sigma_i}^r, v_{\sigma_i}^r) \right) + \tilde{u}(t) \end{aligned} \tag{11.34}$$

where

$$\tilde{u}(t) = \sum_{i=1}^{N} s_i \left(\rho_i u_i(t) + r_i(t) + \overline{p}_i \overline{\psi}_i(x_i, v_i) + \hat{\vartheta}_i(t) \right) + \sum_{i=1}^{N} \frac{1}{\iota_i} \tilde{\eta}_i \dot{\tilde{\eta}}_i.$$

In view of $\dot{\hat{\eta}}_i(t) + \rho_i \hat{\eta}_i(t) \geq 0$ in (11.25), one has

$$\hat{\eta}_i(t) \geq e^{-\rho_i t} \hat{\eta}_i(0)$$

which implies $\hat{\eta}_i(t) \geq 0$ if $\hat{\eta}_i(0) \geq 0$. Then, in view of the definition of η_i in (11.33), we have

$$\begin{aligned} \tilde{u}(t) &= \sum_{i=1}^{N} s_i \left(\rho_i u_i(t) + r_i(t) + \overline{p}_i \overline{\psi}_i(x_i, v_i) + \hat{\vartheta}_i(t) \right) + \sum_{i=1}^{N} \frac{1}{\iota_i} \tilde{\eta}_i \dot{\tilde{\eta}}_i \\ &\leq \sum_{i=1}^{N} s_i \left(\rho_i u_i(t) + \Pi + \vartheta_i^*(t) \right) + \sum_{i=1}^{N} \frac{1}{\iota_i} \tilde{\eta}_i \dot{\tilde{\eta}}_i \end{aligned} \tag{11.35}$$

where

$$
\begin{aligned}
\Pi &= \eta_i\big(1+\overline{\psi}_i(x_i, v_i)\big) \\
\vartheta_i^*(t) &= \sum_{j\in N_i^*}(\gamma_{ij}(t)-1)\vartheta_j(t).
\end{aligned}
$$

In view of Lemma 2.6, it is deduced that

$$
s_i\eta_i\big(1+\overline{\psi}_i(x_i, v_i)\big) \le \frac{\eta_i s_i^2\big(1+\overline{\psi}_i(x_i, v_i)\big)^2}{\sqrt{s_i^2\big(1+\overline{\psi}_i(x_i, v_i)\big)^2+\zeta_i^2}} + \eta_i\zeta_i \tag{11.36}
$$

where $\zeta_i(t) = h_{i1}e^{-h_{i2}t}$. Then, by utilizing (11.21), (11.35) can be rewritten as

$$
\tilde{u}(t) \le \sum_{i=1}^{N}\left(\frac{1}{d_i}(\rho_i\mathscr{N}_i(\kappa_i)-1)\dot{\kappa}_i + \eta_i\zeta_i\right) + \tilde{u}_1(t) + \tilde{u}_2(t) - \tilde{u}_3(t) \tag{11.37}
$$

where

$$
\begin{aligned}
\tilde{u}_1(t) &= \sum_{i=1}^{N}s_i\Big(-c\sum_{j=1}^{N}l_{ij}x_j(t_{k_j}^j) - c\sum_{j=1}^{N}l_{ij}v_j(t_{k_j}^j) + \mathscr{P}_{1i}(t) + \mathscr{P}_{3i}(t)\Big) \\
\tilde{u}_2(t) &= \sum_{i=1}^{N}s_i\sum_{j\in N^*}(\gamma_{ij}(t)-1)\vartheta_j(t) \\
\tilde{u}_3(t) &= \sum_{i=1}^{N}(\rho_i/2\iota_i)\big(\tilde{\eta}_i^2-\eta_i^2\big).
\end{aligned}
$$

Then,

$$
\begin{aligned}
\tilde{u}_1(t) &= \sum_{i=1}^{N}s_i\Big(-c\sum_{j=1}^{N}l_{ij}x_j(t_{k_j}^j) - c\sum_{j=1}^{N}l_{ij}v_j(t_{k_j}^j) - b_is_i(t) + c\sum_{j=1}^{N}l_{ij}x_{\sigma_j}^r(t) + c\sum_{j=1}^{N}l_{ij}v_{\sigma_j}^r(t)\Big) \\
&= \sum_{i=1}^{N}s_i\Big[-c\sum_{j=1}^{N}l_{ij}\Big(e_{x_j}(t)+\tilde{x}_j(t)+e_{v_j}(t)+\tilde{v}_j(t)\Big)-b_is_i(t)\Big] \\
&= \sum_{i=1}^{N}s_i\Big(-c\sum_{j=1}^{N}l_{ij}s_j(t) - b_is_i(t)\Big) + \sum_{i=1}^{N}s_i\Big[-c\sum_{j=1}^{N}l_{ij}\Big(e_{x_j}(t)+e_{v_j}(t)\Big)\Big]
\end{aligned} \tag{11.38}
$$

where

$$\sum_{i=1}^{N} s_i\left[-c\sum_{j=1}^{N} l_{ij}\left(e_{x_j}(t)+e_{v_j}(t)\right)\right] = \begin{bmatrix}\tilde{x}\\ \tilde{v}\end{bmatrix}^T \begin{bmatrix}-c\mathscr{L} & -c\mathscr{L}\\ -c\mathscr{L} & -c\mathscr{L}\end{bmatrix}\begin{bmatrix}e_x\\ e_v\end{bmatrix}$$
$$\leq 2c\|\mathscr{L}\|\|\varsigma(t)\|\|e(t)\| \tag{11.39}$$

with

$$e_x^T(t) = [e_{x_1}^T(t), e_{x_2}^T(t), \cdots, e_{x_N}^T(t)], \quad e(t) = [e_x^T(t), e_v^T(t)]^T$$
$$e_v^T(t) = [e_{v_1}^T(t), e_{v_2}^T(t), \cdots, e_{v_N}^T(t)]$$

and $\varsigma(t)$ is defined in the section of control objective. When the event is not triggered, from (11.26), it can be verified that

$$\|e(t)\|^2 \triangleq \sum_{i=1}^{N}\left(|e_{x_i}|^2+|e_{v_i}|^2\right) \leq \sum_{i=1}^{N}\theta\mu_i(t) \leq \theta\sum_{i=1}^{N}\overline{\mu}_i. \tag{11.40}$$

Thus, one has

$$\|e(t)\| \leq \sqrt{\theta}\sqrt{\sum_{i=1}^{N}\overline{\mu}_i}, \quad \theta = \max_{1\leq i\leq N}\{\theta_i(0)\}.$$

Then, using Lemma 2.7, (11.39) becomes

$$2c\|\mathscr{L}\|\|\varsigma(t)\|\|e(t)\|$$
$$\leq 2c\|\mathscr{L}\|\left(\frac{\sum_{i=1}^{N}\overline{\mu}_i}{2}+\frac{\theta\|\varsigma(t)\|^2}{2}\right) = c\theta\|\mathscr{L}\|\|\varsigma(t)\|^2 + c\|\mathscr{L}\|\sum_{i=1}^{N}\overline{\mu}_i. \tag{11.41}$$

Then, substituting (11.41) into (11.38) gives

$$\tilde{u}_1(t)\leq\sum_{i=1}^{N}s_i\left(-c\sum_{j=1}^{N}l_{ij}s_j(t)-b_is_i(t)\right)+\Lambda+c\|\mathscr{L}\|\sum_{i=1}^{N}\overline{\mu}_i \tag{11.42}$$

where $\Lambda = c\theta\|\mathscr{L}\|\|\varsigma(t)\|^2$. In addition, from (11.37), we have

$$\tilde{u}_2(t)\leq\frac{1}{2}\sum_{i=1}^{N}s_i^2+\frac{1}{2}\sum_{i=1}^{N}\sum_{j=1}^{N}(\gamma_{ij}(t)-1)^2\vartheta_j^2(t). \tag{11.43}$$

It can be obtained from (11.37), (11.42) and (11.43) that:

$$\begin{aligned}\tilde{u}(t) \leq & \sum_{i=1}^{N}\Big(\frac{1}{d_i}(\rho_i\mathcal{N}_i(\kappa_i)-1)\dot{\kappa}_i+\eta_i\zeta_i\Big)+\frac{1}{2}\sum_{i=1}^{N}\sum_{j=1}^{N}(\gamma_{ij}(t)-1)^2\vartheta_j^2(t)\\ & -\sum_{i=1}^{N}\frac{\rho_i}{2\iota_i}\tilde{\eta}_i^2+\sum_{i=1}^{N}\frac{\rho_i}{2\iota_i}\eta_i^2+\sum_{i=1}^{N}\Big(\tilde{x}_i^2+\tilde{v}_i^2\Big)+c\|\mathscr{L}\|\sum_{i=1}^{N}\overline{\mu}_i\\ & +\sum_{i=1}^{N}s_i\Big(-c\sum_{j=1}^{N}l_{ij}s_j(t)-b_is_i(t)\Big)+c\theta\|\mathscr{L}\|\|\varsigma(t)\|^2.\end{aligned} \tag{11.44}$$

According to Assumption 11.2, it can be arrived that

$$\begin{aligned}&\sum_{i=1}^{N}(\tilde{x}_i+\tilde{v}_i)^T(f_{\sigma_i}(t,x_i,v_i)-f_{\sigma_i}(t,x_{n_i}^r,v_{n_i}^r))\\ &\leq\sum_{i=1}^{N}\Big(\frac{3\alpha_i+\beta_i}{2}\tilde{x}_i^T\tilde{x}_i+\frac{3\beta_i+\alpha_i}{2}\tilde{v}_i^T\tilde{v}_i\Big)\leq m_1\tilde{x}^T\tilde{x}+m_2\tilde{v}^T\tilde{v}\end{aligned} \tag{11.45}$$

where

$$\begin{aligned}m_1 &= \max_{1\leq i\leq N}\{(3\alpha_i+\beta_i)/2\}, \quad m_3=\max\{m_1,m_2+1\}\\ m_2 &= \max_{1\leq i\leq N}\{(3\beta_i+\alpha_i)/2\}.\end{aligned}$$

By utilizing (11.34), (11.44), and (11.45), one has

$$\begin{aligned}\mathscr{I}V(t)\leq\, & \tilde{x}^T[c(L+L^T)+2D]\tilde{v}+\tilde{v}^T\tilde{v}+c\|L\|\sum_{i=1}^{N}\overline{\mu}_i\\ & +\sum_{i=1}^{N}\Big(\frac{1}{d_i}(\rho_i\mathcal{N}_i(\kappa_i)-1)\dot{\kappa}_i+\eta_i\zeta_i\Big)+\sum_{i=1}^{N}s_i\Big(-c\sum_{j=1}^{N}l_{ij}s_j(t)-d_is_i(t)\Big)\\ & -\sum_{i=1}^{N}\frac{\rho_i}{2\iota_i}\tilde{\eta}_i^2+\sum_{i=1}^{N}\frac{\rho_i}{2\iota_i}\eta_i^2+\sum_{i=1}^{N}\Big(\tilde{x}_i^2+\tilde{v}_i^2\Big)+\frac{1}{2}\sum_{i=1}^{N}\sum_{j=1}^{N}\overline{\vartheta}(\gamma_{ij}(t)-1)^2\\ & +m_1\tilde{x}^T\tilde{x}+m_2\tilde{v}^T\tilde{v}+c\theta\|L\|\|\varsigma(t)\|^2\\ \leq & \sum_{i=1}^{N}\Big(\frac{1}{d_i}(\rho_i\mathcal{N}_i(\kappa_i)-1)\dot{\kappa}_i+\eta_i\zeta_i\Big)+\varsigma^T(t)\hat{\Phi}\varsigma(t)+\sum_{i=1}^{N}\frac{\rho_i}{2\iota_i}\eta_i^2\\ & +c\|\mathscr{L}\|\sum_{i=1}^{N}\overline{\mu}_i+\frac{1}{2}\sum_{i=1}^{N}\sum_{j=1}^{N}\overline{\vartheta}(\gamma_{ij}(t)-1)^2-\sum_{i=1}^{N}\frac{\rho_i}{2\iota_i}\tilde{\eta}_i^2\end{aligned} \tag{11.46}$$

where

$$\hat{\Phi} = \text{diag}\{\chi I_N - c\mathscr{L} - D, \chi I_N - c\mathscr{L} - D\}$$
$$\chi = m_3 + 1 + c\theta\|\mathscr{L}\|.$$

By defining $\tilde{L} = (-\mathscr{L} - \mathscr{L}^T)/2$, the above inequality can be transformed into

$$\mathscr{I}V(t) \leq \sum_{i=1}^{N}\left(\frac{1}{d_i}\big(\rho_i\mathscr{N}_i(\kappa_i) - 1\big)\dot{\kappa}_i + \eta_i\zeta_i\right) + \frac{1}{2}\sum_{i=1}^{N}\sum_{j=1}^{N}\overline{\vartheta}\big(\gamma_{ij}(t) - 1\big)^2$$
$$+ c\|\mathscr{L}\|\sum_{i=1}^{N}\overline{\mu}_i + \varsigma^T(t)\Phi\varsigma(t) - \sum_{i=1}^{N}\frac{\rho_i}{2\iota_i}\tilde{\eta}_i^2 + \sum_{i=1}^{N}\frac{\rho_i}{2\iota_i}\eta_i^2 \tag{11.47}$$

where

$$\Phi = \text{diag}\{\chi I_N + c\tilde{L} - D, \chi I_N + c\tilde{L} - D\}.$$

When $\Phi < 0$, the following inequality holds:

$$\mathscr{I}V(t) \leq \sum_{i=1}^{N}\left(\frac{1}{d_i}\big(\rho_i\mathscr{N}_i(\kappa_i) - 1\big)\dot{\kappa}_i + \eta_i\zeta_i\right) - \varpi_1 V(t) + \varpi_2 \tag{11.48}$$

where

$$\varpi_1 = \min\{(-\lambda_{\min}(\Phi))/\lambda_{\max}(\Omega), \rho\}, \quad \rho = \min_{1\leq i\leq N}\{\rho_i\}.$$

Since $\gamma_{ij}(t) \in \{0, 1\}$, it is implied that

$$\varpi_2 = c\|\mathscr{L}\|\sum_{i=1}^{N}\overline{\mu}_i + \frac{1}{2}\sum_{i=1}^{N}\sum_{j=1}^{N}\overline{\vartheta}(\gamma_{ij}(t) - 1)^2 + \sum_{i=1}^{N}\frac{\rho_i}{2\iota_i}\eta_i^2$$

is a finite constant. In addition, $\Omega > 0$ if $\Phi < 0$. Integrating both sides of (11.48), it is derived that

$$V(t) \leq \sum_{i=1}^{N}\frac{1}{d_i}\int_0^t \big(\rho_i(\tau)\mathscr{N}_i(\kappa_i) - 1\big)\dot{\kappa}_i(\tau)\,d\tau + \phi_4 \tag{11.49}$$

where

$$\phi_4 = \phi_5 + V(0) + \varpi_2/\varpi_1$$
$$\phi_5 = \sum_{i=1}^{N} \eta_i \int_0^{+\infty} \zeta_i(\tau)\, d\tau > \sum_{i=1}^{N} \eta_i \int_0^{t} \zeta_i(\tau) d\tau$$

are finite constants.

It then follows from (11.49) that

$$\mathbb{E}\{V(t)\} \leq \mathbb{E}\left\{\sum_{i=1}^{N} \int_0^t \frac{1}{d_i} \big(\rho_i(\tau)\mathscr{N}_i(\kappa_i) - 1\big)\dot{\kappa}_i(\tau)\, d\tau\right\} + \phi_6 \tag{11.50}$$

where

$$\phi_6 = V(0) + \frac{\frac{1}{2}\sum_{i=1}^{N}\sum_{j=1}^{N} \overline{\vartheta}\Upsilon_{ij}}{\varpi_1} + \frac{c\|\mathscr{L}\| \sum_{i=1}^{N} \overline{\mu}_i}{\varpi_1} + \frac{\sum_{i=1}^{N} \frac{\rho_i}{2\iota_i}\eta_i^2}{\varpi_1} + \phi_5$$

is a finite constant.

By using Theorems 11.1 and (11.50), it can be concluded that $\mathbb{E}\{\mathscr{N}_i(t)\}$, $\mathbb{E}\{\int_0^t d_i\big(\rho_i(\tau)\mathscr{N}_i(\kappa_i) - 1\big)\dot{\kappa}_i(\tau)\, d\tau\}$, and $\mathbb{E}\{V(t)\}$ are bounded. Based on (11.33), we can obtain

$$\mathbb{E}\left\{\frac{1}{2}\varsigma^T(t)\Omega\varsigma(t)\right\} \leq \mathbb{E}\{V(t)\}, \quad \mathbb{E}\left\{\sum_{i=1}^{N} \frac{1}{2\iota_i}\tilde{\eta}_i^2\right\} \leq \mathbb{E}\{V(t)\} \tag{11.51}$$

namely

$$\mathbb{E}\big\{\|\varsigma(t)\|\big\} \leq \mathbb{E}\big\{\sqrt{2\nu_1 V(t)}\big\} \leq \sqrt{2\nu_1 V_m}$$
$$\mathbb{E}\big\{\|\eta(t)\|\big\} \leq \mathbb{E}\big\{\sqrt{2\nu_2 V(t)}\big\} \leq \sqrt{2\nu_2 V_m}$$

where

$$\nu_1 = \frac{1}{\lambda_{\min}(\Omega)}, \quad \nu_2 = \min_{1\leq i\leq N}\{\iota_i\}$$
$$\eta(t) = [\eta_1(t), \eta_2(t), \cdots, \eta_N(t)]^T.$$

In addition, V_m is the upper bound of $\mathbb{E}\{V(t)\}$. The signals of the error system converge to a compact set

$$\Xi = \{\varsigma(t), \eta(t) | \mathbb{E}\{\|\varsigma(t)\|\} \leq \sqrt{2\nu_1 V_m}, \mathbb{E}\{\|\eta(t)\|\} \leq \sqrt{2\nu_2 V_m}\}.$$

Then, $\mathbb{E}\{\tilde{x}_i\}$ and $\mathbb{E}\{\tilde{v}_i\}$ also converge to a compact set. In addition, in view of ϕ_5 in (11.49) and ϕ_6 in (11.50), it can be obtained that the upper bound of the set is positively correlated with the magnitude of $\zeta_i(t)$, the probability of attack, and the limitation of attack strength $\overline{\vartheta}$. The proof is completed. □

Remark 11.7 The *cluster consensus* problem with known fault direction and upper bound of the scaling factors under denial-of-service attack was investigated in [1], in which the ETM with a fixed threshold was adopted. The *fault-tolerant* analysis for the TVUND faults in this chapter is more difficult than the case with known fault direction and upper bound of the scaling factors in [1]. Instead of using the ETM with a fixed threshold as in [1], the ETM with both adaptive and memory characteristics is adopted in Theorem 11.2. It can be observed that the contribution of this chapter is not obtained from [1] through a trivial extension.

11.4 Simulation Examples

In this section, first, we show the damage of the FDI attacks to the *cluster consensus* performance. Then, the effectiveness and superiority of the control scheme proposed in this chapter are illustrated with two fault cases.

In the simulation, a system of five heterogeneous second-order agents, partitioned into two subgroups, is analyzed. The directed communication topology among these agents is presented in Fig. 8.2.

The dynamic acceleration of the heterogeneous agents is set as

$$\begin{aligned} f_{\sigma_1}(t, x_i(t), v_i(t)) &= 27\cos(2.5t) - \sin(x_i) - 0.25v_i + 1,\ i \in \{1, 3, 4\} \\ f_{\sigma_2}(t, x_i(t), v_i(t)) &= 21\sin(2t) - 0.5v_i, i \in \{2, 5\}. \end{aligned} \tag{11.52}$$

The nonlinear uncertainties are $w_i(t) = \sin(0.1i \times t)$ and it follows from Assumption 11.3 that $\overline{\psi}_i(x_i, v_i) = 1$. The false data of the FDIAs are generated by $\vartheta_j(t) = 5 + 2\sin(t)$, and the probabilities of the successful FDIAs in different communication channels are set as $\gamma_{21}(t) = 0.03$, $\gamma_{31}(t) = 0.34$, $\gamma_{41}(t) = 0.1$, $\gamma_{43}(t) = 0.15$, and $\gamma_{52}(t) = 0.02$. The initial position and speed states of the agents are given by $[1, 1, -1, 0, -1, 2, 3]^T$ and $[1, 1, 0, 1, 0, -1, 2]^T$, respectively. In addition, the parameters of the *Nussbaum function*, ETM, and input signal are listed in Table 11.1. Define the state consensus error index as follows:

$$J(t) = \frac{1}{5}\sqrt{\sum_{i=1}^{5}\left(v_i(t) - v^r_{\sigma_i}(t)\right)^2 + \sum_{i=1}^{5}\left(x_i(t) - x^r_{\sigma_i}(t)\right)^2}.$$

In accordance with the pinning control law (11.22), the agents numbered 1 and 2 are designated as the pinning nodes. Additionally, based on Theorem 11.2, the pinning gains and coupling coefficients are chosen as $c = 10$ and $b_i = 110$, $i = \{1, 2\}$, respectively.

Table 11.1 The values of the designed parameters for $i \in \mathscr{V}$

Symbol	Values	Symbol	Values	Symbol	Values
ε_{11}	0.05	$\theta_i(0)$	0.2	b_i	0.0001
ε_{12}	0.001	ε_1	0.001	h_{i1}	3
ε_{13}	0.1	ε_2	0.5	h_{i2}	0.01
ε_{14}	0.1	ε_3	3	ρ_i	0.00001
ε_{15}	0.1	$\overline{\mu}_i$	50	ι_i	5
ε_{2i}	1.4	$\underline{\mu}_i$	40	$\kappa_i(0)$	0.5

11.4.1 Cluster Consensus Without Actuator Fault Under FDI Attacks

Figure 11.2 shows the impact of the FDI attacks on the system group consensus performance. The state consensus error index $J(t)$ is introduced to quantify the group consensus performance of the MAS. The curves in Fig. 11.2 describe the state consensus error of the MAS with or without FDI attacks by using the control scheme in [30]. It can be observed that the attacks result in a decline in the group consensus performance of the system.

11.4.2 Cluster Consensus With Actuator Fault Under FDI Attacks

The *fault-tolerant* control method proposed in Theorem 11.2 (Plan A) is compared with the *fault-tolerant* control method proposed in [12] (Plan B).

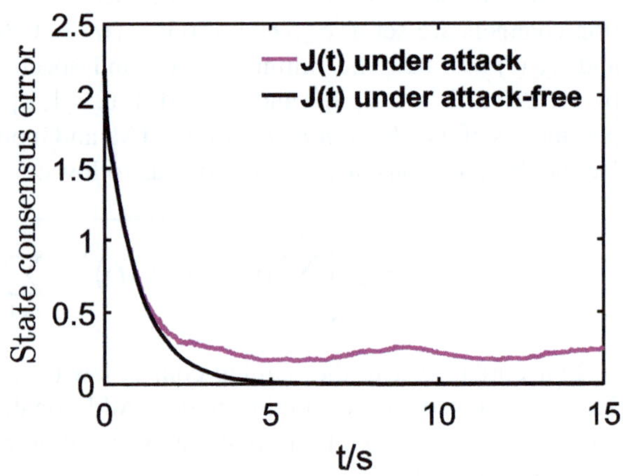

Fig. 11.2 Damage of the FDI attacks to the group consensus performance of the MAS

Case 1 (Loss of effectiveness and bias faults). Consider the loss of effectiveness actuator faults with known direction, i.e, $0 < \rho_i(t) < 1$, and bias fault, and the efficiency factors $\rho_i(t)$ and additive faults $r_i(t)$ are described by

$$\begin{cases} r_i(t) = 2\sin(0.1i \times t), \\ \rho_1(t) = 0.9 + 0.1\cos(0.2t), \\ \rho_2(t) = 0.8 - 0.1\sin(t), \quad i \in \{1, 2, \cdots, 5\}, \\ \rho_3(t) = 0.6 + 0.2\sin(1.1t), \quad 0 \le t < 15, \\ \rho_4(t) = 0.3 + 0.1\sin(1.1t), \\ \rho_5(t) = 0.4 - 0.1\cos(0.6t). \end{cases}$$

Figure 11.3 displays the simulation results of the two *fault-tolerant* control methods in Case 1. It can be observed that both of the two *fault-tolerant* control methods can guarantee bounded cluster consensus in Case 1, and the estimates of their respective scalars tend to be stable. In addition, Fig. 11.4 shows the variation curve of the update gains and the *Nussbaum functions*. It can be obtained that the gains κ_i and functions $\mathcal{N}(\kappa_i)$ tend to stabilize quickly, and the values of $\mathcal{N}(\kappa_i)$ are greater than zero, which is consistent with the fault direction in Case 1. Furthermore, Fig. 11.5 shows the event-triggered instants of the control scheme proposed in this chapter. It can be found that the ETM effectively reduces the communication burden.

Case 2 (TVUND faults). Consider the TVUND faults, and the efficiency factors $\rho_i(t)$ and additive faults $r_i(t)$ are described by

$$\begin{cases} r_i(t) = 2\sin(0.1i \times t) \\ \rho_1(t) = 1.2 + 0.1\cos(0.2t) \\ \rho_2(t) = -1.5 - 0.1\sin(t), \quad i \in \{1, 2, \cdots, 5\} \\ \rho_3(t) = -0.6 + 0.2\sin(1.1t), \quad 0 \le t < 15 \\ \rho_4(t) = 0.7 + 0.1\sin(1.1t) \\ \rho_5(t) = 1.2 - 0.1\cos(0.6t). \end{cases}$$

Figure 11.6 shows the simulation results of the two *fault-tolerant* methods in Case 2. It can be observed that Plan A can achieve *cluster consensus* in Case 2, whereas Plan B can not ensure that the system achieves group consensus, and the system is unstable in Case 2. These findings can be summarized as follows: (1) Plan A successfully bypasses the obstacles caused by the TVUND faults, while Plan B fails to do so; (2) The presence of TVUND faults compromises the stability of the system.

Similarly, Fig. 11.7 displays the curves of the adaptive gains and the Nussbaum functions. It is evident that the gains κ_i and the functions $\mathcal{N}(\kappa_i)$ rapidly stabilize. Moreover, the values of the Nussbaum functions align with the direction of the faults, demonstrating their effectiveness in adapting to fault directions.

In the subsequent analysis, the performance of the designed ETM in this chapter (Plan C) is evaluated by comparing it with the ETM presented in [6] (Plan D). The

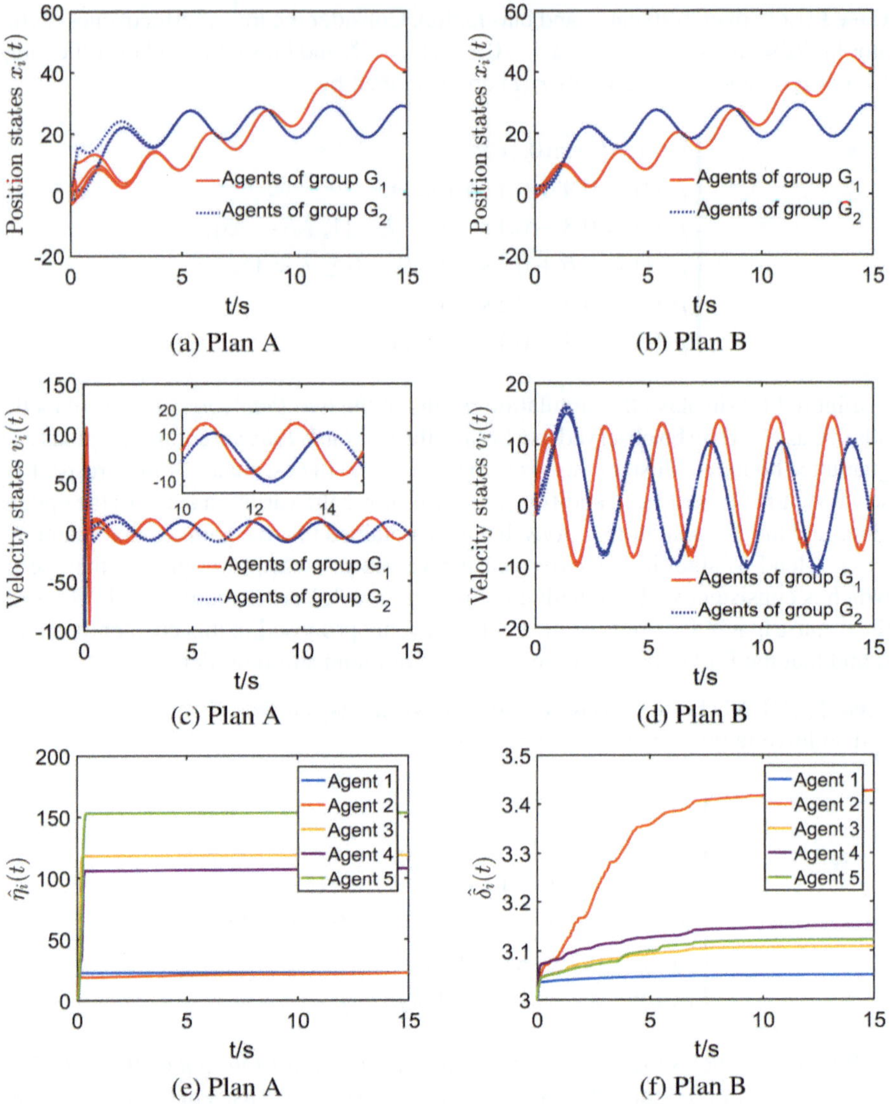

(a) Plan A (b) Plan B

(c) Plan A (d) Plan B

(e) Plan A (f) Plan B

Fig. 11.3 Simulation results in Case 1

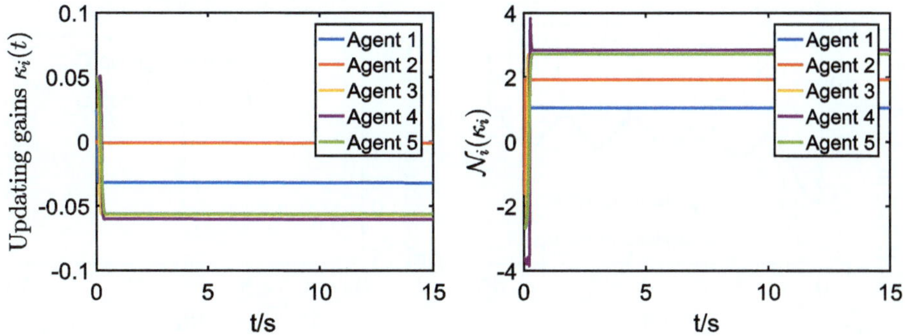

(a) (b)

Fig. 11.4 Update gains and Nussbaum functions in Case 1

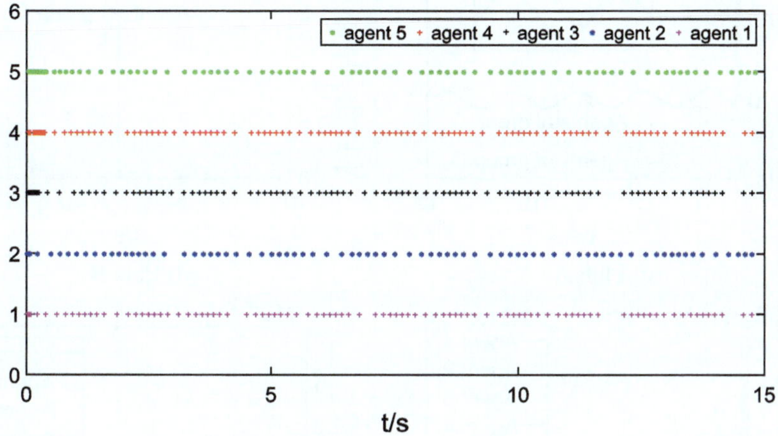

Fig. 11.5 Trigger instants of the agents in Case 1

same factors $\rho_i(t)$ and bias faults $r_i(t)$ are utilized, with the reverse fault being introduced in the simulation at 5 s. The simulation results are illustrated in Fig. 11.8, and Table 11.2 provides the count of triggers for each agent. It is observed that upon the occurrence of the actuator fault, Plan C demonstrates a markedly smaller state error compared to Plan D, and Plan D exhibits a higher number of triggers than Plan C. Additionally, Plan C remains unaffected by the faults, while Plan D is significantly impacted after 5 s. It is also noted that the triggers of Plan D are sparse near the motion trajectory's troughs (or peaks) and more denser in the non-troughs (or non-peaks). This phenomenon is effectively mitigated in Plan C. Therefore, it can be concluded that Plan C can not only effectively ensure the *cluster consensus* performance of the MAS but also substantially alleviates the communication load.

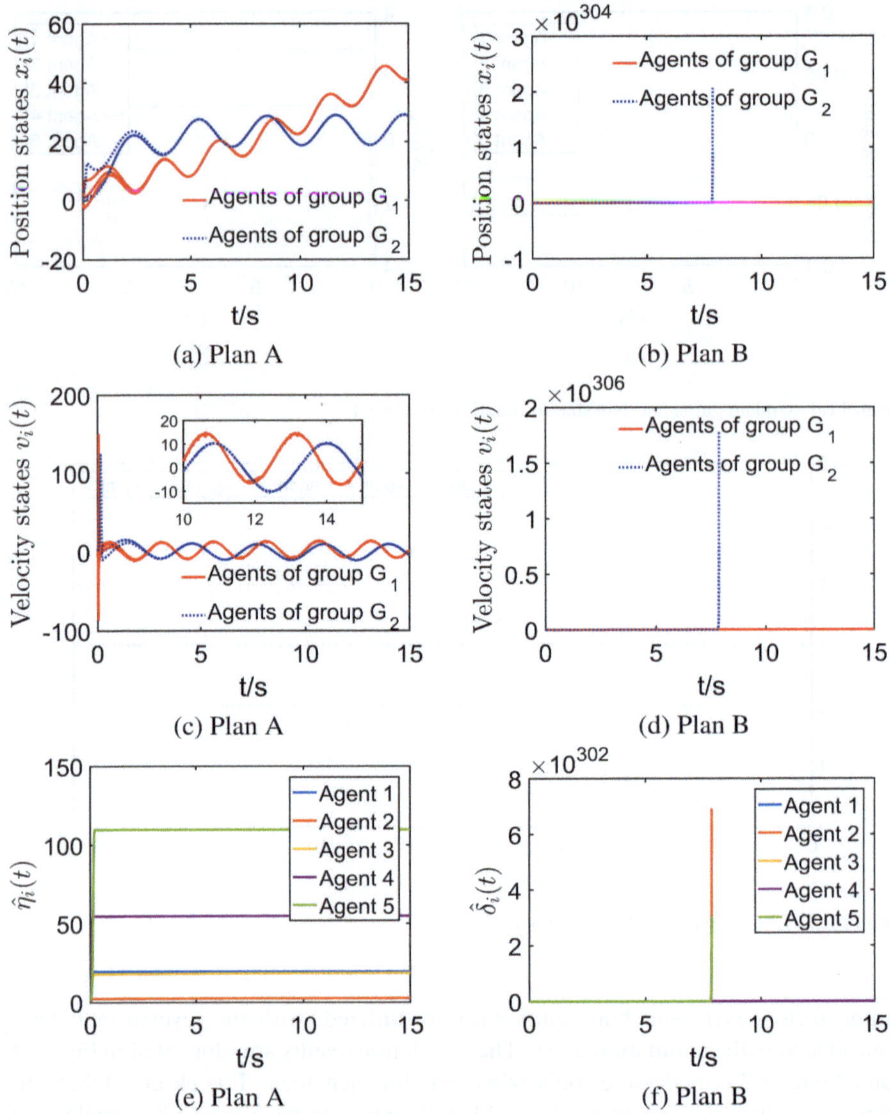

Fig. 11.6 Simulation results of Plan A (left) and Plan B (right)

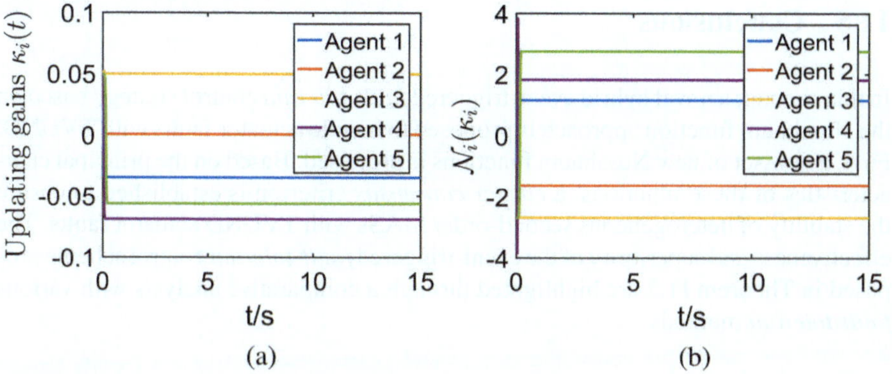

Fig. 11.7 Update gains $\kappa_i(t)$ and Nussbaum functions $\mathcal{N}_i(\kappa_i)$

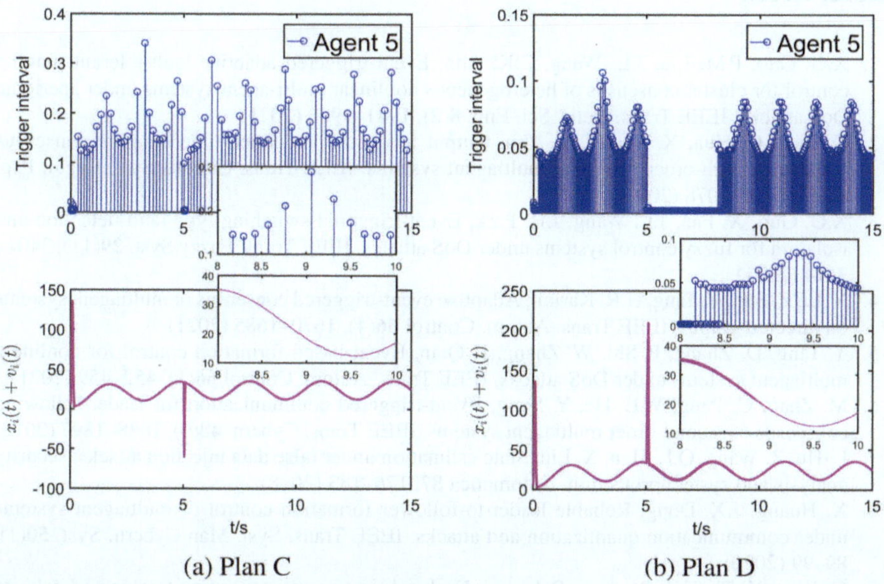

Fig. 11.8 Trigger instants of Plan C (left) and Plan D (right)

Table 11.2 Number of triggers in Plan C and Plan D

Agent	1	2	3	4	5
ETM in Plan C	118	124	222	164	171
ETM in Plan D	363	1329	498	456	1505

11.5 Conclusions

In this chapter, a novel hybrid event-triggered *fault-tolerant* control strategy based on the Nussbaum function approach is proposed to handle actuator faults with TVUND. First, a new set of new Nussbaum functions is proposed. Based on the principal characteristics of these functions, a *cluster consensus* criterion is established to ensure the stability of heterogeneous second-order MASs with TVUND actuator faults. The effectiveness and superiority of the event-triggered *fault-tolerant* control scheme proposed in Theorem 11.2 are highlighted through a comparative analysis with various *fault-tolerant* methods.

References

1. X.G. Guo, P.M. Liu, J.L. Wang, C.K. Ahn, Event-triggered adaptive fault-tolerant pinning control for cluster consensus of heterogeneous nonlinear multi-agent systems under aperiodic DoS attacks. IEEE Trans. Netw. Sci. Eng. **8**(2), 1941–1956 (2021)
2. K. Li, C.C. Hua, X. You, C.K. Ahn, Output feedback predefined-time bipartite consensus control for high-order nonlinear multiagent systems. IEEE Trans. Circuits Syst. Regul. Pap. **68**(7), 3069–3078 (2021)
3. X.G. Guo, X. Fan, J.L. Wang, J.H. Park, Event-triggered switching-type fault detection and isolation for fuzzy control systems under DoS attacks. IEEE Trans. Fuzzy Syst. **29**(11), 3401–3414 (2021)
4. X. Li, Z. Sun, Y. Tang, H.R. Karimi, Adaptive event-triggered consensus of multiagent systems on directed graphs. IEEE Trans. Autom. Control **66**(4), 1670–1685 (2021)
5. Y. Tang, D. Zhang, P. Shi, W. Zhang, F. Qian, Event-based formation control for nonlinear multiagent systems under DoS attacks. IEEE Trans. Autom. Control **66**(1), 452–459 (2021)
6. M. Zhao, C. Peng, W.L. He, Y. Song, Event-triggered communication for leader-following consensus of second-order multiagent systems. IEEE Trans. Cybern. **48**(6), 1888–1897 (2018)
7. L. Hu, Z. Wang, Q.L. Han, X. Liu, State estimation under false data injection attacks: security analysis and system protection. Automatica **87**, 176–183 (2018)
8. X. Huang, J.X. Dong, Reliable leader-to-follower formation control of multiagent systems under communication quantization and attacks. IEEE Trans. Syst. Man Cybern. Syst. **50**(1), 89–99 (2020)
9. G. Wen, X. Zhai, Z. Peng, A. Rahmani. Fault-tolerant secure consensus tracking of delayed nonlinear multi-agent systems with deception attacks and uncertain parameters via impulsive control. Commun. Nonlinear Sci. Numer. Simulat. **82**, Art. No. 105043 (2020)
10. K. Li, C.C. Hua, X. You, C.K. Ahn, Leader-following consensus control for uncertain feedforward stochastic nonlinear multiagent systems. IEEE Trans. Neural Netw. Learn. Syst. **34**(2), 1049–1057 (2023)
11. L. Zhao, G.H. Yang, Cooperative adaptive fault-tolerant control for multi-agent systems with deception attacks. J. Frankl. Inst. **357**(6), 3419–3433 (2020)
12. X. Jin, S. Wang, J. Qin, W.X. Zheng, Y. Kang, Adaptive fault-tolerant consensus for a class of uncertain nonlinear second-order multi-agent systems with circuit implementation. IEEE Trans. Circuits Syst. Regul. Pap. **65**(7), 2243–2255 (2018)
13. Y. Song, X. Huang, C. Wen, Robust adaptive fault-tolerant PID control of MIMO nonlinear systems with unknown control direction. IEEE Trans. Ind. Electron. **64**(6), 4876–4884 (2017)
14. W.D. Xu, X.G. Guo, J.L. Wang, W.W. Che, Z.G. Wu, Nonlinear disturbance observer-based fault-tolerant sliding-mode control for 2-D plane vehicular platoon with utvfd and anas. IEEE Trans. Cybern. **54**(4), 2050–2061 (2024)

15. X.G. Guo, D.C. Tan, C.K. Ahn, J.L. Wang, Fully distributed adaptive fault-tolerant sliding-mode control for nonlinear leader-following multiagent systems with ANASs and IQCs. IEEE Trans. Cybern. **52**(5), 2763–2774 (2022)
16. E. Ribeiro, A.J.M. Cardoso, C. Boccaletti, Fault-tolerant strategy for a photovoltaic DC-DC converter. IEEE Trans. Power Electron. **28**(6), 3008–3018 (2013)
17. A. Theissler, Detecting known and unknown faults in automotive systems using ensemble-based anomaly detection. Knowl.-Based Syst. **123**, 163–173 (2017)
18. X.G. Guo, W.D. Xu, J.L. Wang, J.H. Park, Distributed neuroadaptive fault-tolerant sliding-mode control for 2-D plane vehicular platoon systems with spacing constraints and unknown direction faults. Automatica **129**, Art. No. 109675 (2021)
19. X.G. Guo, W.D. Xu, J.L. Wang, J.H. Park, H. Yan, BLF-based neuroadaptive fault-tolerant control for nonlinear vehicular platoon with time-varying fault directions and distance restrictions. IEEE Trans. Intell. Transp. Syst. **23**(8), 12388–12398 (2022)
20. X.Y. Guo, C.L. Wang, Z. Dong, Z.T. Ding, Adaptive containment control for heterogeneous MIMO nonlinear multiagent systems with unknown direction actuator faults. IEEE Trans. Autom. Control **68**(9), 5783–5790 (2023)
21. K. Zhao, Y.D. Song, C.L.P. Chen, L. Chen, Adaptive asymptotic tracking with global performance for nonlinear systems with unknown control directions. IEEE Trans. Autom. Control **67**(3), 1566–1573 (2022)
22. B. Fan, Q. Yang, S. Jagannathan, Y. Sun, Output-constrained control of nonaffine multiagent systems with partially unknown control directions. IEEE Trans. Autom. Control **64**(9), 3936–3942 (2019)
23. Q. Wang, C. Sun, Adaptive consensus of multiagent systems with unknown high-frequency gain signs under directed graphs. IEEE Trans. Syst. Man Cybern. Syst. **50**(6), 2181–2186 (2020)
24. C. Chen, C. Wen, Z. Liu, K. Xie, Y. Zhang, C.L.P. Chen, Adaptive consensus of nonlinear multi-agent systems with non-identical partially unknown control directions and bounded modelling errors. IEEE Trans. Autom. Control **62**(9), 4654–4659 (2017)
25. J. Huang, Y. Song, W. Wang, C. Wen, G. Li, Fully distributed adaptive consensus control of a class of high-order nonlinear systems with a directed topology and unknown control directions. IEEE Trans. Cybern. **48**(8), 2349–2356 (2018)
26. C.L. Wang, C.Y. Wen, L. Guo, Adaptive consensus control for nonlinear multi-agent systems with unknown control directions and time-varying actuator faults. IEEE Trans. Autom. Control **66**(9), 4222–4229 (2021)
27. X.M. Li, Q. Zhou, P. Li, H.Y. Li, R.Q. Lu, Event-triggered consensus control for multi-agent systems against false data-injection attacks. IEEE Trans. Cybern. **50**(5), 1856–1866 (2020)
28. M. Meng, G. Xiao, B. Li, Adaptive consensus for heterogeneous multi-agent systems under sensor and actuator attacks. Automatica **122**, Art. No. 109242 (2020)
29. X. Liu, T. Chen, Finite-time and fixed-time cluster synchronization with or without pinning control. IEEE Trans. Cybern. **48**(1), 240–252 (2019)
30. Q. Ma, Z. Wang, G.Y. Miao, Second-order group consensus for multi-agent systems via pinning leader-following approach. J. Frankl. Inst. **351**(3), 1288–1300 (2014)

Chapter 12
Intrusion- and Fault-Tolerant Cluster Consensus of MASs with Multiple Attacks

Abstract This chapter investigates the *cluster consensus* tracking problem of the high-order *multi-agent systems* (MASs) that are beset by time-varying unknown non-identical direction (TVUND) faults under multiple *cyber attacks*, including denial-of-service (DoS) attacks and false data injection attacks (FDIAs). Prompted by this, a fully distributed hierarchical event-triggered (ET) *intrusion-* and *fault-tolerant* controller is introduced. This controller operates across the cyber layer and the physical layer and is based on the *Nussbaum function* technique. It incorporates a state-dependent asynchronous ET mechanism (ETM) that ensures there is a positive time interval between triggers. The created cyber layer facilitates communication between various agents, preventing direct physical interaction and thereby minimizing the spread of errors. This dual-layer control frame significantly enhances the flexibility of control design when compared to single-layer frame. Furthermore, a ground-breaking stability lemma based on the Nussbaum function (Lemma 12.1) has been provided for the first time. Similar to Chap. 11, the control scheme can align the input direction with the fault direction by utilizing the Nussbaum function.

12.1 Introduction

With the growing exploration of *complete consensus* in *multi-agent systems* (MASs), increasing attention has shifted toward multi-objective *cluster consensus*, such as the topological conditions for achieving *cluster consensus* [1] and the *cluster consensus* under signed graphs [2]. The communication among agents in MASs relies heavily on cyber mediums [3]. In Chaps. 9–11, the *cluster consensus* control strategies of MASs under common attacks such as denial-of-service (DoS) attacks and false data injection attacks (FDIAs) are studied respectively. It is noted that the previous chapters only consider cases involving a single type of attack. However, in practice, the coordination of multiple *cyber attacks* is common [4, 5]. For handling multiple cyber attacks, studies [5, 6] have employed linear matrix inequality (LMI) techniques to realize *intrusion-tolerant* in MASs and networked control systems, respectively. However, LMI-based controllers face scalability challenges due to the dimensional

X.-G. Guo et al., *Anti-disturbance Intrusion- and Fault-Tolerant Control of Multi-agent Systems*, https://doi.org/10.1007/978-981-96-9614-7_12

explosion, making them impractical for large-scale MASs. Consequently, developing a *intrusion-tolerant cluster consensus* approach for MASs under multiple *cyber attacks* remains an open and pressing research problem.

On the other hand, the aforementioned results concerning multiple *cyber attacks* assume fault-free physical machines. However, in practice, actuator faults frequently arise during physical processes, often leading to performance degradation or even severe accidents [7–11]. As discussed in Chap. 11, excess faults and reverse faults are prevalent in chemical and industrial processes [12], as well as in circuits and sensors, manifesting as excess fault currents, excessive electrical or thermal stresses [13, 14]. Additionally, Nussbaum-type gain methods have been extensively applied to address challenges associated with unknown control directions [13, 15]. Recently, [16, 17] have proposed Nussbaum functions with exponentially increasing frequencies. However, these Nussbaum function-based methods are not directly applicable to address the unknown direction fault problem considered in this chapter, due to the time-varying nature of the unknown direction faults. Moreover, the Nussbaum function design in [17] relies on prior knowledge of global information about the MAS, including agent number, which limits its applicability. Consequently, for MASs with time-varying unknown nonidentical direction (TVUND) faults subjected to multiple *cyber attacks*, developing fully distributed Nussbaum-based *fault-tolerant* control methods is both a practical and challenging research endeavor.

Motivated by the above discussions, this chapter investigates the *cluster consensus* tracking problem for MASs with TVUND faults in the presence of multiple *cyber attacks*. Two primary challenges are identified.

(1) The absence of global information in MASs subjected to TVUND faults and multiple *cyber attacks* poses a significant challenge for designing fully distributed control strategies.
(2) The intricate mutual coupling among TVUND faults, multiple *cyber attacks*, and mismatched nonlinearities further complicates the control design and stability analysis.

The main contributions are as follows.

(1) To achieve *intrusion-* and *fault-tolerant* control in MASs, a hierarchical framework consisting of a cyber layer and a physical layer is developed. In the cyber layer, a switching input control mechanism is incorporated into a virtual node to facilitate agent interaction and mitigate the effects of DoS attacks. Unlike the zero-input (open-loop) control approach adopted during attack periods in [18], this method employs a non-zero input control in the virtual node, ensuring continuous operation during attacks. Furthermore, the proposed control scheme operates in a fully distributed manner, eliminating the need for *a priori* knowledge of the MASs' global information, which distinguishes it from the strategies in [16, 17]. Compared to the single-layer control approach outlined in Chap. 11, the dual-layer framework introduced here prevents direct interactions among agents in the physical layer, effectively suppressing error propagation and enhancing system robustness.

(2) Stability lemma in a random framework : A novel *Nussbaum function* stability lemma (Lemma 12.1) is introduced in this chapter to address the challenges posed by TVUND faults within a random framework. This lemma not only eliminates the requirement for identical or partially known unknown fault directions as in [19], but it also serves as a powerful analytical tool for assessing the stability of agents experiencing time-varying fault directions under random conditions. Furthermore, unlike the methods in [16, 17], the frequency of the proposed *Nussbaum function* is independent of the agents serial numbers, setting it apart from the function used in Chap. 11.

(3) New ETM: A new event-triggered mechanism (ETM) is also proposed, which relies on a state-dependent asynchronous framework and incorporates a simplified analysis of the Zeno phenomenon. This approach differs from the state-independent constant thresholds used in [20], which fail to capture the system's dynamic nature. It also improves upon the state-dependent dynamic thresholds in [18], which involve complex Zeno behavior analysis. Additionally, unlike [21], the proposed ETM eliminates the need for continuous monitoring of neighbors' states during event condition detection, reducing computational overhead and improving system efficiency.

12.2 Problem Formulation and Preliminaries

12.2.1 High-Order MAS Model

Consider a nonlinear MAS composed of N agents. Each agent has an n integrator dynamics as follows

$$\begin{aligned} \dot{x}_{i,q}(t) &= x_{i,q+1}(t) + \theta_i^T f_{i,q}(\overline{x}_{i,q}), \quad i = 1, 2, \ldots, N \\ \dot{x}_{i,n}(t) &= u_i^F(t) + \theta_i^T f_{i,n}(x_i) + w_i(t, x_i) \\ y_i(t) &= x_{i,1}(t), \quad q = 1, 2, \ldots, n-1 \end{aligned} \tag{12.1}$$

where $x_i(t) = [x_{i,1}(t), \ldots, x_{i,n}(t)]^T$, $y_i(t) \in \mathbb{R}$ and $u_i^F(t) \in \mathbb{R}$ are the states, output, and faulty control input, respectively; $f_{i,q}(\overline{x}_{i,q}) \in \mathbb{R}^{p_i}$ with $\overline{x}_{i,q} = [x_{i,1}(t), \ldots, x_{i,q}(t)]^T \in \mathbb{R}^q$ and $f_{i,n}(x_i) \in \mathbb{R}^{p_i}$ are known nonlinear functions; $\theta_i \in \mathbb{R}^{p_i}$ is the unknown parameter vector, and $w_i(t, x_i) \in \mathbb{R}$ represents the unknown disturbances. In addition, similar to Chap. 11, the faulty control input signal $u_i^F(t)$ is modeled as

$$u_i^F(t) = \rho_i(t) u_i(t) + r_i(t) \tag{12.2}$$

in which $u_i(t)$ is the nominal input signal to be designed, $\rho_i(t) \neq 0$ and $r_i(t)$ are referred to as the efficiency of the actuator and the bias faults, respectively. In addition, $\rho_i(t)$ and $r_i(t)$ satisfy Assumption 11.2.

In this chapter, each subgroup σ_i is assigned a distinct trajectory $\hat{y}_{\sigma_i}(t) \in \mathbb{R}$, representing the virtual leader's path to be followed by the agents in that subgroup. It is assumed that $\hat{y}_{\sigma_i}(t)$ and its finite nth derivatives are piecewise continuous and bounded. This is a common assumption in the existing literature, such as [17] and related works. Typically, $\hat{y}_{\sigma_i}(t)$ is produced by a user-designed dynamic system, ensuring that both the trajectory $\hat{y}_{\sigma_i}(t)$ and its derivatives remain bounded and accessible for control purposes.

Remark 12.1 The dynamic model (12.1) studied in this chapter can capture the behavior of numerous real-world systems, including spring-mass-damper systems [17], ship dynamics [22], robot manipulators [9, 23], and quarter-car active suspension systems [24]. In the model, the known nonlinear functions $f_{i,q}(\cdot)$ can describe established mechanical structures or physical relationships. For instance, in the dynamics of a single-link manipulator (as illustrated in the Simulation Examples), the interaction between the link's acceleration $\ddot{q}$, velocity $\dot{q}$, and angular position q is defined by the system's mechanical structure. However, parameters like the damping coefficient are considered unknown due to measurement difficulties. Unlike simpler first-order systems [1], second-order systems [18], and nonlinear uncertain systems [25], the higher-order system model introduced in this paper offers broader applicability. Furthermore, reverse and excess faults frequently arise in various applications, such as chemical and industrial processes [12], circuits, and sensors. Examples include excessive fault currents, overloading from electrical or thermal stresses [13, 14], and negative sensor offset faults [26]. Similarly, in the single-link manipulator example, the motor voltage input could experience reversals or excessive values.

12.2.2 *Multiple Cyber Attack Model*

In an open network environment, agents are vulnerable to *cyber attacks*. This chapter considers a variety of *cyber attacks* targeting the system, encompassing DoS attacks and FDIAs. Here, the sequence of data crafted by the attacker is injected into the control channel of agent, and the attack model is formulated as

$$\hat{\vartheta}_i(t) = \sum_{j \in \mathcal{N}_i^*} \rho_{ij}(t)\vartheta_j(t) \tag{12.3}$$

where $\hat{\vartheta}_i(t)$ denotes the control deviation produced by FDIAs, and the false data $\vartheta_j(t)$ is a function carefully designed by the attackers, which can usually avoid being blocked by the defense mechanism of the system. The weighting $\rho_{ij}(t)$ is an independent Bernoulli variable that satisfies $\mathbb{E}\{\rho_{ij}(t)\} = p_{ij}$ and is defined as

$$\rho_{ij}(t) = \begin{cases} 1, & \text{if data is injected through the edge } (j, i) \\ 0, & \text{otherwise.} \end{cases} \tag{12.4}$$

Combining (12.2) with (12.3), the real faulty control input can be expressed as

$$u_i^F(t) = \rho_i(t)\big(u_i(t) + \hat{\vartheta}_i(t)\big) + r_i(t). \tag{12.5}$$

In addition, this chapter also considers the damage from DoS attacks on the communication channels among different agents. Due to the limited energy, the model of DoS attack is described by

$$\mathscr{S}(t) = \begin{cases} 1, & t \in [t_m, t_m + \nu_m) \\ 0, & t \in [t_m + \nu_m, t_{m+1}) \end{cases} \tag{12.6}$$

where t_{m+1} and $t_m + \nu_m$ with $m \in \mathbb{N}$ denote the ending and starting instants of the mth attacks, respectively. Therefore, the communication between different agents is down and interrupted due to the attack when $\mathscr{S}(t) = 0$, and the communication is up and normal when $\mathscr{S}(t) = 1$. In addition, it is assumed that the system has a communication recovery mechanism with periodic attempts. Define $\wp_1 > 0$ as the inter-attempt time.

Assumption 12.1 ([18]) Let $N_a(t_1, t_2)$ and $\Im_a(t_1, t_2)$ be the number and duration of DoS attacks in $[t_1, t_2)$, respectively. Then, there exist constants $F_0 \geq 0$, $F_1 > 1$, $T_0 \geq 0$ and $T_1 > 1$ that satisfy the following conditions:

$$N_a(t_1, t_2) \leq F_0 + \frac{t_2 - t_1}{F_1} \tag{12.7}$$

$$\Im_a(t_1, t_2) \leq T_0 + \frac{t_2 - t_1}{T_1}. \tag{12.8}$$

Remark 12.2 Unlike disturbances, actuator faults, and parameter uncertainties, random FDIAs are deliberate and carefully crafted by attackers to manipulate the control or measurement channels [27]. These attacks are often designed to bypass detection mechanisms. In contrast to scenarios involving a single FDIA or a single denial-of-service (DoS) attack, which may target both control and communication channels simultaneously [18, 27], or focus on just one channel [28, 29], this chapter considers the combined impact of FDIAs and DoS attacks. The coordinated efforts of FDIAs on control channels and DoS attacks on communication channels present a significantly more dangerous threat.

Remark 12.3 Both FDIAs and DoS attacks are subject to specific constraints. For DoS attacks, Assumption 12.1 aligns with common practices in [18, 30], which typically include periodic sleep intervals $[t_m, t_m + \nu_m)$ to restore energy. For FDIA, the energy constraint is modeled by an unknown upper bound $\overline{\vartheta}$, satisfying

$$\|\vartheta\|^2 \leq \overline{\vartheta}, \quad \vartheta = [\vartheta_1, \vartheta_2, \cdots, \vartheta_N]^T.$$

As noted in [27, 29], FDIA energy is typically limited. Furthermore, to avoid detection, attackers often mimic actual system signals, which are inherently bounded.

These constraints are critical for ensuring the control system's ability to mitigate or resist such attacks and are widely adopted in existing studies, such as [18, 27, 29].

12.2.3 Control Objective and Preliminaries

Before proceeding, the following definition is provided:

Definition 12.1 Given the high-order MAS (12.1) with any initial states $x_i(0)$, the agent's output $y_i(t)$ reaches *cluster consensus* if

$$\lim_{t\to\infty} \mathbb{E}\{\|y_i(t) - \hat{y}_{\sigma_i}(t)\|\} = 0, \quad i \in \mathscr{V}, \quad \sigma_i = 1, \ldots, s. \tag{12.9}$$

In this chapter, the control objective is to develop a fully distributed hierarchical ET *intrusion-* and *fault-tolerant* control strategy, ensuring that the output signal $y_i(t)$ of the agents in MAS (12.1) achieves bounded *cluster consensus* despite the presence of multiple *cyber attacks*, that is,

$$\lim_{t\to\infty} \mathbb{E}\{\|y_i(t) - \hat{y}_{\sigma_i}(t)\|\} < \overline{y} < \infty$$

for all agents, where $\overline{y}$ is an unknown positive bounded constant independent of t.

Furthermore, several key assumptions and a lemma are presented below.

Assumption 12.2 The directed communication graph $\mathscr{G}$ of high-order MAS (12.1) has a directed spanning tree. In addition, the root node should be able to receive information from one of the virtual leaders $\hat{y}_{\sigma_i}(t)$, $\sigma_i \in \{1, \cdots, s\}$.

Assumption 12.3 For the unknown disturbances $w_i(t, x_i)$ for each agent in (12.1), there is a positive unknown constant $\overline{q}_i$ and a locally Lipschitz positive function $\overline{w}_i(x_i)$ such that

$$|w_i(t, x_i)| \le \overline{q}_i \overline{w}_i(x_i), \quad i \in \mathscr{V}. \tag{12.10}$$

Remark 12.4 In contrast to the assumptions of strongly connected or undirected graphs in [17, 28], Assumption 12.2 is a milder assumption, ensuring that there are no isolated agents in the MASs. Moreover, the consideration of unknown disturbances $w_i(t, x_i)$ for each agent in (12.1) signifies the presence of bounded *disturbances* or bounded time-varying parameters. Then, Assumption 12.3 is relevant and applicable.

Lemma 12.1 *Given Lyapunov function $V(t) \ge 0$, and functions $\kappa(t)$ and $\rho(t) \ne 0$ defined on $[0, t_f)$ with $t_f > 0$, if the following condition holds:*

$$\mathbb{E}\{V(t)\} \le \mathbb{E}\left\{ \int_0^t \frac{\dot{\kappa}}{\phi} \left[\rho(\tau)\mathscr{N}\big(\kappa(\tau)\big) + 1\right] d\tau + \varphi \right\} \tag{12.11}$$

where

$$\mathcal{N}(\kappa) = 2\varepsilon_1\kappa \sinh(\varepsilon_1\kappa^2)\cos(\varepsilon_2\kappa) - \varepsilon_2 \cosh(\varepsilon_1\kappa^2)\sin(\varepsilon_2\kappa)$$

is a Nussbaum function; $b > 0$ and φ are constants. Then, $\kappa(t)$, $\mathbb{E}\{V(t)\}$, and $\mathbb{E}\{\int_0^t (\dot{\kappa}/\phi)\left[\rho(\tau)\mathcal{N}\big(\kappa(\tau)\big) + 1\right] d\tau\}$ are bounded on $[0, t_f)$.

Proof First, from the given *Nussbaum function* in Lemma 12.1, one can obtain that

$$H(\kappa) \triangleq \int_0^{\kappa} \mathcal{N}(\tau)\, d\tau = \cosh(\varepsilon_1\kappa^2)\cos(\varepsilon_2\kappa). \tag{12.12}$$

According to the characteristics of trigonometric functions, there exists $\kappa \in [\underline{a}_q, \overline{a}_q]$, $q \in \mathbb{Z}$ such that

$$\text{sign}\{\rho\}\cos(\varepsilon_2\kappa) < -0.5 \tag{12.13}$$

where

$$\begin{cases} \underline{a}_q = \frac{2q\pi + \frac{5\pi}{3}}{\varepsilon_2}, \\ \overline{a}_q = \frac{2q\pi + \frac{7\pi}{3}}{\varepsilon_2}, \end{cases} \quad \text{if } \text{sign}(\rho) = -1$$

or

$$\begin{cases} \underline{a}_q = \frac{2q\pi + \frac{2\pi}{3}}{\varepsilon_2}, \\ \overline{a}_q = \frac{2q\pi + \frac{4\pi}{3}}{\varepsilon_2}, \end{cases} \quad \text{if } \text{sign}(\rho) = 1.$$

In the following, complete the proof by seeking contradictions. Suppose that $\kappa(t)$ is unbounded. For the case $\kappa \in [\underline{a}_q, \overline{a}_q]$, $q = 1, 2, \ldots$, it can be concluded from (12.11) that

$$\begin{aligned} \mathbb{E}\{V(t)\} &\le \mathbb{E}\bigg\{\int_0^t \frac{\dot{\kappa}}{\phi}\rho(\tau)\mathcal{N}\big(\kappa(\tau)\big)\, d\tau + \frac{1}{\phi}\kappa(t)\bigg\} + \varphi_1 \\ &\le \mathbb{E}\bigg\{ - \frac{1}{2\phi}\underline{\rho}\cosh(\varepsilon_1\kappa^2) + \frac{1}{\phi}\kappa(t)\bigg\} + \varphi_2 \end{aligned} \tag{12.14}$$

where

$$\varphi_1 = \varphi - \kappa(0)/\phi, \quad \varphi_2 = \varphi_1 - \int_0^{\kappa(0)} (\underline{\rho}/\phi)\mathcal{N}\big(\kappa(\tau)\big)\, d\kappa(\tau).$$

When $q \to \infty$, $\kappa(t) \to \infty$ holds; thus, one has

$$-(1/2\phi)\underline{\rho}\cosh(\varepsilon_1\kappa^2) + \frac{1}{\phi}\kappa(t) \to -\infty$$

since the growth rate of $\cosh(\varepsilon_1\kappa^2)$ is higher than that of κ. From (12.14), $\mathbb{E}\{V(t)\} < -\infty$ holds, which appears to be a contradiction since $V(t) \geq 0$. Therefore, $\kappa(t)$ is bounded, which implies that $\mathbb{E}\{V(t)\}$ and $\mathbb{E}\{\int_0^t (\dot{\kappa}/\phi)\left[\rho(\tau)\mathscr{N}\left(\kappa(\tau)\right) + 1\right] d\tau\}$ are bounded. This completes the proof. □

Remark 12.5 To achieve the control objectives that MAS realizes *cluster consensus*, and meanwhile ensures the applicability of the *fault-tolerant* method in the random framework, the above new Lemma 12.1 is developed. Since the random FDIAs are considered in this chapter, the existing lemmas in [25] cannot be applied directly.

12.3 Fully Distributed Hierarchical Cluster Consensus Control Strategy

In this section, a fully distributed hierarchical ET *intrusion-* and *fault-tolerant* control scheme is designed. A schematic representation of the control scheme is depicted in Fig. 12.1. In addition, the *cluster consensus* and the Zeno phenomenon of MAS (12.1) is analyzed.

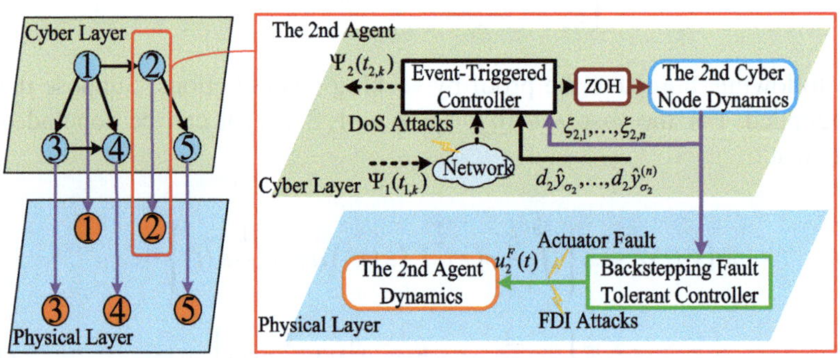

Fig. 12.1 Control system schematic

12.3.1 Design of Intrusion- and Fault-Tolerant Controller

Drawing inspiration from [25], this section proposes a fully distributed hierarchical ET *intrusion-* and *fault-tolerant* control scheme, which consists of a virtual cyber layer and a physical layer as illustrated in Fig. 12.1. Except for the virtual leader, there is a one-to-one correspondence between the nodes in the cyber layer and the agents in the physical layer. The virtual cyber layer addresses the issues of *cluster consensus* and DoS attacks, supplying a local reference signal for the corresponding agents in the physical layer. Meanwhile, the physical layer tackles unknown direction faults and FDIAs, and is responsible for tracking the local reference signals. The design of the hierarchical controller is bifurcated into two main components: (1) the virtual cyber layer, and (2) the actual controller to the agents.

12.3.1.1 Design of The Virtual Cyber Layer

The dynamics of the nodes in the virtual cyber layer are designed as

$$\begin{aligned} \dot{\xi}_{i,q} &= \xi_{i,q+1}, \quad i \in \mathscr{V} \\ \dot{\xi}_{i,n} &= v_i, \quad q = 1, 2, \ldots, n-1 \end{aligned} \tag{12.15}$$

where $\xi_{i,q}$ and $\xi_{i,n}$ are local reference states and v_i denotes the input of virtual nodes. For $i \in \mathscr{V}$, define

$$\varepsilon_i = \xi_{i,1} - \hat{y}_{\sigma_i} \tag{12.16}$$

$$\overline{\varepsilon}_i = \left(1 + \frac{d}{dt}\right)^{n-1} \varepsilon_i = \sum_{q=0}^{n-1} \varepsilon_q \varepsilon_i^{(q)} \tag{12.17}$$

$$\Psi_i = \sum_{q=0}^{n-1} \varepsilon_q \xi_{i,q+1}, \quad \overline{\Psi}_i = \sum_{q=0}^{n-1} \varepsilon_q \hat{y}_{\sigma_i}^{(q)} \tag{12.18}$$

where

$$\varepsilon_q = C_{n-1}^q = \frac{(n-1)!}{(n-1-q)!q!}.$$

Furthermore, ε_i and $\overline{\varepsilon}_i$ are auxiliary variables for analysis. Although the information of leaders is only available for some agents (i.e., the pinning agents), tracking error ε_i is the direct indicator to measure system performance. Therefore, auxiliary variables ε_i and $\overline{\varepsilon}_i$ are essential in the theoretical analysis. Then, differentiating (12.17), one has

$$\dot{\overline{\varepsilon}}_i = v_i + \sum_{q=0}^{n-2} \varepsilon_q \xi_{i,q+2} - \sum_{q=0}^{n-1} \varepsilon_q \hat{y}_{\sigma_i}^{(q+1)}. \tag{12.19}$$

In order to reduce the redundant communication between different agents, the following ETM is introduced in the cyber layer:

$$t_{j,k+1} = \inf\{t | t > t_{j,k}, g_j(t) \geq 0\} \tag{12.20}$$

$$g_j(t) = |\Psi_j(t) - \Psi_j(t_{j,k})| - \frac{\eta_j(t)}{c} \tag{12.21}$$

where

$$\eta_j(t) = \underline{\eta}_j + (\overline{\eta}_j - \underline{\eta}_j)e^{-|\Psi_j(t) - \Psi_j(t_{j,k})|}$$

with $\overline{\eta}_j > 0$ and $\underline{\eta}_j > 0$ being constants, $k \in \mathbb{N}$, and $c > 0$ is the coupling coefficient.

In order to reduce resource consumption, pinning control technology is introduced into the virtual node input v_i. Now, v_i is proposed as

$$v_i = \begin{cases} -c \sum\limits_{j=1}^{N} a_{ij} F_{ij} + \mathscr{P}_i - \sum\limits_{q=0}^{n-2} \varepsilon_q \xi_{i,q+2}, & t \in [t_m, t_m + \nu_m) \\ -\sum\limits_{q=0}^{n-2} \varepsilon_q \xi_{i,q+2}, & t \in [t_m + \nu_m, t_{m+1}) \end{cases} \tag{12.22}$$

where $m \in \mathbb{N}$, $F_{ij} = \Psi_i(t) - \Psi_j(t_{j,k})$, c is coupling coefficient, and

$$\mathscr{P}_i = \begin{cases} -cb_i\left(\Psi_i - \overline{\Psi}_i\right) + c \sum\limits_{j=1}^{N} l_{ij} \overline{\Psi}_i + b_i \sum\limits_{q=0}^{n-1} \varepsilon_q \hat{y}_{\sigma_i}^{(q+1)}, & i \in \tilde{\mathscr{V}}_{\sigma_i} \\ -cb_i\left(\Psi_i - \overline{\Psi}_i\right) + b_i \sum\limits_{q=0}^{n-1} \varepsilon_q \hat{y}_{\sigma_i}^{(q+1)}, & \mathscr{D}_i = 0 \text{ or } i \in \hat{\mathscr{V}}_{\sigma_i} \\ 0, & \text{otherwise} \end{cases} \tag{12.23}$$

where $b_i > 0$ represents the pinning gain.

As depicted in Fig. 12.1, in the virtual cyber layer, the problems of *cluster consensus*, DoS attacks, and redundant data transmission are addressed through pinning control (12.23), switching input method (12.22), and ETM (12.20), respectively. The pinning control method allows a subset of agents to access the leader's information, reducing communication burden and achieving *cluster consensus* by communication. The switching input method rescues the impact of DoS attacks on agent performance, using limited information to maintain system performance during attacks. The ETM reduces network burden by reducing redundant data transfers.

Remark 12.6 In the designed cyber layer, a state-dependent asynchronous ETM (12.20) is proposed for each node, eliminating the need for continuous communication among nodes. Each node's triggering condition relies solely on its local information $\Psi_i(\cdot)$, contrasting with the dependency on real-time neighbor data seen in [21]. This reduces the communication burden significantly. Additionally, the trigger threshold $\eta_j(t)/c$ is maximized at trigger time $t_{j,k}$ and dynamically adjusts between $(t_{j,k}, t_{j,k+1})$. Unlike fixed-threshold ETMs designed in [20], the state-dependent-based time-varying threshold further reduces redundant transmissions. Moreover, each node maintains a strictly positive inter-event time, enhancing the practicality of the ETM, as illustrated in Sect. 12.4.

12.3.1.2 Design The Actual Input of The Agent

In this part, the backstepping method including n steps is used to construct the actual input u_i of the agent i. First, the following notations are defined

$$z_{i,1} = y_i - \xi_{i,1} \tag{12.24}$$

$$z_{i,q} = x_{i,q} - \kappa_{i,q-1} \tag{12.25}$$

$$\Upsilon_{i,q} = \sum_{h=2}^{q} \frac{\partial \kappa_{i,h-1}}{\partial \hat{\theta}_i} z_{i,h}, \quad \Pi_{i,q} = \sum_{p=1}^{q-1} \frac{\partial \kappa_{i,q-1}}{\partial x_{i,p}} \tag{12.26}$$

where $q = 2, 3, \ldots, n$ and $\kappa_{i,q-1}$ as an intermediate variable is designed in Step $(q-1)$.

Step 1: Taking the time derivative of $z_{i,1}$ and recalling (12.1), one can obtain

$$\dot{z}_{i,1} = \dot{x}_{i,1} - \dot{\xi}_{i,1} = z_{i,2} + \kappa_{i,1} + \theta_i^T f_{i,1}(\overline{x}_{i,1}) - \xi_{i,2}. \tag{12.27}$$

Consider the following Lyapunov function:

$$V_{i,1} = \frac{1}{2} z_{i,1}^2 + \frac{1}{2} \tilde{\theta}_i^T \Gamma_i^{-1} \tilde{\theta}_i \tag{12.28}$$

where $\tilde{\theta}_i = \hat{\theta}_i - \theta_i$ with $\hat{\theta}_i$ being the estimate of θ_i and Γ_i is a positive definite matrix. Then, taking the infinitesimal operator of $V_{i,1}$ and noting the time derivative of $z_{i,1}$ in (12.27), one has

$$\begin{aligned} \mathscr{I} V_{i,1} &= z_{i,1} \dot{z}_{i,1} + \tilde{\theta}_i^T \Gamma_i^{-1} \dot{\hat{\theta}}_i \\ &= z_{i,1} [z_{i,2} + \kappa_{i,1} + \hat{\theta}_i^T f_{i,1} - \xi_{i,2}] + \tilde{\theta}_i^T \Gamma_i^{-1} [\dot{\hat{\theta}}_i - z_{i,1} \Gamma_i f_{i,1}] \end{aligned} \tag{12.29}$$

where θ_i has been replaced with $\hat{\theta}_i - \tilde{\theta}_i$.

Let

$$\kappa_{i,1} = -z_{i,1} - \hat{\theta}_i^T f_{i,1} + \xi_{i,2} \tag{12.30}$$

$$\psi_{i,1} = z_{i,1} \Gamma_i f_{i,1} \tag{12.31}$$

where $\psi_{i,1}$, as a component of the adaptive update law in the final step, is a tuning function for $\hat{\theta}_i$. Then, substituting (12.30) and (12.31) into (12.29) gives

$$\mathscr{I} V_{i,1} = -z_{i,1}^2 + z_{i,1}z_{i,2} + \tilde{\theta}_i^T \Gamma_i^{-1}(\dot{\hat{\theta}}_i - \psi_{i,1}). \tag{12.32}$$

Now, the state $x_{i,1}$ can track the generated local reference $\xi_{i,1}$ with the help of the virtual control law (12.30) and tuning function (12.31). In addition, the item $z_{i,1}z_{i,2}$ is left to the Step 2, which does not exist when the model (12.1) is a first-order system.

Step 2: In view of the fact that $\kappa_{i,1}$ is a function of $\xi_{i,1}$, $\xi_{i,2}$, $x_{i,1}$, and $\hat{\theta}_i$, the time derivative of $z_{i,2}$ is

$$\dot{z}_{i,2} = z_{i,3} + \kappa_{i,2} + \theta_i^T f_{i,2} - \frac{\partial \kappa_{i,1}}{\partial x_{i,1}}\dot{x}_{i,1} - \sum_{h=1}^{2} \frac{\partial \kappa_{i,1}}{\partial \xi_{i,q}}\dot{\xi}_{i,q} - \frac{\partial \kappa_{i,1}}{\partial \hat{\theta}_i}\dot{\hat{\theta}}_i. \tag{12.33}$$

Similar to Step 1, by using the following Lyapunov function candidate

$$V_{i,2} = V_{i,1} + \frac{1}{2}z_{i,2}^2 \tag{12.34}$$

it follows from (12.32) and (12.33) that

$$\begin{aligned} \mathscr{I} V_{i,2} &= \mathscr{I} V_{i,1} + z_{i,2}\dot{z}_{i,2} \\ &= z_{i,2}\Bigg[z_{i,3} + \kappa_{i,2} + \hat{\theta}_i^T f_{i,2} - \tilde{\theta}_i^T f_{i,2} - \frac{\partial \kappa_{i,1}}{\partial \hat{\theta}_i}\dot{\hat{\theta}}_i - \sum_{o=1}^{2} \frac{\partial \kappa_{i,1}}{\partial \xi_{i,o}}\dot{\xi}_{i,o} \\ &\quad - \Pi_{i,2}\left(x_{i,2} + (\hat{\theta}_t^T - \tilde{\theta}_i^T) f_{i,1}\right)\Bigg] - z_{i,1}^2 + z_{i,1}z_{i,2} + \tilde{\theta}_i^T \Gamma_i^{-1}(\dot{\hat{\theta}}_i - \psi_{i,1}). \end{aligned} \tag{12.35}$$

Let

$$\kappa_{i,2} = -z_{i,1} - z_{i,2} - \hat{\theta}_i^T\left(f_{i,2} - \Pi_{i,2}f_{i,1}\right) + \Pi_{i,2}x_{i,2} + \sum_{o=1}^{2} \frac{\partial \kappa_{i,1}}{\partial \xi_{i,o}}\dot{\xi}_{i,o} + \frac{\partial \kappa_{i,1}}{\partial \hat{\theta}_i}\psi_{i,2} \tag{12.36}$$

$$\psi_{i,2} = \psi_{i,1} + z_{i,2}\Gamma_i\left(f_{i,2} - \Pi_{i,2}f_{i,1}\right) \tag{12.37}$$

then it can be arrived that

$$\mathscr{I} V_{i,2} = -\sum_{h=1}^{2} z_{i,q}^2 + z_{i,2}z_{i,3} + \left[\tilde{\theta}_i^T \Gamma_i^{-1} - \Upsilon_{i,2}\right](\dot{\hat{\theta}}_i - \psi_{i,2}). \tag{12.38}$$

In this step, the state $x_{i,2}$ can track the intermediate variable $\kappa_{i,1}$ with the help of the virtual control law (12.36) and tuning function (12.37).

Step q ($3 \leq q \leq n-1$)*:* Similarly, $\kappa_{i,q-1}$ is a smooth function of $\xi_{i,1}$, $\xi_{i,2}$, ..., $\xi_{i,q}$, $x_{i,1}$, $x_{i,2}$,..., $x_{i,q-1}$, and $\hat{\theta}_i$, and one has

$$\dot{z}_{i,q} = z_{i,q+1} + \kappa_{i,q} + \theta_i^T f_{i,q} - \Pi_{i,q}\dot{x}_{i,p} - \sum_{o=1}^{q} \frac{\partial \kappa_{i,q-1}}{\partial \xi_{i,o}}\dot{\xi}_{i,o} - \frac{\partial \kappa_{i,q-1}}{\partial \hat{\theta}_i}\dot{\hat{\theta}}_i. \quad (12.39)$$

Consider the following Lyapunov function:

$$V_{i,q} = V_{i,q-1} + \frac{1}{2}z_{i,q}^2.$$

Let

$$\kappa_{i,q} = \left(\Upsilon_{i,q-1}\Gamma_i - \hat{\theta}_i^T\right)\left(f_{i,q} - \Pi_{i,q}f_{i,p}\right) + \sum_{o=1}^{q} \frac{\partial \kappa_{i,q-1}}{\partial \xi_{i,o}}\dot{\xi}_{i,o}$$
$$- z_{i,q-1} - z_{i,q} + \frac{\partial \kappa_{i,q-1}}{\partial \hat{\theta}_i}\psi_{i,q} + \Pi_{i,q}x_{i,p+1} \quad (12.40)$$
$$\psi_{i,q} = \psi_{i,q-1} + z_{i,q}\Gamma_i\left(f_{i,q} - \Pi_{i,q}f_{i,p}\right). \quad (12.41)$$

Then, taking the infinitesimal operator of $V_{i,q}$ and using the facts (12.39), (12.40), and (12.41) yields

$$\mathscr{I}V_{i,q} = -\sum_{h=1}^{q} z_{i,h}^2 + z_{i,q}z_{i,q+1} + \left[\tilde{\theta}_i^T\Gamma_i^{-1} - \Upsilon_{i,q}\right](\dot{\hat{\theta}}_i - \psi_{i,q}) \quad (12.42)$$

where the fact

$$\mathscr{I}V_{i,q-1} = [\tilde{\theta}_i^T\Gamma_i^{-1} - \Upsilon_{i,q-1}](\dot{\hat{\theta}}_i - \psi_{i,q-1}) - \sum_{h=1}^{q-1} z_{i,h}^2 + z_{i,q-1}z_{i,q}$$

has been used. Based on the recursive idea, $x_{i,q}$ can track the intermediate signal $\xi_{i,q-1}$ generated in the previous step.

Step n: Recalling that $\kappa_{i,n-1}$ is a function of $\xi_{i,1}, \xi_{i,2}, \ldots, \xi_{i,n}, x_{i,1}, x_{i,2}, \ldots, x_{i,n-1}$, and $\hat{\theta}_i$, it can be obtained that

$$\mathscr{I} z_{i,n} = \rho_i(t)\Big(u_i(t) + \sum_{j\in\mathscr{N}*} \rho_{ij}(t)\vartheta_j(t)\Big) + r_i(t) + \theta_i^T f_{i,n} + w_i - \Pi_{i,n}\dot{x}_{i,p}$$
$$- \sum_{o=1}^{n} \frac{\partial \kappa_{i,n-1}}{\partial \xi_{i,o}} \dot{\xi}_{i,o} - \frac{\partial \kappa_{i,n-1}}{\partial \hat{\theta}_i} \dot{\hat{\theta}}_i. \tag{12.43}$$

Recall Assumption 12.3, define

$$\gamma_i = \max\{\overline{q}_i, \sup_{t\geq 0} \big(r_i(t) + \sum_{j\in\mathscr{N}*} \rho_i(t)\rho_{ij}(t)\vartheta_j(t)\big)\}$$

and consider the following Lyapunov function:

$$V_{i,n} = V_{i,n-1} + \frac{1}{2}z_{i,n}^2 + \frac{1}{2}\tilde{\gamma}_i^2 \tag{12.44}$$

where $\tilde{\gamma}_i = \hat{\gamma}_i - \gamma_i$ with $\hat{\gamma}_i$ being the estimate of γ_i. Combining (12.42) with (12.43), the infinitesimal operator of $V_{i,n}$ is

$$\mathscr{I} V_{i,n} \leq \mathscr{I} V_{i,n-1} + z_{i,n}\dot{z}_{i,n} + \tilde{\gamma}_i\dot{\hat{\gamma}}_i$$
$$= -\sum_{h=1}^{n-1} z_{i,h}^2 + z_{i,n-1}z_{i,n} + \tilde{\gamma}_i\dot{\hat{\gamma}}_i + \Big[\tilde{\theta}_i^T \Gamma_i^{-1} - \Upsilon_{i,n-1}\Big](\dot{\hat{\theta}}_i - \psi_{i,n-1})$$
$$+ z_{i,n}\gamma_i(1 + \overline{\psi}_i) + z_{i,n}\Big[\rho_i(t)u_i(t) - \Pi_{i,n-1}\big(x_{i,p+1} + (\hat{\theta}_t^T - \tilde{\theta}_i^T) f_{i,p}\big)$$
$$+ \hat{\theta}_i^T f_{i,n} - \tilde{\theta}_i^T f_{i,n} - \sum_{o=1}^{n} \frac{\partial \kappa_{i,n-1}}{\partial \xi_{i,o}} \dot{\xi}_{i,o} - \frac{\partial \kappa_{i,n-1}}{\partial \hat{\theta}_i} \dot{\hat{\theta}}_i\Big]. \tag{12.45}$$

Let

$$\psi_{i,n} = \psi_{i,n-1} + z_{i,n}\Gamma_i\big(f_{i,n} - \sum_{o=1}^{n-1} \frac{\partial \kappa_{i,n-1}}{\partial x_{i,o}} f_{i,o}\big) \tag{12.46}$$

$$\dot{\hat{\theta}}_i = \psi_{i,n} \tag{12.47}$$

and according to Lemma 2.6, it can be proved that

$$z_{i,n}\gamma_i\big(1 + \overline{\psi}_i(x_i)\big) \leq \gamma_i|z_{i,n}||1 + \overline{\psi}_i(x_i)| \leq \frac{\gamma_i z_{i,n}^2\big(1 + \overline{\psi}_i(x_i)\big)^2}{\sqrt{z_{i,n}^2\big(1 + \overline{\psi}_i(x_i)\big)^2 + \zeta_i^2}} + \gamma_i\zeta_i \tag{12.48}$$

where $\zeta_i = \iota_{i1}e^{-\iota_{i2}t}$, which ensures the non-singularity of the fractional term in (12.48).

Now, the actual input $u_i(t)$ of the agent i is designed as

$$u_i(t) = \mathcal{N}(\kappa_{i,n})\overline{\kappa}_{i,n} \tag{12.49}$$

$$\dot{\kappa}_{i,n} = \phi_i z_{i,n}\overline{\kappa}_{i,n} \tag{12.50}$$

$$\begin{aligned}\overline{\kappa}_{i,n} &= z_{i,n-1} + z_{i,n} + \left(\hat{\theta}_i^T - \Upsilon_{i,n-1}\Gamma_i\right)\left(f_{i,n} - \Pi_{i,n} f_{i,p}\right) - \Pi_{i,n} x_{i,p+1}\\ &\quad - \sum_{o=1}^{n} \frac{\partial \kappa_{i,n-1}}{\partial \xi_{i,o}}\dot{\xi}_{i,o} - \frac{\partial \kappa_{i,n-1}}{\partial \hat{\theta}_i}\psi_{i,n} + \frac{\hat{\gamma}_i z_{i,n}\left(1+\overline{\psi}_i\right)^2}{\sqrt{z_{i,n}^2\left(1+\overline{\psi}_i\right)^2 + \zeta_i^2}}\end{aligned} \tag{12.51}$$

where $\phi_i > 0$ is a parameter to be designed. Substituting (12.46)–(12.51) into (12.45) yields

$$\begin{aligned}\mathscr{I} V_{i,n} \le &- \sum_{h=1}^{n} z_{i,h}^2 + z_{i,n}\rho_i(t)\mathcal{N}(\kappa_{i,n})\overline{\kappa}_{i,n} + \frac{1}{\phi_i}\dot{\kappa}_{i,n} + \gamma_i\zeta_i\\ &+ \tilde{\gamma}_i\left(\dot{\hat{\gamma}}_i - \frac{z_{i,n}^2\left(1+\overline{\psi}_i\right)^2}{\sqrt{z_{i,n}^2\left(1+\overline{\psi}_i\right)^2 + \zeta_i^2}}\right).\end{aligned} \tag{12.52}$$

Let

$$\dot{\hat{\gamma}}_i = \frac{z_{i,n}^2\left(1+\overline{\psi}_i\right)^2}{\sqrt{z_{i,n}^2\left(1+\overline{\psi}_i\right)^2 + \zeta_i^2}}. \tag{12.53}$$

Then, it further yields

$$\mathscr{I} V_{i,n} \le -\sum_{h=1}^{n} z_{i,h}^2 + \frac{\dot{\kappa}_{i,n}}{\phi_i}\left(\rho_i(t)\mathcal{N}(\kappa_{i,n})+1\right) + \gamma_i\zeta_i. \tag{12.54}$$

Hence, the final control law (12.49) and adaptive update law (12.51) and (12.53) are obtained.

Remark 12.7 To regulate the agents in a MAS, certain techniques must be implemented to acquire information from neighboring agents. Specifically, for each cyber node, the virtual controller (12.22) relies on the latest local signals $\Psi_j(t_{j,k})$ from its neighbors $j \in \mathcal{N}_i^*$ and continuous data $b_i\hat{y}_{\sigma_i}^{(h)}(t), h = 0, \cdots, n$ provided by the virtual leader. It also utilizes its own local reference signals $\xi_{i,q}, q = 1, \cdots, n$. Similarly, for each physical agent, the controller (12.49), (12.50), and (12.53) requires the agents state information $x_{i,q}(t), q = 1, \cdots, n$ and reference signals $\xi_{i,q}(t), q = 1, \cdots, n$, which are generated by the cyber layer. Some of these signals, such as the latest local signals $\Psi_j(t_{j,k}), j \in \mathcal{N}_i^*$ from neighboring agents, are more readily accessible through communication network, whereas the agent's state information $x_{i,q}, q = 1, \cdots, n$ must be obtained via sensors. Additionally, interactions among nodes are conducted through broadcast communication.

Remark 12.8 In the physical layer, the challenges posed by TVUND faults, FDIAs, and mismatched nonlinearities are addressed using the *Nussbaum function* technique (12.49), an adaptive compensation method (12.53), and an adaptive estimation mechanism (12.47), respectively. Specifically, the *Nussbaum function* technique adjusts the input direction to align with the unknown fault direction. Meanwhile, the adaptive compensation method mitigates the negative impact of FDIAs through real-time adaptive control. Finally, the adaptive estimation mechanism facilitates the identification of unknown system parameters, a crucial step in achieving the objectives. Compared to the single-layer *intrusion-tolerant* approach via LMI in [5, 6], the fully distributed hierarchical ET *fault-tolerant* controller (12.49) provides greater flexibility and is more applicable to large-scale MASs.

Remark 12.9 *Fault-tolerant* pinning control within a *cluster consensus* framework has been explored in [18]. Unlike the *fault-tolerant* control strategy in [18], which relies on virtual leader information (either directly or indirectly), the controller (12.49) proposed in this chapter replaces the virtual leaders data with local references v_i and $\xi_{i,q}, q = 1, \ldots, n$ generated by the cyber layer. While [18] identifies two types of nodes requiring pinning ($\mathscr{D}_i = 0$ and $i \in \tilde{\mathscr{V}}_i$), the fully distributed hierarchical scheme introduces a third type of node, i.e., nodes responsible for transmitting information to other clusters ($i \in \hat{\mathscr{V}}_i$). Although the number of pinned nodes increases, the overall resource cost is reduced due to the reduced transmission of virtual leader data and lower energy consumption from computation compared to communication.

Remark 12.10 In contrast to the signal type attack consideration in [27, 31], the compounded effects of multiple type attack (FDIAs and DoS attacks) present a significant challenge by obstructing communication and injecting false data at the same time. These factors could rapidly degrade *cluster consensus* performance. To tackle this, the hierarchical control strategy effectively separates the impact of different attack types, maintaining *cluster consensus* despite the combined assault. This method surpasses single-attack handling approaches by mitigating the superposition effects of multiple attacks.

12.3.2 Cluster Consensus Analysis

Now, the main theorem is presented here.

Theorem 12.1 *Consider MAS (12.1) with UNDFs (12.2), FDIAs (12.3) and (12.4) and DoS attacks (12.6). Under Assumptions 12.1–12.3 and for a positive symmetric matrix P in (12.67)–(12.68) satisfying Lemma 2.2, if the conditions (10.15)* *and (10.16) in Lemma 10.1 are satisfied, then the fully distributed hierarchical ET intrusion- and fault-tolerant controller (12.49) with cyber layer dynamic (12.15), ETM (12.20), and adaptive update laws (12.50) and (12.53) guarantees that*

(a) all signals of the MAS are globally bounded and the control objective is achieved;
(b) the Zeno behavior is excluded and there exists a lower bound of inter-event time.

Proof First, define the following *cluster consensus* error:

$$\begin{aligned}\varpi_i &= \sum_{j=1}^{N} a_{ij}(\xi_{i,1}-\xi_{j,1})+b_i(\xi_{i,1}-\hat{y}_{\sigma_i})-\sum_{j=1}^{N} a_{ij}(\hat{y}_{\sigma_i}-y_{\sigma_j}) \\ &= \sum_{j=1}^{N} a_{ij}\big[(\xi_{i,1}-\hat{y}_{\sigma_i})-(\xi_{j,1}-y_{\sigma_j})\big]+b_i(\xi_{i,1}-\hat{y}_{\sigma_i})\end{aligned} \tag{12.55}$$

$$\overline{\varpi}_i = \left(1+\frac{d}{dt}\right)^{n-1}\varpi_i = \sum_{q=0}^{n-1}\varepsilon_q \varpi_i^{(q)} \tag{12.56}$$

$$\overline{\varepsilon} = [\overline{\varepsilon}_1, \overline{\varepsilon}_2, \ldots, \overline{\varepsilon}_N]^T, \quad \overline{\varpi} = [\overline{\varpi}_1, \overline{\varpi}_2, \ldots, \overline{\varpi}_N]^T \tag{12.57}$$

where $\varepsilon_i, i = 1, \ldots, N$ is defined in (12.17). Then, it can be checked that

$$\overline{\varpi} = (\mathscr{L} + \mathscr{B})\overline{\varepsilon}. \tag{12.58}$$

Substituting (12.55) into (12.56) gives

$$\overline{\varpi}_i = \sum_{j=1}^{N} a_{ij}(\Psi_i - \Psi_j) + b_i\left(\Psi_i - \sum_{q=0}^{n-1}\varepsilon_q \hat{y}_{\sigma_i}^{(q)}\right) - \sum_{j=1}^{N} a_{ij}\left[\sum_{q=0}^{n-1}\varepsilon_q\left(\hat{y}_{\sigma_i}^{(q)} - y_{\sigma_j}^{(q)}\right)\right]. \tag{12.59}$$

Noting (12.19) and (12.22), it can be checked that

$$\begin{aligned}\dot{\overline{\varepsilon}}_i &= -c\sum_{j=1}^{N} a_{ij}F_{ij} - cb_i\left(\Psi_i - \sum_{q=0}^{n-1}\varepsilon_q \hat{y}_{\sigma_i}^{(q)}\right) + c\sum_{j=1}^{N} l_{ij}\left(\sum_{q=0}^{n-1}\varepsilon_q y_{\sigma_j}^{(q)}\right) + (b_i - 1)\sum_{q=0}^{n-1}\varepsilon_q \hat{y}_{\sigma_i}^{(q+1)} \\ &= -c\sum_{j=1}^{N} a_{ij}F_{ij} - cb_i\left(\Psi_i - \sum_{q=0}^{n-1}\varepsilon_q \hat{y}_{\sigma_i}^{(q)}\right) + c\sum_{j=1}^{N} l_{ij}\left(\sum_{q=0}^{n-1}\varepsilon_q y_{\sigma_j}^{(q)}\right) + \Xi_i\end{aligned} \tag{12.60}$$

where

$$\begin{aligned}
F_{ij} &= \Psi_i(t) - \Psi_j(t_{j,k}) \\
&= [\Psi_i(t) - \Psi_j(t)] + [\Psi_j(t) - \Psi_j(t_{j,k})] \\
&= [\Psi_i(t) - \Psi_j(t)] + \Theta_j(t), \\
\Theta_j(t) &= \Psi_j(t) - \Psi_j(t_{j,k}) \\
\Xi_i &= (b_i - 1)\sum_{q=0}^{n-1} \varepsilon_q \hat{y}_{\sigma_i}^{(q+1)}.
\end{aligned}$$

Combining (12.59) with (12.60), one can get

$$\dot{\bar{\varepsilon}}_i = -c\overline{\varpi}_i - c\sum_{j=1}^{N} a_{ij}\Theta_j + \Xi_i = -c\overline{\varpi}_i + \Delta_i + \Xi_i \tag{12.61}$$

where $\Delta_i = -c\sum_{j=1}^{N} a_{ij}\Theta_j$.

Consider the following Lyapunov function:

$$V = \frac{1}{2}\overline{\varpi}^T P \overline{\varpi} \tag{12.62}$$

where P is a positive definite matrix satisfying Lemma 2.2.

When $t \in [t_m, t_m + \nu_m)$, taking the derivative of V yields

$$\begin{aligned}
\dot{V} &= \overline{\varpi}^T P(\mathscr{L} + \mathscr{B})\dot{\bar{\varepsilon}} \\
&= -\frac{c}{2}\overline{\varpi}^T Q\overline{\varpi} + \overline{\varpi}^T P(\mathscr{L} + \mathscr{B})(\Delta + \Xi)
\end{aligned} \tag{12.63}$$

where

$$\begin{aligned}
&\Delta = [\Delta_1, \Delta_2, \ldots, \Delta_N]^T, \quad \Xi = [\Xi_1, \Xi_2, \ldots, \Xi_N]^T \\
&Q = P(\mathscr{L} + \mathscr{B}) + (\mathscr{L} + \mathscr{B})^T P.
\end{aligned}$$

Recalling the ETM (12.20), one can obtain

$$\Delta_i^2 \le N\sum_{j=1}^{N} a_{ij}\overline{\eta}_j^2, \quad \|\Delta\|^2 \le \sum_{i=1}^{N}\sum_{j=1}^{N} a_{ij}N\overline{\eta}_j^2. \tag{12.64}$$

By using Lemma 2.7, it can be proved that

$$\begin{aligned}\varpi^T P(\mathscr{L}+\mathscr{B})\Delta &\leq \varpi^T c\lambda_{\min}(Q)\frac{1}{c\lambda_{\min}(Q)}P(\mathscr{L}+\mathscr{B})\Delta \\ &\leq \frac{c\lambda_{\min}(Q)}{4}\varpi^T\varpi + \frac{\|P(\mathscr{L}+\mathscr{B})\Delta\|^2}{c\lambda_{\min}(Q)} \\ &\leq \frac{c}{8}\varpi^T Q\varpi + \frac{\mu_1}{c} \end{aligned} \tag{12.65}$$

where

$$\mu_1 = \frac{2\|P(\mathscr{L}+\mathscr{B})\|^2 \sum_{i=1}^N \sum_{j=1}^N a_{ij} N\overline{\eta}_j^2}{\lambda_{\min}(Q)}.$$

Similarly, one has

$$\begin{aligned}\varpi^T P(\mathscr{L}+\mathscr{B})\Xi &\leq \frac{c}{8}\varpi^T Q\varpi + \frac{2\|P(\mathscr{L}+\mathscr{B})\Xi\|^2}{c\lambda_{\min}(Q)} \\ &\leq \frac{c}{8}\varpi^T Q\varpi + \frac{\mu_2}{c} \end{aligned} \tag{12.66}$$

where

$$\mu_2 = \frac{2\|P(\mathscr{L}+\mathscr{B})\|^2\|\Xi\|^2}{\lambda_{\min}(Q)}$$

is bounded due to the fact that $\hat{y}_{\sigma_i}$ and its finite nth derivatives are bounded.

Then, by substituting the above two inequalities into (12.63), $\dot{V}$ is bounded by

$$\dot{V} \leq -\frac{c}{4}\varpi^T Q\varpi + \frac{\mu_1+\mu_2}{c} \leq -\beta_1 V + \beta_2 \tag{12.67}$$

where

$$\beta_1 = \frac{c\lambda_{\min}(Q)}{2\lambda_{\max}(P)}, \quad \beta_2 = \frac{\mu_1+\mu_2}{c}.$$

When $t \in [t_m + \nu_m, t_{m+1})$, taking the derivative of V yields

$$\begin{aligned}\dot{V} &= \varpi^T P(\mathscr{L}+\mathscr{B})\dot{\bar{\varepsilon}} \\ &= \varpi^T P(\mathscr{L}+\mathscr{B})\bar{\Xi} \\ &\leq \frac{c}{64}\varpi^T Q\varpi + \frac{8\mu_3}{c} \\ &\leq \beta_3 V + \beta_4 \end{aligned} \tag{12.68}$$

where

$$\beta_3 = \frac{c\lambda_{\max}(Q)}{32\lambda_{\min}(P)}, \quad \beta_4 = \frac{\mu_3}{c}, \quad \bar{\Xi} = [\bar{\Xi}_1, \bar{\Xi}_2, \ldots, \bar{\Xi}_N]^T$$
$$\bar{\Xi}_i = -\sum_{q=0}^{n-1} \varepsilon_q \hat{y}_{\sigma_i}^{(q+1)}, \quad \mu_3 = \frac{2\|P(\mathscr{L}+\mathscr{B})\|^2\|\bar{\Xi}\|^2}{\lambda_{\min}(Q)}.$$

Then, from Lemma 10.1, $V(t)$ and $\overline{\varpi}$ are bounded if the conditions (10.15) and (10.16) in Lemma 10.1 are satisfied, and this together with (12.58) gives the boundedness of $\overline{\varpi}_i$ and $\overline{\varepsilon}_i$. In addition, it can be obtained from (12.17) and (12.55) that

$$\varpi_i^{(q)} = \frac{s^q}{(s+1)^{n-1}}\overline{\varpi}_i, \quad \varepsilon_i^{(q)} = \frac{s^q}{(s+1)^{n-1}}\overline{\varepsilon}_i$$

with s being the Laplace operator. By combining the boundedness of $\overline{\varpi}_i$ and $\overline{\varepsilon}_i$ with the stability of $\frac{s^p}{(s+1)^{n-1}}$, it can be obtained that $\varpi_i^{(q)}$ and $\varepsilon_i^{(q)}$, $(q = 0, 1, \ldots, n-1)$ are bounded. Then, one can obtain the boundedness of $\xi_{i,1}, \xi_{i,2}, \ldots, \xi_{i,n}$ by (12.16). As a result, Ψ_i, $F_{i,j}$, and v_i are bounded.

In what follows, the convergence of agents in the physical layer will be proved. First, integrating both sides of (12.54) yields

$$\begin{aligned} V_{i,n}(t) \le& -\sum_{q=1}^{n}\int_0^t z_{i,q}^2(\tau)\,d\tau + \int_0^t \frac{\dot{\kappa}_{i,n}}{\phi_i}\left[\rho_i(\tau)\mathscr{N}\big(\kappa_{i,n}(\tau)\big)+1\right]d\tau \\ &+ V_{i,n}(0) + \gamma_i \int_0^t \zeta_i(\tau)\,d\tau \\ \le& \int_0^t \frac{\dot{\kappa}_{i,n}}{\phi_i}\left[\rho_i(\tau)\mathscr{N}\big(\kappa_{i,n}(\tau)\big)+1\right]d\tau + \varphi \end{aligned} \tag{12.69}$$

where

$$\varphi = V_{i,n}(0) + \gamma_i \int_0^t \zeta_i(\tau)\,d\tau \le V_{i,n}(0) + \gamma_i \int_0^\infty \zeta_i(\tau)\,d\tau = V_{i,n}(0) + \gamma_i \frac{\iota_{i1}}{\iota_{i2}}$$

is a finite constant. In addition, taking expectation of (12.69) yields

$$\mathbb{E}\Big\{V_{i,n}(t)\Big\} \le \mathbb{E}\Big\{\int_0^t \frac{\dot{\kappa}_{i,n}}{\phi_i}\left[\rho_i(\tau)\mathscr{N}\big(\kappa_{i,n}(\tau)\big)+1\right]d\tau\Big\} + \mathbb{E}\{\varphi\}. \tag{12.70}$$

From Lemma 12.1, the boundedness of $\mathbb{E}\{\int_0^t \frac{\dot{\kappa}_{i,n}}{\phi_i}\left[\rho_i(\tau)\mathscr{N}\left(\kappa_{i,n}(\tau)\right)+1\right]d\tau\}$, $\mathbb{E}\{\kappa_{i,n}\}$, and $\mathbb{E}\{V_{i,n}\}$ is obtained. As a result, $z_{i,q}$, $\tilde{\theta}_i$, $\hat{\theta}_i$, $\tilde{\gamma}_i$, and $\hat{\gamma}_i$ are bounded. Since $z_{i,1}$ and $\xi_{i,1}$ are bounded, one can obtain the boundedness of $x_{i,1}$ by (12.24). Noting that $\xi_{i,1}$, $\xi_{i,2}$, ..., $\xi_{i,n}$, and v_i are bounded and using the fact that $\kappa_{i,1}$ is a function of $x_{i,1}$, $\xi_{i,1}$, $\xi_{i,2}$, and $\hat{\theta}_i$, one can conclude that $\kappa_{i,1}$ is bounded. It is further obtained that $x_{i,2}$ in (12.25) is bounded. By using similar methods, the boundedness of $x_{i,q}$, $\kappa_{i,q-1}$ $(q=2,\ldots,n)$, and $u_i(t)$ can be obtained. In addition, since the value of γ_i is positively related to $\sum_{j\in N^*}\rho_{ij}(t)\vartheta_j(t)$, the value of $\mathbb{E}\{\varphi\}$ is positively related to the probability p_{ij} and energy upper bound $\overline{\vartheta}$ of the FDIAs. Up to now, the result (a) in Theorem 12.1 has been proved.

Next, the inter-event time greater than 0 will be proved. Recalling the definition of $\Theta_j(t)$ and taking the time derivative of $\Psi_j(t)$ yield

$$\Theta_j(t)=\Psi_j(t)-\Psi_j(t_{j,k}) \tag{12.71}$$

$$\dot{\Psi}_j(t)=\sum_{q=0}^{n-2}\varepsilon_q\xi_{j,q+2}+\varepsilon_{n-1}v_j \tag{12.72}$$

where $\dot{\Psi}_j(t)$ is bounded. Then, using the boundedness of $v_j, \xi_{j,2},\ldots,\xi_{j,n}$, $j=1,\ldots,N$, one can get

$$|\dot{\Theta}_j(t)|=|\dot{\Psi}_j(t)|\le\varsigma_j \tag{12.73}$$

where ς_j is the upper bound of $|\dot{\Psi}_j(t)|$ obtained from (12.72). In addition, base on the ETM (12.20), $\Theta_j(t_{j,k})=0$ and

$$\lim_{t\to t_{j,k+1}^-}|\Theta_j(t)|=\frac{\underline{\eta}_j+(\overline{\eta}_j-\underline{\eta}_j)\hbar}{c}$$

with $0<\hbar=e^{-|\Theta_j(t)|}<1$. Therefore, one has

$$t_{j,k+1}-t_{j,k}\ge\frac{\underline{\eta}_j}{c\varsigma_j}. \tag{12.74}$$

Then, *Zeno behavior* is ruled out. The proof is completed. □

For DoS attacks, Lemma 10.1 and Theorem 12.1 provide sufficient conditions for the upper bounds on the attack frequency and duration under which the control strategy (12.49) ensures that all system signals remain globally bounded in an expected sense. Specifically, Theorem 12.1 determines the maximum permissible duration of DoS attacks, while Lemma 10.1 establishes conditions (10.15)-(10.15) for the allowable attack frequency. Under these constraints and the attack model (12.6), the system benefits from periods without attacks, which help mitigate the impact of DoS attacks and maintain bounded system stability. Additionally, for FDIAs, the adaptive compensation method (12.53) effectively addresses the issue as long as the injected false data $\vartheta_j(t)$ in (12.3) remains bounded. It is worth noting that as highlighted in Remark 12.1, assuming an upper bound on $\vartheta_j(t)$ is both reasonable and common, as demonstrated in [27, 29]. Furthermore, the random FDIA model (12.3) adopted in this chapter is more general, accommodating a wider range of attack scenarios. Notably, the proposed control strategy does not impose any restrictions on the duration of FDIAs or TVUND faults, allowing these faults and attacks to persist throughout the MAS's operation.

Remark 12.11 Compared with existing researchs, the proposed fully distributed hierarchical ET *intrusion-tolerant* and *fault-tolerant* control strategy offers three key advantages. First, unlike [17, 27, 32], the TVUND faults, FDIAs, and mismatched nonlinearities of an individual agent are confined to that agent, as communication is restricted to the cyber layer. This containment prevents fault propagation across agents. Second, the strategy enables heterogeneous agents to employ different types of Nussbaum functions for addressing NUDF, enhancing flexibility in controller design. This stands in contrast to the homogeneous Nussbaum function approaches in [16, 17]. Third, the control strategy is fully distributed. Each agent relies solely on its own information and that of its immediate neighbors, without requiring global network data, distinguishing it from approaches in [17, 28]. These features significantly enhance the practicality and scalability of the proposed control method.

Remark 12.12 The *complete consensus* problem for MASs with nonidentical unknown control directions in a secure cyber environment, as addressed in [25], uses a hierarchical control scheme. However, *cluster consensus* introduces additional complexity due to the need to manage both intra-cluster and inter-cluster communications. This makes the design of *cluster consensus* protocols inherently more challenging. The *intrusion-tolerant* and *fault-tolerant cluster consensus* analysis presented in this chapter is more complex than the scenario in [25], which assumes a secure cyber environment without actuator faults or mismatched nonlinearities. Additionally, the higher-order MASs considered in this chapter have greater practical relevance than the nonlinear MASs in [25]. It is important to emphasize that the contributions of this chapter are not straightforward extensions of [25] but involve novel developments tailored to more challenging scenarios.

12.4 Simulation Examples

In this section, a practical example is presented to demonstrate the practical benefits and advantages of the hierarchical ET *intrusion-* and *fault-tolerant* control strategy designed in Theorem 12.1.

In this simulation, the considered MAS is composed of five individual single-link manipulators, each driven by a brush DC motor. These manipulators are organized into two distinct clusters, each tasked with achieving separate goals. The connectivity structure among these clusters is depicted in Fig. 8.2. The dynamic of the single-link manipulator borrowed from [23] is described by

$$\begin{cases} H\ddot{q} + W\dot{q} + B\sin(q) = I \\ L\dot{I} = -R\dot{q} - K_B I + U \end{cases} \tag{12.75}$$

where U and I denote the input voltage and current of motor, respectively. $\ddot{q}$, $\dot{q}$, and q represent the link acceleration, velocity, and angular position, respectively. In addition, H, W, B, L, R, and K_B are system parameters, and their physical meanings can be found in [23], hence, omitted here.

Setting $x_1 = q, x_2 = \dot{q}$, and $x_3 = I/H$ and considering the unknown disturbances $w_i(t, x_i)$, then dynamic (12.75) is rewritten in the form of (12.1) as

$$\begin{aligned} \dot{x}_{i,1} &= x_{i,2} + \theta_i^T f_{i,1}(\overline{x}_{i,1}) \\ \dot{x}_{i,2} &= x_{i,3} + \theta_i^T f_{i,2}(\overline{x}_{i,2}) \\ \dot{x}_{i,3} &= u_i + \theta_i^T f_{i,3}(x_i) + w_i(t, x_i), \quad i = 1, 2, \ldots, 5 \\ y_i &= x_{i,1} \end{aligned}$$

where

$$\theta_i = -\frac{1}{H}[W, B, \frac{R}{L}, \frac{K_B H}{L}]^T, \quad u_i = \frac{U}{LH}, \quad f_{i,1}(\overline{x}_{i,1}) = \mathbf{0}$$
$$f_{i,2}(\overline{x}_{i,2}) = [x_{i,2}, \sin(x_{i,1}), 0, 0]^T, \quad f_{i,3}(x_i) = [0, 0, x_{i,2}, x_{i,3}]^T.$$

Furthermore, in this example, the unknown disturbances are defined as $w_i(t, x_i) = \sin(0.1i \times t)$, which leads to $\overline{w}_i(x_i) = 1$ in accordance with Assumption 12.3. Additionally, the parameters of the systems are detailed in Table 12.1.

The FDIA generator in (12.3) is given as $\vartheta_i(t) = 0.5\sin(t)$, and the probabilities of injecting false data into different channels are set to $\text{prob}\{\gamma_{21} = 1\} = 0.03$, $\text{prob}\{\gamma_{31} = 1\} = 0.04$, $\text{prob}\{\gamma_{41} = 1\} = 0.1$, $\text{prob}\{\gamma_{43} = 1\} = 0.15$, and $\text{prob}\{\gamma_{52} = 1\} = 0.02$. Meanwhile, in this example, for given parameters $F_0 = 1$, $T_0 = 2$ and $\beta^* = 0.05$, the number and duration of DoS attacks are set as $N_a(0, 30) = 11$ and $\Im_a(0, 30) = 8.5$, respectively, which satisfy the upper bound of the number $F_0 + 30/F_1 = 17.1180$ and duration $T_0 + 30/T_1 = 8.8739$ obtained in Theorem 12.1, respectively.

Table 12.1 Parameters in agent i of single-link manipulators

Symbol	Agent 1	Agent 2	Agent 3	Agent 4	Agent 5
W/H	1	1.1	1	1	1
B/H	0.4	0.1	0.3	0.1	0.3
$R/(LH)$	0.2	0.3	0.1	0.3	0.2
K_B/L	2	2	2	3	2

Table 12.2 Initial states of virtual nodes and physical agent

Symbol	$i = 1$	$i = 2$	$i = 3$	$i = 4$	$i = 5$
$\xi_{i,1}(0)$	3	2	–3	2	–1
$\xi_{i,2}(0)$	1	–1	2	3	2
$\xi_{i,3}(0)$	2	2	5	–3	3
$x_{i,1}(0)$	3	5	1	4	2
$x_{i,2}(0)$	–2	4	3	–1	6
$x_{i,3}(0)$	1	5	4	1	3

Based on the dynamics of the single-link manipulator, a cyber layer with third-order dynamics has been proposed. The initial states of the virtual nodes and physical agents are provided in Table 12.2. Moreover, the trajectories for the virtual leaders of the two clusters are specified as $\hat{y}_1 = 2\cos(t)$ and $\hat{y}_2 = 5 + 2\sin(t)$.

It should be noted that the directions of actuator faults for all single-link manipulators are different and unknown. In order to better demonstrate the controller effect, in this example, the time-varying parameters within the fault model (12.2) are set as follows:

$$\begin{cases} r_i(t) = 2\sin(0.1i \times t), \quad i \in \{1, 2, \cdots, 5\} \\ \rho_1(t) = 1.1 + 0.1\cos(0.2t) \\ \rho_2(t) = -1.5 - 0.2\sin(t) \\ \rho_3(t) = -0.7 + 0.3\sin(1.1t) \\ \rho_4(t) = 0.8 + 0.1\sin(1.3t) \\ \rho_5(t) = 1.2 - 0.2\cos(0.5t). \end{cases}$$

In addition, according to Theorem 12.1, nodes 1 and 2 are pinning nodes. Moreover, the parameters of the controller in Theorem 12.1 are given in Table 12.3.

The simulation results are illustrated in Figs. 12.2, 12.3, 12.4, 12.5, 12.6 and 12.7, confirming that the hierarchical ET *intrusion-* and *fault-tolerant* control scheme (12.49) provided in Theorem 12.1 successfully achieves the control objective. Figure 12.2 presents the trajectories of two states ($x_{i,1}$ and $x_{i,2}$, $i = 1, \ldots, 5$) for the five single-link manipulators. It is evident that the angular positions and velocities of all manipulators achieve *cluster consensus* under multiple *cyber attacks*. Figure 12.3 displays the displacements of leaders $\hat{y}\sigma_i$ and virtual node states ξi, 1 for $i = 1, ..., 5$ in the cyber layer, which also achieve *cluster consensus*. Notably, due to the adoption of the switching input mechanism in (12.22), the cyber layer agents exhibit smooth transitions even during DoS attack periods. Figure 12.4 highlights the trigger instants for each agent, illustrating the discontinuous communication between the single-link manipulators. This demonstrates the efficacy of the event-triggered mechanism in reducing communication frequency while maintaining system stability.

Table 12.3 The values of the designed parameters for $i \in \mathscr{V}$

Symbol	Values	Symbol	Values	Symbol	Values
ε_{11}	0.05	ε_{21}	1.4	$\kappa_i(0)$	0.5
ε_{12}	0.00001	ε_{22}	0.4	$\theta_i(0)$	$[0.1, 0.1, 0.1, 0.1]^T$
ε_{13}	0.1	ε_{23}	1.4	Γ_1	diag$\{9, 9, 9, 9\}$
ε_{14}	0.1	ε_{24}	1.4	Γ_2	diag$\{3, 17, 15, 9\}$
ε_{15}	0.1	ε_{25}	1.4	Γ_3	diag$\{3, 9, 9, 9\}$
ι_{i1}	3	$\underline{\eta}_i$	10	Γ_4	diag$\{3, 12, 9, 9\}$
ι_{i2}	0.001	$\overline{\eta}_i$	14	Γ_5	diag$\{3, 13, 11, 11\}$
d_1	30	ϕ_i	0.00001	c	55
d_2	30	$\eta_i(0)$	1		

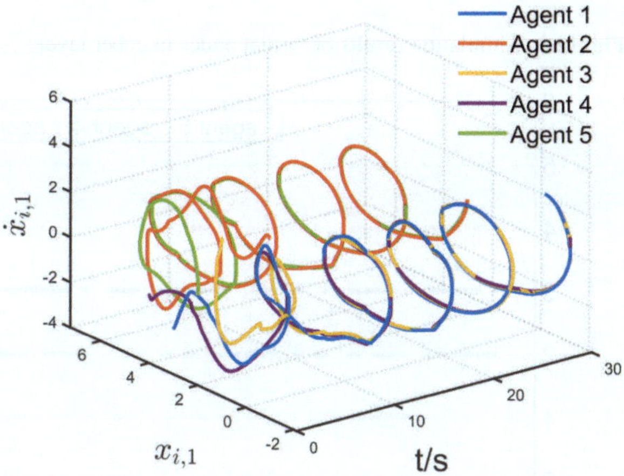

Fig. 12.2 Evolution of $[x_{i,1}, \dot{x}_{i,1}]$

At the physical layer, the outputs y_i of the agents ($i = 1, \ldots, 5$) successfully track the local references ξ_i generated by the cyber layer, as illustrated in Fig. 12.3. The tracking performance of the angular positions y_i for each single-link manipulator is shown in Fig. 12.5, where it is evident that all manipulators effectively counteract the adverse effects of NUDF and accurately follow the local references ξ_i provided by the cyber layer. Figure 12.6 depicts the evolution of the controller gains $\kappa_i(t)$ and the outputs of the *Nussbaum functions* $\mathscr{N}_i(\kappa_i(t))$, $i = 1, \ldots, 5$. These gains converge to constant values, allowing the hierarchical controller to correctly determine the motor voltage directions. In this simulation, each single-link manipulator requires the estimation of four parameters, represented as $\hat{\theta}_i^{4\times 1}$ for $i = 1, \ldots, 5$, resulting in a total of 20 estimators across the system. For simplicity, Fig. 12.7 only displays the gain estimates $\hat{\theta}_1$ of the first agent.

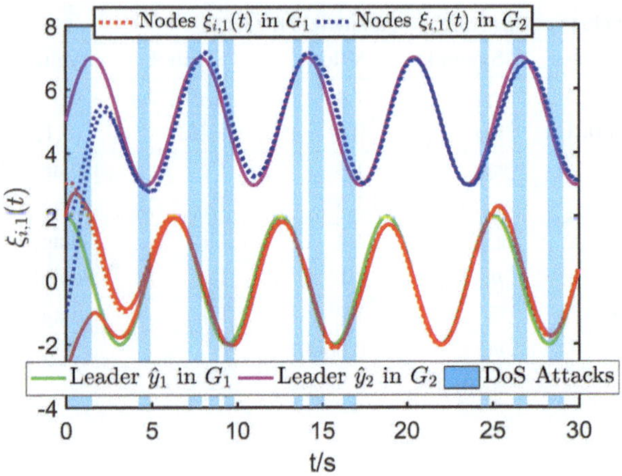

Fig. 12.3 Simulation results of virtual nodes in cyber layer

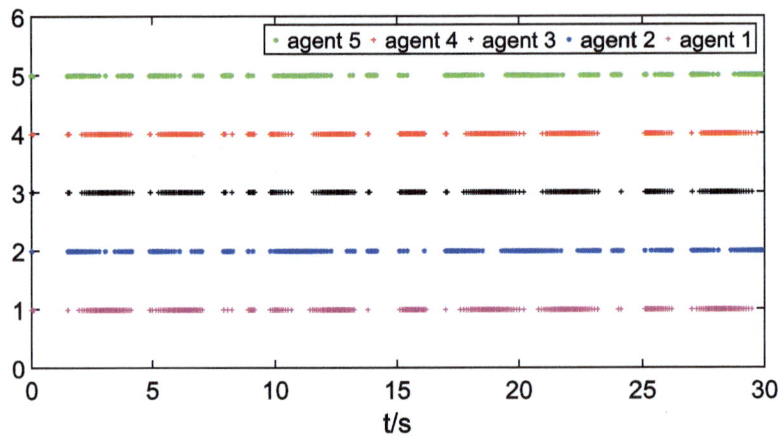

Fig. 12.4 Trigger instant for each agents

To evaluate the *cluster consensus* performance of the hierarchical control strategy under multiple *cyber attacks*, define output error χ and its Euclidean norm as

$$\chi \triangleq [\chi_1, \chi_2, \chi_3, \chi_4, \chi_5]^T, \quad \chi_i = y_i - \hat{y}_{\sigma_i}, \quad \|\chi\| = \sqrt{\sum_{i=1}^{N} \chi_i^2}.$$

Figure 12.8 presents the Euclidean norm of the output error χ for the proposed hierarchical *intrusion-* and *fault-tolerant* control scheme in Theorem 12.1, compared with several alternative strategies: the hierarchical control strategy without switching input in (12.22) for the cyber layer, the state-independent ETM from [20], and the

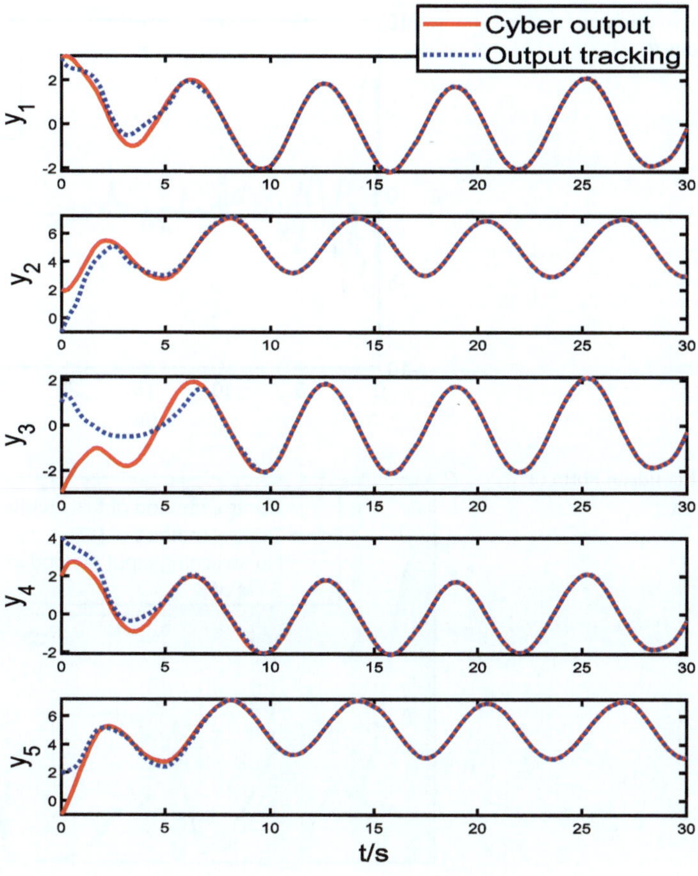

Fig. 12.5 Evolution of output y_i

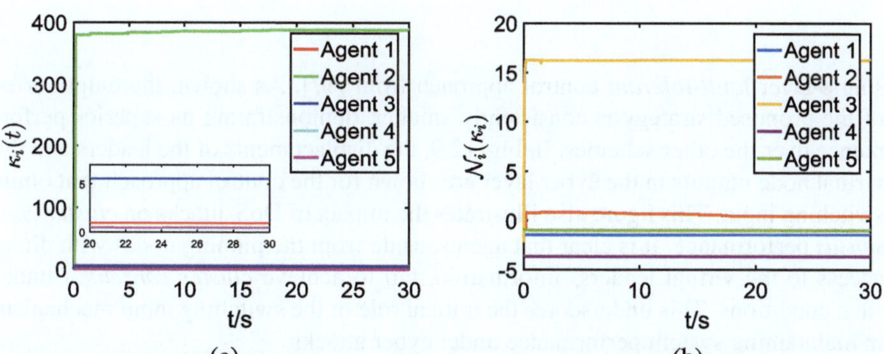

Fig. 12.6 Evolution of gains κ_i and Nussbaum function $\mathcal{N}_i(\kappa_i)$

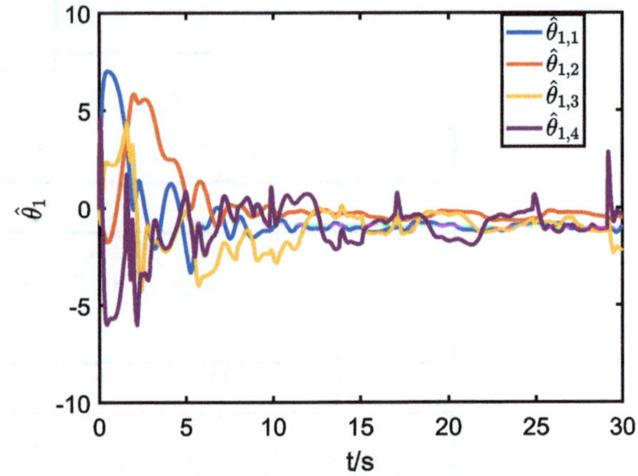

Fig. 12.7 Evolution of adaptive gains θ_1

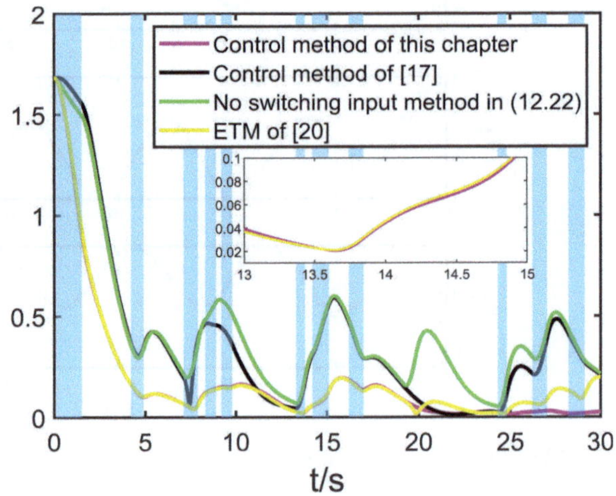

Fig. 12.8 Euclidean norm of output error χ

single-layer *fault-tolerant* control approach from [17]. As shown, the output error of the proposed strategy is consistently smaller, demonstrating its superior performance over the other schemes. In Fig. 12.9, the displacements of the leaders and the virtual node outputs in the cyber layer are shown for the control approach that omits switching input. This figure also illustrates the impact of DoS attacks on *cluster consensus* performance. It is clear that agents, aside from the pinning nodes with direct access to the virtual leaders' information, fail to achieve *cluster consensus* under such conditions. This underscores the critical role of the switching input mechanism in maintaining system performance under cyber attacks.

Figure 12.10 illustrates the trajectories of two states ($x_{i,1}$ and $x_{i,2}$, $i = 1, \ldots, 5$) under the control strategy from [17], where the *Nussbaum function* design relies on global information. By comparing Fig. 12.10 with Fig. 12.2, it becomes evident that the hierarchical ET *fault-tolerant* control strategy proposed in this chapter delivers

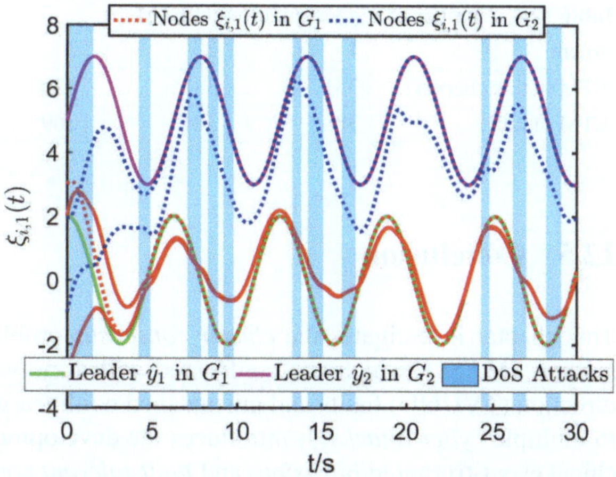

Fig. 12.9 Simulation results of no switching input method in (12.22)

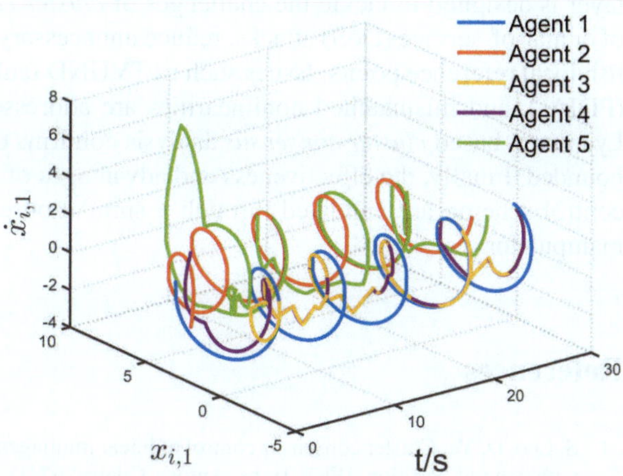

Fig. 12.10 Evolution of $[x_{i,1}, \dot{x}_{i,1}]$ in [17]

superior *cluster consensus* performance than [17]. This observation aligns with the conclusions drawn from Fig. 12.8, further validating the effectiveness of the proposed approach in practical scenarios.

To highlight the benefits of the proposed state-dependent ETM with a time-varying threshold, Table 12.4 presents the number of triggers for both the state-independent ETM with a fixed threshold from [20] and the mechanism proposed in this chapter. As shown in Fig. 12.8, the output error achieved by the proposed ETM is comparable to that of [20]. However, the data in Table 12.4 indicate that the proposed ETM results in fewer trigger events. This demonstrates that the state-dependent ETM in this chapter can conserve communication resources while maintaining similar *cluster consensus* performance.

Table 12.4 Number of triggers in different ETM

Agent	1	2	3	4	5
ETM in this chapter	331	331	320	339	294
ETM in [20]	388	384	369	398	326

12.5 Conclusion

This chapter investigates the *cluster consensus* problem of high-order *multi-agent systems* (MASs), which are confronted with time-varying unknown nonidentical direction (TVUND) faults and mismatched nonlinearities, all while being subjected to multiple *cyber attacks*. It introduces the development of fully distributed hierarchical event-triggered *intrusion-* and *fault-tolerant* control schemes. A virtual cyber layer is designed to tackle the challenges of *cluster consensus*, mitigate the effects of denial-of-service (DoS) attacks, reduce unnecessary data transmission, and establish local reference points. Issues such as TVUND faults, false data injection attacks (FDIAs), and mismatched nonlinearities are addressed in the physical layer. The Lyapunov-based *cluster consensus* analysis confirms that all signals of the MAS are bounded. Finally, the effectiveness and advantages of the hierarchical *fault-tolerant* control scheme are validated through a simulation example involving single-link manipulators.

References

1. S. Luo, D. Ye, Cluster consensus control of linear multiagents systems under directed topology with general partition. IEEE Trans. Autom. Control **67**(4), 1929–1936 (2022)
2. D.G. Pasquale, M.E. Valcher, Consensus for clusters of agents with cooperative and antagonistic relationships. Automatica **135**, Art. No. 110002 (2022)
3. D. Zhang, G. Feng, Y. Shi, D. Srinivasan, Physical safety and cyber security analysis of multi-agent systems: a survey of recent advances. IEEE-CAA J. Automatica Sin. **8**(2), 319–333 (2021)
4. X.G. Guo, B.Q. Wang, J.L. Wang, C.K. Ahn, Z.G. Wu, Edge-event-triggered encryption-decryption observer-based control of multiagent systems for privacy protection under multiple cyber attacks. Inf. Sci. **642**, Art. No.119128 (2023)
5. J. Liu, T. Yin, D. Yue, H.R. Karimi, J. Cao, Event-based secure leader-following consensus control for multiagent systems with multiple cyber attacks. IEEE Trans. Cybern. **51**(1), 162–173 (2021)
6. L. Zha, R. Liao, J. Liu, X. Xie, E. Tian, J. Cao, Dynamic event-triggered output feedback control for networked systems subject to multiple cyber attacks. IEEE Trans. Cybern. **52**(12), 13800–13808 (2022)
7. X.G. Guo, P.M. Liu, Z.G. Wu, D. Zhang, C.K. Ahn, Hybrid event-triggered group consensus control for heterogeneous multi-agent systems with TVNUD faults and stochastic FDI attacks. IEEE Trans. Autom. Control **68**(12), 8013–8020 (2023)

8. P.M. Liu, X.G. Guo, J.L. Wang, X.P. Xie, F.W. Yang, Fully distributed hierarchical ET intrusion- and fault-tolerant group control for MASs with application to robotic manipulators. IEEE Trans. Autom. Sci. Eng. **21**(3), 2868–2881 (2024)
9. S.C. Tong, Y.M. Li, Adaptive fuzzy output feedback tracking backstepping control of strict-feedback nonlinear systems with unknown dead zones. IEEE Trans. Fuzzy Syst. **20**(1), 168–180 (2012)
10. S.Y. Xiao, J.X. Dong, Distributed fault-tolerant tracking control for uncertain homogeneous and heterogeneous MASs under asynchronous sampled communications. IEEE Trans. Netw. Sci. Eng. **9**(3), 1866–1879 (2022)
11. C.H. Xie, G.H. Yang, Decentralized adaptive fault-tolerant control for large-scale systems with external disturbances and actuator faults. Automatica **85**, 83–90 (2017)
12. Q.P. He, S.J. Qin, J. Wang, A new fault diagnosis method using fault directions in fisher discriminant analysis. AlChE J. **51**(2), 555–571 (2005)
13. X.G. Guo, W.D. Xu, J.L. Wang, J.H. Park, Distributed neuroadaptive fault-tolerant sliding-mode control for 2-D plane vehicular platoon systems with spacing constraints and unknown direction faults. Automatica **129**, Art. No. 109675 (2021)
14. E. Ribeiro, A.J.M. Cardoso, C. Boccaletti, Fault-tolerant strategy for a photovoltaic DC-DC converter. IEEE Trans. Power Electron. **28**(6), 3008–3018 (2013)
15. N. Ji, J. Liu, Vibration control for a three-dimensional variable length flexible string with time-varying actuator faults and unknown control directions. IEEE Trans. Autom. Sci. Eng. **20**(4), 2761–2771 (2023)
16. J. Huang, Y. Song, W. Wang, C. Wen, G. Li, Fully distributed adaptive consensus control of a class of high-order nonlinear systems with a directed topology and unknown control directions. IEEE Trans. Cybern. **48**(8), 2349–2356 (2018)
17. C.L. Wang, C.Y. Wen, L. Guo, Adaptive consensus control for nonlinear multi-agent systems with unknown control directions and time-varying actuator faults. IEEE Trans. Autom. Control **66**(9), 4222–4229 (2021)
18. X.G. Guo, P.M. Liu, J.L. Wang, C.K. Ahn, Event-triggered adaptive fault-tolerant pinning control for cluster consensus of heterogeneous nonlinear multi-agent systems under aperiodic DoS attacks. IEEE Trans. Netw. Sci. Eng. **8**(2), 1941–1956 (2021)
19. Q. Wang, C. Sun, Adaptive consensus of multiagent systems with unknown high-frequency gain signs under directed graphs. IEEE Trans. Syst. Man Cybern. Syst. **50**(6), 2181–2186 (2020)
20. W. Wang, C. Wen, J. Huang, J. Zhou, Adaptive consensus of uncertain nonlinear systems with event triggered communication and intermittent actuator faults. Automatica **111**, Art. No. 108667 (2020)
21. X.W. Li, Y. Tang, H.R. Karimi, Consensus of multi-agent systems via fully distributed event-triggered control. Automatica **116**, Art. No. 108898 (2020)
22. H. Kaufman, I. Barkana, K. Sobel, *Direct Adaptive Control Algorithms: Theory and Applications*, 2nd edn., Springer, New York (1998)
23. J.J. Carroll, D.M. Dawson, Integrator backstepping techniques for the tracking control of permanent magnet brush DC motors. IEEE Trans. Ind. Appl. **31**(2), 248–255 (1995)
24. Y.M. Li, T.C. Wang, W. Liu, S.C. Tong, Neural network adaptive output-feedback optimal control for active suspension systems. IEEE Trans. Syst. Man Cybern. Syst. **52**(6), 4021–4032 (2022)
25. Y.W. Wang, Y. Lei, T. Bian, Z.H. Guan, Distributed control of nonlinear multiagent systems with unknown and nonidentical control directions via event-triggered communication. IEEE Trans. Cybern. **50**(5), 1820–1832 (2020)
26. A. Theissler, Detecting known and unknown faults in automotive systems using ensemble-based anomaly detection. Knowl. Based Syst. **123**, 163–173 (2017)
27. X.M. Li, Q. Zhou, P. Li, H.Y. Li, R.Q. Lu, Event-triggered consensus control for multi-agent systems against false data-injection attacks. IEEE Trans. Cybern. **50**(5), 1856–1866 (2020)
28. X.G. Guo, D.Y. Zhang, J.L. Wang, J.H. Park, L. Guo, Observer-based event-triggered composite anti-disturbance control for multi-agent systems under multiple disturbances and stochastic FDIAs. IEEE Trans. Autom. Sci. Eng. **20**(1), 528–540 (2023)

29. W.L. He, X.Y. Gao, W.M. Zhong, F. Qian, Secure impulsive synchronization control of multi-agent systems under deception attacks. Inf. Sci. **459**, 354–368 (2018)
30. Z. Feng, G. Hu, Secure cooperative event-triggered control of linear multiagent systems under DoS attacks. IEEE Trans. Control Syst. Technol. **28**(3), 741–752 (2020)
31. A.Y. Lu, G.H. Yang, Stability analysis for cyber-physical systems under denial-of-service attacks. IEEE Trans. Cybern. **51**(11), 5304–5313 (2021)
32. C. Chen, C. Wen, Z. Liu, K. Xie, Y. Zhang, C.L.P. Chen, Adaptive consensus of nonlinear multi-agent systems with non-identical partially unknown control directions and bounded modelling errors. IEEE Trans. Autom. Control **62**(9), 4654–4659 (2017)

Chapter 13
Concluding Remarks

Abstract This chapter summarizes the key findings and contributions of this book, discusses the challenges of *intrusion-tolerant*, *fault-tolerant*, and anti-disturbance control, and outlines potential future research directions.

13.1 Summary of the Book

This book revolves around the immune intelligence of multi-agent systems (MASs) for effective anti-disturbance, fault tolerance, and intrusion tolerance. Beginning with the basic control tasks for an MAS including the complete consensus control and cluster consensus control, we explored the challenges of integrating immune capabilities into intelligent systems impacted by various unfavorable factors. We gradually touched onto the immune coordination problems that arise when disturbance, fault, and attack occur simultaneously and presented effective and inspiring self-healing solutions. Throughout the book, we discussed key design considerations, and supported these discussions by rigorous analyses and mathematical proofs.

Chapter 3 presented two distinct anti-disturbance control strategies based on the mechanisms of *disturbance* observation and *disturbance* compensation to achieve *complete consensus* of the nonlinear MASs satisfying Lipschitz functions under norm-bounded external *disturbance* using state feedback. In addition, an adaptive memory event-triggered mechanism (AMETM) was proposed to reduce the transmission of redundant data via adjusting the threshold dynamically with the change of the system state.

Based on the anti-disturbance strategies proposed in Chap. 3, a novel adaptive memory state observer-based anti-disturbance scheme under DoS attacks was developed in Chap. 4 to improve the observer accuracy by adding a buffer for the system output measurements. The designed control strategy not only realizes the anti-disturbance *complete consensus* objective, but also provides more reasonable control signals when DoS attacks occur. The assumption that the state is measurable was removed. Moreover, global information was not required in the updates of the proposed AMETM.

X.-G. Guo et al., *Anti-disturbance Intrusion- and Fault-Tolerant Control of Multi-agent Systems*, https://doi.org/10.1007/978-981-96-9614-7_13

Chapter 5 explored an observer-based event-triggered H_∞ *complete consensus* control strategy for nonlinear MASs under norm-bounded external disturbances and stochastic FDIAs over a directed communication network. Different from the previous result in Chap. 4, the observer errors were regarded as *disturbance* and then attenuated by H_∞ norm bounds, together with the external disturbances. In addition, a fault detection scheme was designed by constructing a residual signal with adaptive threshold and the performance of the fault detection mechanism was improved by introducing a *disturbance* compensation mechanism.

Based on Chaps. 3–6 presented a CHADC strategy for a class of nonlinear MASs subjected to stochastic FDIAs and multiple disturbances under a directed communication topology. The *disturbance* observer was designed to counteract the *disturbance* produced by an exogenous system. Based on the H_∞ control technique, the bounded errors and variables caused by the other type of disturbances and FDIAs were further attenuated simultaneously. In addition, a novel adaptive compensation technique was proposed to ensure the *complete consensus* performance of MASs by evaluating the estimate of the upper bounds of the unknown but bounded disturbance.

Chapter 7 proposed two novel fully distributed adaptive *fault-tolerant* control strategies for nonlinear MASs with or without asymmetric nonlinear actuator saturation (ANASs), respectively. Unlike Chaps. 3–6, this chapter was used IQC to describe the characteristics and sources of the considered disturbances. At the same time, by combining the pseudo-PID sliding-mode control (SMC) method with adaptive control techniques, the leader-following *complete consensus* was achieved in *finite time*. In addition, the designed controllers only need local neighbor information instead of global topology information.

Chapter 8 investigated the *cluster consensus* control for heterogeneous MAS with nonlinear dynamics, actuator faults, and IQCs under a directed topology. Based on the pinning control method, two adaptive *fault-tolerant* control strategies with fixed and adaptive pinning gains respectively were proposed to guarantee *cluster consensus* in *finite time*. To attenuate the adverse effects of actuator faults and IQCs, an adaptive input compensation scheme was developed via the estimation of unknown parameters.

Chapter 9 studied the *cluster consensus* problem for second-order nonlinear MASs with directed communication graph under periodic DoS attacks. By using the input switching technology, the state tracking error systems under periodic DoS attacks were established. On this basis, an event-triggered control strategy was designed, which does not require the MASs to satisfy the in-degree balance condition.

Based on the cluster consensus control scheme proposed in Chap. 9 under periodic DoS attacks, Chap. 10 developed a *cluster consensus* scheme for heterogeneous nonlinear second-order MASs under aperiodic DoS attacks. By using the input switching method and adaptive compensation mechanism, an event-triggered adaptive *fault-tolerant* pinning control strategy was designed to achieve *cluster consensus* under simultaneous aperiodic DoS attacks, actuator faults, and IQCs. Unlike the centralized ETM in Chap. 9, the trigger instants of the ETM designed in this chapter were asynchronous.

Chapter 11 devoted the *fault-tolerant cluster consensus* problems for heterogeneous MASs with time-varying unknown nonidentical direction (TVUND) faults under stochastic FDIAs. In order to overcome the challenge brought by the summation of multiple Nussbaum functions, a group of new Nussbaum functions was proposed. At the same time, a *fault-tolerant* strategy based on the new group of *Nussbaum functions* was designed that require neither the upper and lower bounds of the fault scaling factors nor fault directions. In addition, a hybrid ETM with both adaptive and memory characteristics was proposed to improve the tracking performance of the system state at the inflection points.

Chapter 12 investigated *cluster consensus* tracking problem for a class of high-order MASs with TVUND faults under multiple *cyber attacks*. Based on the *Nussbaum function* technique, a fully distributed hierarchical (cyber layer and physical layer) *intrusion-* and *fault-tolerant* control strategy was presented. Unlike the challenge resulting from the summation of multiple Nussbaum functions in Chap. 11, this chapter avoided the problem of summation of multiple Nussbaum functions by introducing a virtual cyber layer. In addition, the constructed virtual cyber layer avoided the interaction in the physical processes and thus reduced the error propagation. This two layer controller structure greatly increased the flexibility of the controller design.

13.2 Future Missions

The work detailed in this book focuses on the design of anti-disturbance, intrusion-tolerant, and fault-tolerant control strategies to address the consensus control problems in MASs. It also explores the development of efficient controllers for implementing these tolerance strategies. The results provide valuable insights into the application of immune control to MASs, with the theoretical findings being validated through extensive simulations. However, as immune solutions are still a relatively novel approach for MASs, several interesting and challenging open problems remain, which warrant further exploration and research.

- **Disturbances utilization**. In MASs, disturbances are an inevitable consequence of model uncertainties and the complexities of the operating environment. These disturbances can adversely affect the performance of individual agents as well as the overall system. To counteract the negative effects of disturbances, various control strategies have been proposed to enhance the robustness and performance of systems. These strategies include H_∞ control [1, 2], disturbance observer-based controls [3–5], compensation methods based on estimating disturbance bounds [6, 7], active disturbance rejection [8, 9], and composite hierarchical anti-disturbance approaches for handling multiple homogeneous disturbances [10]. However, most of these solutions focus solely on mitigating the negative impact of disturbances on system performance, without considering the potential benefits that disturbances might offer. In practical scenarios, well-managed beneficial disturbances can contribute to energy savings, reduced steady-state errors, and faster convergence, providing a more efficient system overall. Therefore, there are numerous intriguing research avenues that can be pursued in this field.

- **Active fault-tolerant control**. For faults with unknown time-varying directions, different adaptive *fault-tolerant* control strategies has proposed in this book to ensure the performance of MASs by gradually adjusting control parameters and control inputs. However, its reliance on slow adaptation mechanisms often delays the response to the occur of faults, especially small or incipient ones. This limitation can hinder timely adjustment of input signals and compromise overall system performance. Active *fault-tolerant* control, in contrast, is equipped with fault diagnosis capabilities that allow it to detect and identify faults as they occur, enabling a more immediate and targeted response [11, 12]. This proactive approach significantly enhances the system's resilience and operational efficiency. Future research could explore the integration of advanced diagnostic techniques within active *fault-tolerant* control frameworks to achieve faster and more accurate fault detection. Furthermore, integrating active *fault-tolerant* control with distributed optimization and learning-based methods could unlock new possibilities for achieving resilient and efficient coordination in complex, dynamic environments.
- **Intrusion-tolerant control against stealthy attack**. The intrusion-tolerant control methods provided in this monograph assume that attacks can be detected by the system. However, attackers often design stealthy attacks that evade conventional detection mechanisms [13, 14], posing significant challenges to system security and resilience. Designing intrusion-tolerant control strategies capable of addressing stealthy attacks is an interesting and pressing research direction. It is interesting and meaningful to focus on developing advanced detection frameworks that can identify subtle anomalies indicative of stealthy attacks, as well as robust control schemes that maintain system stability and performance even under undetectable conditions. Incorporating techniques such as adaptive estimation, data-driven analysis, encryption/decryption algorithm, and game-theoretic approaches could provide a solid foundation for tackling this challenge in complex networked systems.
- **Preset-time and preset-accuracy control**. Convergence time and convergence accuracy are two critical metrics for evaluating system performance [15, 16]. Unlike uniformly ultimately bounded stability, prescribed accuracy control enables the predefinition of the steady-state accuracy of a system, offering a highly effective approach to precision control. Regarding convergence time, finite-time control, fixed-time control, and preset-time control are commonly employed methods [17]. Among these, prescribed-time control demonstrates significant advantages by allowing the convergence time to be predefined, making it highly desirable for practical applications. Combining preset-time and preset-accuracy control to ensure both convergence time and steady-state accuracy in the presence of disturbances, attacks, and faults is a highly meaningful research direction. The research mentioned is intriguing and meritorious due to its focus on creating robust

control schemes that maintain system performance and resilience in challenging conditions. In addition, leveraging techniques such as advanced optimization and reinforcement learning algorithms could further enhance the applicability and effectiveness of such control strategies in complex, real-world scenarios.

References

1. L.W. An, G.H. Yang, Data-based distributed sensor scheduling for multiple linear systems with H_∞ performance preservation. IEEE Trans. Autom. Control **67**(12), 6834–6841 (2022)
2. M. Rigotti Thompson, M. Torres-Torriti, F.A. Auat Cheein, G. Troni, H_∞-based terrain disturbance rejection for hydraulically actuated mobile manipulators with a nonrigid link. IEEE/ASME Trans. Mechatron. **25**(5), 2523–2533 (2020)
3. Z. Bao, Y. Zhu, H. Teng, Y. Cui, L. Guo, Deep-coupled sliding mode disturbance observer-based composite attitude control for flexible communication spacecrafts. IEEE Trans. Ind. Electron. **72**(1), 833–844 (2025)
4. X. Fang, F. Liu, X. Gao, Composite learning control of overactuated manned submersible vehicle with disturbance/uncertainty and measurement noise. IEEE Trans. Neural Netw. Learn. Syst. **32**(12), 5575–5583 (2021)
5. Z. Huang, M. Chen, Augmented disturbance observer-based appointed-time tracking control of UAVs under exogenous disturbance. IEEE Trans. Intell. Veh. **9**(1), 2822–2835 (2024)
6. N. Ji, J. Liu, Vibration control for a three-dimensional variable length flexible string with time-varying actuator faults and unknown control directions. IEEE Trans. Autom. Sci. Eng. **20**(4), 2761–2771 (2023)
7. J. Wu, W. Sun, S.F. Su, Y. Wu, Adaptive asymptotic tracking control for input-quantized nonlinear systems with multiple unknown control directions. IEEE Trans. Cybern. **53**(8), 5216–5225 (2023)
8. J. Han, From PID to active disturbance rejection control. IEEE Trans. Ind. Electron. **56**(3), 900–906 (2009)
9. S. Shen, J. Xu, P. Chen, Q. Xia, An intelligence attitude controller based on active disturbance rejection control technology for an unmanned helicopter. IEEE Trans. Veh. Technol. **72**(3), 2936–2946 (2023)
10. L. Guo, W. Li, Y. Zhu, X. Yu, Z. Wang, Composite disturbance filtering: a novel state estimation scheme for systems with multisource, heterogeneous, and isomeric disturbances. IEEE Open J. Ind. Electron. Soc. **4**, 387–400 (2023)
11. R.Z. Chen, X.G. Guo, Q. Li, J.L. Wang, L. Guo, *Disturbance-Observer-Based Model-Free Adaptive Active Fault-Tolerant Consensus Control for MASs with TVDAF*. Automatica. **173**, Art. No. 112114 (2025)
12. J. Wang, J. Zhang, B. Qu, H. Wu, J. Zhou, Unified architecture of active fault detection and partial active fault-tolerant control for incipient faults. IEEE Trans. Syst. Man Cybern. Syst. **47**(7), 1688–1700 (2017)
13. S.X. Ding, L. Li, D. Zhao, C. Louen, T. Liu, Application of the unified control and detection framework to detecting stealthy integrity cyber-attacks on feedback control systems. Automatica **142**, Art. No. 110352 (2022)
14. F. Miao, Q. Zhu, M. Pajic, G.J. Pappas, Coding schemes for securing cyber-physical systems against stealthy data injection attacks. IEEE Trans. Control Netw. Syst. **4**(1), 106–117 (2017)
15. P.M. Liu, X.G. Guo, J.L. Wang, D. Coutinho, Z.G. Wu, Preset-time and preset-accuracy human-in-the-loop cluster consensus control for MASs under stochastic actuation attacks. IEEE Trans. Autom. Control **69**(3), 1675–1688 (2024)

16. P.M. Liu, X.G. Guo, J.L. Wang, D. Coutinho, L.H. Xie, Chattering-free fault-tolerant cluster control and fault direction identification for HIL UAV swarm with pre-specified performance. IEEE-CAA J. Automatica Sin. **12**(1), 183–197 (2025)
17. W.D. Xu, X.G. Guo, J.L. Wang, Z.G. Wu, X.P. Xie, Distributed fuzzy leader-follower vehicular formation control with appointed-time performances and obstacle avoidance. IEEE Trans. Veh. Technol. **74**(1), 152–165 (2024)

Index

X.-G. Guo et al., *Anti-disturbance Intrusion- and Fault-Tolerant Control of Multi-agent Systems*, https://doi.org/10.1007/978-981-96-9614-7

MIX
Papier aus verantwortungsvollen Quellen
Paper from responsible sources
FSC® C105338

If you have any concerns about our products,
you can contact us on
ProductSafety@springernature.com

In case Publisher is established outside the EU,
the EU authorized representative is:
Springer Nature Customer Service Center GmbH
Europaplatz 3, 69115 Heidelberg, Germany

Printed by Libri Plureos GmbH
in Hamburg, Germany